Additional Praise for *The Sense of the Universe*

"*The Sense of the Universe* represents the project on phenomenological cosmology. His main stance is that the universe as a whole cannot be an object in the natural attitude of consciousness and is comprehended as a 'saturated phenomenon,' a concept developed by contemporary French philosopher Jean-Luc Marion. This concept allows one to respond to the question on the sense and origin of the universe not in terms of an objective foundation that could be discovered by theoretical thinking but rather to treat the universe as the ultimate existential background of this thinking. Correspondingly, the universe cannot be considered anymore as a phenomenon without human beings articulating it. Such an approach to cosmology opens a new avenue to its dialogue with theology by relating the articulation of the world with experience of the Divine in the one and same human subject."

Ruslan Loshakov
Luleå University of Technology

"*The Sense of the Universe* by Alexei Nesteruk is a deep and well-structured book about science and theology that focuses in particular on a study on the philosophical foundations of cosmology. Nesteruk's central tenet is that a phenomenological analysis of the foundations of physics reveals that in doing physics we also reveal the nature of our humanity. He sets out, in convincing detail, that such analysis of cosmology in particular can reveal the hidden psychological and spiritual aspirations that guide the study of the world. Nesteruk interprets ideas concerning the universe as a whole, and its origin, existentially as reflecting the basic anxieties of human existence in the vast cosmos. In this view, the study of cosmological is interpreted as an inevitable part of the teleology pertaining to all human acts. The universe as a whole, which is the inexhaustible context of the living experience, is then seen as 'the infinite created unknowable' which, from an epistemological point of view, is similar to that of the Divine. Nesteruk unfolds, through the analysis of ideas of the universe, a hidden theological commitment in cosmology related to the metaphysical and theological mystery of the human existence."

Chris Dewdney
University of Portsmouth

The Sense of the Universe

The Sense of the Universe

Philosophical Explication of Theological Commitment in Modern Cosmology

Alexei V. Nesteruk

Fortress Press
Minneapolis

THE SENSE OF THE UNIVERSE
Philosophical Explication of Theological Commitment in Modern Cosmology

Cover image: Joe Reinke
Cover design: © Thinkstock: Spiral Galaxy NGC 5584 / Stocktrek

Library of Congress Cataloging-in-Publication Data
Print ISBN: 978-1-4514-7038-3
eBook ISBN: 978-1-4514-9417-4

The paper used in this publication meets the minimum requirements of American National Standard for Information Sciences — Permanence of Paper for Printed Library Materials, ANSI Z329.48-1984.

Manufactured in the U.S.A.

This book was produced using PressBooks.com, and PDF rendering was done by PrinceXML.

Contents

Preface

This book, in a way, represents a further extension of some ideas on the dialogue between science and theology that were formulated in my previous books *Light from the East* and *The Universe as Communion*. My position on the general approach to the dialogue between theology and science has experienced a considerable transformation toward an understanding that theology and science cannot enter this dialogue on the same footing, that is, on equal terms. Orthodox Christianity treats theology as experience, related specifically to communion of human persons with God while being in the physical universe. In fact, life *is* communion, so that all other activities (including a scientific one) originate in this communion. In this sense, to establish the dialogue between theology and science means to appropriate science theologically, that is, experientially or existentially. The symmetry between theology and science is not sustainable simply because existence, that is, life as a center of disclosure and manifestation, precedes its explication through science. This asymmetry reveals itself in the dialogue as a certain (*a priori*) theological commitment. Theological commitment means an intentional approach to science through "the eyes of (existential) faith." Being a commitment, it entails a method: namely a phenomenological explication of those premises in the coherence between human rationality and the rationality of the cosmos that make cosmology possible at all. The aim and the final result of such an analysis is the creation of a solid ground for understanding the sense of cosmology's "dialogue" with theology. This book deals with three principal aspects of explication of the theological commitment in cosmology: 1) the reinstatement of personhood to its central place in the dialogue between theology and cosmology as being a center of disclosure and manifestation in both theology and cosmology; 2) the elucidation of the sense of retaining transcendence while conducting research within the rubrics of intentional immanence, which ultimately elucidates the sense of humanity as not being circumscribed by the necessities of the universe, but carrying in itself the Divine image; and thus 3) the elucidation of cosmology as activity explicating the history of salvation and thus contributing to faith in God.

There are many colleagues and friends who through discussions and indirect support contributed toward the writing of this book. Among many, I would like to express my feelings of gratitude to my sons Dmitri and Mark, as well as my wife Zhanna, for existential support and encouragement. My sister Nina was very helpful in obtaining necessary Russian books while I was outside Russia: my deep gratitude to her. George Horton deserves special thanks as a first reader of the book, contributing a lot to its style and ultimate shape. Among others, my thanks for fruitful conversations and discussions go to Christopher Dewdney, David Matravers, Roy Maartens, David Bacon, David Coule, Joel Matthews, John Bowker, Niels Gregersen, Adrian Lemeni, Michael McCabe, Mogens Wegener, Alfred Osborne, Argyrios Nicolaides, Rev. Christopher Knight, Andrei Pavlenko, Grigory Benevich, Grigory Goutner, Dmitri Biryukov, Marina Vasina, Aleksandr Shevchenko, Alexandr Soldatov, Natalia

Pecherskaya, Tatiana Litvin, Sister Teresa Obolevich, Oksana Kuropatkina, Stoyan Tanev, Ruslan Loshakov, Andrei Grib, Peter Coleman, Antonio Samons, Rev. Konstantin Litvinenko, Rev. Kyrill Kopeikin, Rev. Brian Macdonald-Milne, Brother Christopher Mark (CSWG), Florin Caragiu, Andreas Marcou and Alex Ali. I am particularly grateful to Metropolitan John of Pergamon (Zizioulas) for appreciation of my contribution to the dialogue between Orthodox Christianity and science, as well as for encouragement. The department of mathematics of the University of Portsmouth was very helpful in providing me with the conditions for conducting research and writing: my special thanks to its head Andrew Osbaldestin for comprehension and support. I always used the opportunity to communicate general ideas developed in this book through the courses in theoretical and mathematical physics, as well as history and philosophy of mathematics which I taught an the University of Portsmouth. In this regard I would like to thank all my students who were patient and perceptive in grasping complicated ideas. St. Andrews Biblical and Theological Institute in Moscow was also very helpful for providing me an opportunity to approve ideas developed in this book during its summer institutes in 2005-2014. I am grateful to Alexei Bodrov and Mikhail Tolstoluzhenko for inviting me to collaborate and contribute to various scientific events. Since the project of this book dates back many years, the research and publication of its ideas were indirectly supported by the John Templeton Foundation (JTF) through its grants (#1573 and #11921) as well as through numerous international projects where the ideas of the book were approved. I would like to express my sincere gratitude to the Foundation for this support. Some papers related to the content of this book have been published with the support of JTF through the Copernicus Centre in Krakow. The opinions expressed in those publications as well as in this book are those of the author and do not reflect the views of the JTF.

Introduction: Thinking of The Universe and Theological Commitment

If a human person craves immortality, he must, in his individual and collective life realize the mode of the truly existent, the logic of relations found in cosmic harmony.

–Yannaras, The Meaning of Reality, p. 133

The Universe and the Mystery of Human Existence

This book is not about cosmology as physical research and it is not concerned with the popular interpretations of fashionable cosmological theories. Nor is it about *meta*-cosmology, that is, a metaphysical extension of cosmology, which lags behind cosmological theories and ideas in order to use them as a testing ground for known philosophical ideas. This book is on the sense of modern cosmological ideas as they originate in the being of humanity and the way that ideas about the universe are related to the philosophical and theological mystery of the human condition in the universe. Thus this book positions itself in the field of religion and science or, more precisely, Christian theology and cosmology.

It does not, however, aim to compare contemporary cosmological theories and observations with the ideas of the world in different philosophical or theological systems. We believe that it would be incongruous to bring into correlation the cosmological views of the Fathers of the Early Christian Church (which historically had been rooted in ancient Greek philosophy and astronomy) with the experimental and theoretical results of modern cosmology. Similar to this, it seems doubtful to conduct a comparative hermeneutics of the scriptural texts with modern writings on cosmology in an attempt to reveal some linguistic parallels: such a comparison would exhibit an arbitrary approach that is dictated neither by the needs of theology nor the logic of science.

Instead, the argument starts from the premise that there is a fundamental asymmetry between cosmology, as a definite form of activity and thinking, and that philosophico-theological consciousness which exercises its reflection upon cosmology. This asymmetry consists in a simple fact that although philosophical and theological motives enter implicitly any speculations on the universe, cosmology as a scientific discipline cannot explicate these motives. The motives we imply here enter our discussion as a certain attitude of consciousness that is determined by an ambivalent position of humanity in the universe, that is, on the one hand, being included in or contained by the universe, and, on the other hand, containing the universe as a representation and articulated reality within consciousness.

The implicitly present philosophy is not a "neutral" form of thought, but is imbued with existential meaning that has theological connotations, in the sense that any philosophical reflection as well as scientific theories are "inserted" (bracketed) in the experience of existence, that is, the experience of communion with God. In other words, the aim of this book is to conduct the philosophical analysis of those logical operations of the human mind in research of the universe from within a hidden philosophico-theological "obviousness" that is essential to all acts of consciousness, including scientific ones. From this obviousness cosmology is explicated by us as a certain way of interrogating the reality of the world as well as that of human beings themselves.

Such a philosophically and theologically "enlightened" treatment of cosmology, despite its sheer deviation from mainstream science, is in our opinion very timely because it elucidates not only an existential sense of what cosmologists are speaking of the universe, but also the sense of what they are speaking of themselves, that is, of human beings incarnate in this universe and capable of speculating about it. Thus the main interest of this book is not so much in the sense of physical realities that cosmology attempts to constitute, but in the ways this constitution originates in those anthropological and psychological aspects of humanity's existence that express basic anxieties of existence and represent a theological mystery. Our interest is not in describing that which is in the universe as if this description would be self-evident and not needing any further analysis, but in investigating how this very description became possible. This is a philosophical objective, but one that cannot be fulfilled without recourse to theology.

Correspondingly, the search for the ultimate foundations of cosmological knowledge cannot avoid a certain "theological commitment" related to the stance on the nature and essence of the knowing subject.[1] At the same time, the enquiry into those original

1. Jean Ladrière expressed a thought that in order to explicate the analogy between the deep structure of nature and the structure of human existence as openness, creativity, possibility, etc., one needs to enter what he called the "domain of the word," which, in our parlance, would correspond to thought within the "theological commitment": "The problematic of nature can thus be linked with the problematic of human existence. Still, there is no continuity between these two domains. There are perhaps indications pointing in a certain direction, but it is not within the power of cosmological thought, even when developed, to become a consideration

conditions in the study of the universe without which this study would not be possible explicates this hidden theological commitment.

The analysis of the conditions of knowledge is called in philosophy "transcendental." This analysis deals with two fundamental issues: 1) the intrinsic interlink between human consciousness and the possibility of sensing, judging, and reasoning about the universe; in short: the universe can be presented in thought and knowledge only as constituted within certain transcendental delimiters related to the structures of embodied subjectivity; 2) it is because of the physical and epistemological incommensurability between the universe and human beings, that the universe always remains a *transcendent* background of any *transcendental* knowledge. The "relationship" between the universe and human beings is established on the principles of freedom, that is, free-thinking (related to what Kant called the faculty of reflective judgment, and theologians call the free will of humanity made in the Divine image). This freedom implies that the universe and humanity interact in ways that reflect their mutual constitution: the universe is a never-accomplished mental creation, whereas human subjectivity is the self-correcting structural unity of apperception, the unity of which originates in the thought (intuition and imagination) of the infinity of the universe.

The theological upshot in this transcendental analysis is that humanity remains free and responsible in its thinking of the universe, because this thinking implies free action, free judgment, and choice of theoretical options, which is not subordinated either to the rigidity of the structures of subjectivity, or to the material content of the universe. A theological stance is the possibility of transcendence in cognitive actions, the transcendence either as longing for the

of finality, to enter the domain of the word. Only by meditation on what properly belongs to the word can one open another way of understanding (if one exists), leading towards . . . faith." J. Ladrière, *Language and Belief* (Dublin: Gill & Macmillan, 1972), 186.

incommensurable content of the universe, or as a resistance to any forms of thought that position humanity as part of the cosmic determinism, denying its ability to avoid the dissolution and crush by the mounting number of facts about the universe.[2] Finally, a theological stance in the transcendental analysis of cosmology is the commitment to the view that the very facticity (that is the very possibility and actual fact of existence) of the subject of transcendental knowledge, that is, a human person, originates in and through communion with the divine, as the giver of life and provider of its image.

The study of cosmology through the prism of the philosophically and theologically shaped mind is not in tune with the modern way of treating the real in terms of scientifically representable matter. In this sense such a study is untimely, that is, out of tune with the present, in the same way as philosophy, which deals with the phenomena (in our case the universe) that cannot encounter any immediate response from wider humanity, is untimely. Thus philosophical enquiry in cosmology imbued with a theological commitment reveals itself in an autonomous existence such that it makes things more difficult and complicated. However, here lies the advantage of a philosophical interrogation of cosmology as an autonomously functioning consciousness above and beyond that mass-consciousness which functions in the natural attitude. Skeptics and nihilists, whose

2. This is a different way of expressing that which Gabriel Marcel asserted in 1940 in his book *Du Refus à l'Invocation* (Paris: Gallimard, 1940), when he discussed a paradox related to the representation of the universe as an object: "The more I insist on the objectivity of things, thus cutting off the umbilical cord which binds them to my existence, that one which I call my organo-physical presence to myself, the more I affirm the independence of the world from me, its radical indifference to my destiny, my goals; the more the world thus proclaimed as the only real one, is converted into an illusory spectacle, a great documentary film offered for my curiosity, but which is ultimately abolished because of a simple fact that it ignores me. I mean that the universe tends to be annihilated in the measure that it overwhelms me. And this, I believe, is that which is forgotten whenever one attempts to crush man under the weight of astronomical facts" (32).

presence among intellectuals bears a sign of our times, can raise a disarming question as to whether it is worth doing at all: "What for to study the foundations of the universe?," or, correspondingly, "What for to understand the sense of humanity?"

The response to these questions comes from the definition of philosophy as *love for wisdom* (*philo-sophia*) and *truth* (*aletheia*), which implies *love* in general as a major characteristic of the human condition understood theologically. To enquire into the sense of the universe means not only to know it, but to be in communion with it, to love it. Philosophically and theologically oriented cosmology is not "knowledge" achievable and ready to use. Rather, philosophical cosmology belongs to the realm of those perennial aspects of the human quest for the sense of being that can be addressed only in the rubrics of the so-called *negative certitude*[3] pertaining to the long-lived traditional theology which does not provide us with a definite discursive judgment on the existence of God and what God is; this question drives the human reason only to one possible answer: it is *certain*, but this certainty is *negative*, so that one cannot answer this question in rubrics of reason alone. In similarity with theology when cosmology dares to predicate the "universe as a whole," or "multiverse" (the plurality of the worlds), the outcome of this predication does not resolve the present scientific uncertainty about their actual existence, rather it brings us back to the same "negative

3. *Positive incertitude* is typical for the sciences dealing with knowledge of objects, which can be described as science that operates with some precarious and incomplete data about these objects, which are amended and corrected in the course of science's progress. The paradox of science is exactly in that this uncertainty and corrigibility of its results is the condition for science to function at all. Another aspect of science is that it cannot know things in the context of the wholeness of the world. By contrast in philosophy, in what concerns its perennial questions about the world as a whole, there is no visible progress, so that it is able to speculate about the world only in rubrics of what are called by Jean-Luc Marion *negative certitudes*. See details of this concept in J.-L. Marion, *Certitudes négatives* (Paris: Bernard Grasset, 2010).

certitude" in which no answer to the question of "What?," "Why?" and so forth related to the universe as a whole is possible.

Correspondingly, a philosophical enquiry into cosmology within a theological commitment cannot be judged on the grounds of simplified scientific or commonsense criteria. Philosophical cosmology within a theological commitment characteristically contributes to the understanding and formation of humanity through its interaction with the universe. It represents cosmology as a general strategy of acquisition of the world, a strategy that as such manifests the ongoing incarnation of humanity in the universe, or in rather theological terms, the "humanization" of the cosmos. In this sense, philosophical cosmology within a theological commitment is directly related to philosophical anthropology as well as to the discourse of personhood. Both of them are concerned with the ancient question raised in Greek philosophy, "Why is there existence rather than nothing?" Contemporary physical cosmology attempts to respond to this question; however, its forms of thought remain intrinsically unadjusted to this type of interrogation. Said differently, cosmology is content with what it says in physical terms and what one says about it as it exists.

However, to understand the sense of cosmology one needs to establish a new type of "questioning of cosmology" in which thinking evolves beyond what was stated by cosmology itself. Here one needs an "enlightened" reason, or, as it was expressed by Nietzsche, a "great reason" that, on the one hand, is associated with the embodiment in flesh of the universe and would represent cosmology as a specific way of appropriation of the world. On the other hand this "great reason" is related to the Divine image in humanity, which humanity attempts to restore and fulfill, thus making the process of the humanization of the universe its communion with the Divine. In this sense any philosophical cosmology confesses a free type of thinking not constrained by the findings of the scientific and thus transcendent of physical cosmology by bringing it to the next circle of understanding the essence of being and humanity. The issue is not to think of the essence of cosmology, which would be equivalent to being restricted to its contemporary forms, regardless as to whether we judge it positively or negatively. It is important to realize that by questioning cosmology philosophically and theologically we overcome its seeming neutrality with respect to us, thus advancing our understanding of the very being of cosmology as being in us. Cosmology acts in producing its theories, but it does not think in a philosophical sense (compare with Heidegger's famous assertion that science does not think). The sense of cosmology can become enlightened only when the gulf between its particular theories and human thinking in general is realized.

To establish the sense of cosmology starting from cosmology itself, this cosmology must evolve in a radically reflective or transcendental mode, that is, in fact, to become philosophy. The sense of its theories can be grasped only within a critique originating in experience. This is the realm of transcendental self-experience that can be established through a method of phenomenological reduction. Such reduction aims to overcome a "natural naïveté," that is, a belief that cosmology deals with the things of the outer world. Its ultimate objective could be seen as questioning the neutrality of cosmological propositions (their invariance) with respect to specific historically contingent events of knowing. To remove the elements in this contingency would imply the return to those irreducible certainties that would represent the universe as pertaining to the essence of one's conscious life. It is from this life, with its mundane experiences, that the universe is constituted. Life is understood here not anymore as an empirical psycho-physiological life that belongs to the universe, but as the transcendental self-apprehension that comes forth and from within which the universe emerges as its intentional correlate. By inverting this last proposition, one can assert that it is through cosmology that transcendental subjectivity is revealed as overcoming its own incarnate boundaries. Indeed, by stripping off the layers of the physical and biological, one comes to discover that the universe as a whole appears as an intentional correlate of transcendental consciousness. Thus "putting out of play" the contingent aspects of the universe brings cosmology to a discourse of the transcendental subject, as that center of disclosure and manifestation of the universe through which the latter acquires its own "voice."[4]

However, even this transcendental reduction does not guarantee that we do not fall into a "transcendental naïveté." Such naïveté amounts to thinking that reality presupposes the transcendental subject as that pre-given context-horizon within which reality unfolds. But this transcendental subject still functions as an embodied creature, that is, in the world of physical things. However, the very physical things exist for this subject only as constituted by the thinking subject. With regard to the universe as a whole the situation is different: its alleged totality cannot be constituted by the subject but, vice versa, the subject itself is being constituted by the universe (not in a trivial physical sense).

In order to clarify this thought one must remind the reader that cosmology, as a historically concrete science, is capable of making its claims on the structure and evolution of the universe within the limits of what can be called "positive incertitude," that is, that certainty which is local in time and is subject to amendment and falsification. This can be expressed as those scientific conceptual signifiers that never exhaust the content of that which is supposed to be signified.

4. Cf. T. Torrance, *The Grammar of Theology: Consonance between Theology and Science* (Edinburgh: T. & T. Clark, 2001), 2. See also O. Clément, *Le Christ, Terre des vivants. Essais théologiques*, Spiritualité Orientale n. 17 (Bégrolles-en-Mauges: Abbaye de Bellefontaine, 1976), 102–3.

"Positive incertitude" in science can also be described in terms of so-called *apophaticism* (well known in patristic theology), asserting a simple truth that the appearances of things and constitutions by the finite consciousness deal with a particular, incomplete phenomenality that pertains to objects. With regard to things beyond simple perception and nominations that exceed the capacity of constitution and phenomenality, one can conjecture only in terms of aberrations and approximations. The fact that we can see and speculate about some aspects of the universe does not entail that there are no other aspects of existence than those that are present and perceived by us, but whose presence cannot be affirmed in terms of consciousness and knowledge.

A simple *physical* example of such a hidden aspect of the universe is its dark matter and dark energy, which according to theory constitute 96 percent of the overall matter of the universe. However, the phenomenality of these theoretical constructs is poor: physics does not know what particular particles and fields stand behind these constructs. A *philosophical* example of concealment related to the universe as a whole can be taken as its own contingent facticity, the sense of which cannot probably be disclosed to humanity at all. Indeed, the notion of the universe as a whole, which is claimed to be a subject matter of cosmology, allows one only to have some precarious and incomplete definitions related to the fundamental finitude (spatial, temporal, historical, etc.) of the subject of knowledge.

However, this "positive incertitude" of cosmology does not mean that from a philosophical point of view one must disdain cosmology as irrelevant to any perennial questions. It just implies that the cosmological research has to proceed along the lines of the scientific method in clear understanding that the universe as a whole will never be constituted at all. Then the persistence of cosmology exhibits the courage and heroism of scientists in following their quest for the universe despite the ultimate futility of any hope to have this universe as an object of science. The same takes place in theology when believers explicate their experience of God as an open-ended process in a clear consciousness that the true names of the Divine are beyond this age and any denominations. Correspondingly, in cosmology the persistence of research as a purposive activity of humanity is pointing toward its *telos*, that is, the *telos* of research, which as such is also beyond this age and any denominations. Here is a fundamental paradox of cosmology, as well as any other science, namely, that its incertitude is that condition of its progress consisting in the unceasing correction and amendment of its results and theories. However, in spite of the fact that a human person cannot constitute the universe, so that the universe saturates its intuition and blocks the reason, this person remains an independent center of disclosure and manifestation of the universe, resisting any attempt to be crushed by the grandeur of being. In this sense the "negative certitude" in relation to knowledge of the universe turns out to be a constructive certitude of constituting the human subject.

By interacting with the infinity of the universe human persons form themselves: in the measure that humanity is incapable of constituting the universe as a whole, the human person is constituted by the universe as an "object" of humanity's constant interest and anxiety of its own position in it. This means that the transcendental subject that appropriates the universe into the sphere of its own subjectivity, and is destined to carry out the phenomenological reduction with the goal of revealing the immanent belonging of the universe to consciousness of the subject, is the forming and changing subject who is formed and changed through this very appropriation. One can summarize by saying that the understanding of the sense of cosmology implies the understanding of the formation of the self-consciousness of humanity's position in the universe subject to one important condition: the cosmological picture does not diminish the place of the human in the universe as the center of its disclosure and manifestation. The more cosmology proves that human beings are no more than a speck of dust in the universe, the more the human person resists this by defending the sense of its existence. That which has been said partially explains the sense of what we asserted earlier—that the human "I" is constituted by the universe.

But the shift of the center of cosmological enquiry into the life of transcendental subjectivity still retains the same perennial question of the facticity of this very subjectivity. If it is claimed that the characteristic feature of personal existence, that is, human hypostatic being, is its ability to resist scientific tendencies to denigrate humanity by dissolving it in the natural and cosmic, and, hence, to understand the cosmic conditions of human existence as the only *necessary* ones, then the question arises as to what is the ground and foundation of the contingent facticity of hypostatic existents, that is, of persons? Where do the sufficient conditions of human existence come from? Certainly one can take a classical existentialist position that makes this last question devoid of any sense, for one cannot abstract from the already-present event of life. However, this stance is unsatisfactory for a theologically inclined mind who wants to see in the very fact of conscious existence the manifestation of *truth* (*aletheia*) in an absolute sense, such that the acceptance of conscious existence as an absolute reference point of any further philosophizing implies *belief in the truth of existence.* Thus the knowledge of the universe as unfolded from within human subjectivity is by its essence committed to a simple *existential faith.* For a skeptical scientist or for a modern atheist, it would be problematic to proceed from existential faith to religious faith, that is, to the conviction that truth has foundation in God, for any reference to the Divine would imply transcendence, principally impossible in science and prohibited by the very essence of the phenomenological reduction.[5] However, and this is our main point, the very reduction as well as the functioning of consciousness will be impossible at all if the reference to the source of its contingent facticity would be eidetically removed. In such a case the removal of God as the foundation of consciousness would lead inevitably to a suggestion that there must be another non-worldly foundation of this consciousness, which would be analogous to the idea of God, that God which was previously bracketed out. Correspondingly, we return to the assertion that any hypothetical reduction of God would imply the cessation of functioning of consciousness itself. This is one of the motivations of contemporary phenomenology—to argue that even if the facticity of consciousness cannot be justified, it can at least be explicated through dealing with the saturated phenomena that, in a way, constitute this consciousness.[6]

To understand cosmology within a theological commitment is thus to understand the existential sense of the universe, or, to be more precise, to "understand" what it means to think *of* or commune *with* the universe. What could it mean—the thinking *of* or communion *with* the universe within the conditions of a scientific and technological age in order to avoid such thinking being enslaved by the sphere in which knowledge operates according to some social, but still historically contingent standards? Correspondingly, how could we dress this thinking in words while avoiding all cultural superstitions that engulf our language? And even in the case where

5. One can point to Husserl, who in his *Ideas I* (§58) subjected God to reduction, bracketing it and depriving it of any transconscious status. See E. Husserl, *Ideas Pertaining to a Pure Phenomenology and to a Phenomenological Philosophy. First Book: General Introduction to a Pure Phenomenology*, trans. F. Kersten (Dordrecht: Kluwer, 1998).

6. Later in this book we will discuss the idea of the saturated phenomena in application to cosmology. For a systematic approach to such phenomena, see J.-L. Marion, *In Excess: Studies of Saturated Phenomena* (New York: Fordham University Press, 2002).

we believe that we have achieved such a goal, could we expect any recognition of that form of thinking which intentionally extends beyond the view of the universe framed by varieties of scientific projects, conference discussions, and numerous publications? All these questions implicitly presuppose that the scientific way of thinking of the universe does not cover the fullness of our communion with the universe, which is concealed in the very fact of our existence. This concealment follows, for example, from the fact that humanity is able to interact not only with the physical world of corporeal objects, but also with the realm of intelligible forms, to which cosmology can attest only indirectly. To think of the universe is thus to explicate the sense of the universe on existential grounds, where our understanding of the adjective "existential" follows from the sense that was asserted by existentialists in the twentieth century, namely, that human life and existence is the primary and unquestionable metaphysical fact from which the whole reality is unfolded.[7] And this, as we have mentioned above, contributes to the perennial issue of how to think of humanity. Thinking of the universe in existential categories thus implies the extended vision and perception of the universe, which, in the words of a seventh-century Byzantine, Saint Maximus the Confessor, is the *makro-anthropos*, that is, that which was created in order to be humanized.

To think of the universe on the grounds of existential communion entails freedom of such thinking. It does not necessarily imply the overthrowing of scientific authority in the questions of physical cosmology: it implies that cosmological theories and hypotheses can be interpreted not as propositions about outer realities but as

7. Our usage of the adjective "existential" must be carefully distinguished from the same adjective that is sometimes used by cosmologists in the context of the stated smallness and insignificance of humanity in the universe; see, for example, J. Primack and N. E. Abrams, *The View from the Centre of the Universe: Discovering Our Extraordinary Place in the Cosmos* (London: Fourth Estate, 2006), 273–78.

movements of the human heart and spirit that reflect a fundamental anxiety of existence. In this case the universe is perceived as a certain *whole*, whose *partial* phenomenality is explicated by science. This *whole* includes not only the physically fragmented or united cosmos, but it includes the infinity of human life (the infinity of relations of human beings to created existents) in the universe. Correspondingly, all accumulated forms of knowledge, established in history to this very date, are merely pieces and moments, temporary and provisional sketches of the immensely mysterious phenomenon of personal beings. The "nontechnological" thinking of the universe, even if it will not be able to reproduce the "whole of the universe" (which was, however, attempted in works of art and poetry) and hence will remain no more than a symbol rather than reality, can receive its justification in a deep hope, that through this thinking we learn something of ourselves that has never been present in our vision of all. Being an intentional thinking, thinking of the universe as a whole brings the one who thinks beyond any conditional objectification and positivity. In a way, thinking of the universe is transcending the limits of thought, which requires from the enquirer exceptional discipline, courage, and humility in face of the fact that the task will never be fulfilled, and that they are ready to learn of themselves something that could shatter the image of their own "I."[8]

By thinking of the universe as a whole, we attempt to explicate our intrinsically ambivalent existential situation, being a part of the universe, in the particularities of time and space, and at the same time being at "that" paradoxically central "nowhere" from which the wholeness of the universe is unfolded. Some cosmologists object to this by saying that in terms of time we are living in a very special era in the universe, that it is only now possible to detect the

8. Cf. Primack and Abrams, op. cit., 282; Ladrière, *Language and Belief*, 150.

universe's evolution, its origin in the Big Bang, etc.[9] The universe as described by specific cosmological theories is not contingent from the point of view of these models. However, from the point of view of the very possibility of such a description, that is, from the point of view of the contingent facticity of life of knowing persons, it is still contingent. The pole of "nowhere" remains intact simply because cosmology, which deals with the physical background for existence of embodied human persons (that is, its necessary conditions), is not able to shed light on the nature of the sufficient conditions of existence of intelligent observers and theoreticians of the universe.

It is this pole of "nowhere" in thinking of the universe that deprives this thinking of any essential historical goal, which could be placed at the service of any intellectual or social-political economy, if it is not related to the saving ideals of Christianity. Being engaged in thinking of the universe as a whole, we are immersed not so much in the present of the scientific discourse of the universe but in the present of thinking itself. And this present is dictated not only by the advance of contemporary physical theories of the universe but to a great extent by the advance of thinking *per se*, that is, its free philosophical mode, which is not subjected to the logic of the already known but follows that which Husserl called humanity's "infinite tasks."[10] Here it is appropriate to quote K. Jaspers, slightly rephrasing his text, that our historical consciousness of the universe, in spite of being a temporal phenomenon, is a "free-flying" consciousness without "any ground and original point accessible to knowledge, ultimately rooted in that source which is always and necessarily present in ourselves."[11] This type of thinking, flying away from mundane realities and technological delimiters, will reveal more

9. See, for example, L. M. Krauss, *A Universe from Nothing* (New York: Free Press, 2012), 118–19.
10. E. Husserl, *The Crisis of European Sciences and Transcendental Phenomenology* (Evanston, IL: Northwestern University Press, 1970), 279.
11. K. Jaspers, *Weltgeschichte der Philosophie. Einleitung* (München: R. Piper, 1982), 77.

deeply and clearly the fact of our, as Heidegger termed it, "planetary homelessness" (but still centrality), which pertains to the present intellectual, social, and political unpredictability of the human condition. One must, perhaps, amplify this point by using the term "cosmic homelessness," implying the lack of understanding of the human place in the whole universe. We are homeless because the universe is infinite, and in spite of some claims of our centrality in the universe, we still do not know our place in it, that is, we do not know scientifically the grounds of our facticity in it. What we know for sure, however, is that it is we who articulate the universe, so that, perhaps, as some claim, we are in the center of the universe, but the question of "where" this very center ultimately is, remains in the field of perennial *certitudes négatives*.

While Jaspers could say that the realization of "cosmic homelessness" (as the denial of historical consciousness) becomes "the metaphysical consciousness of being (*Sein*), which being constantly present, must become evident in true being (*Dasein*), as if eternally present,"[12] according to Heidegger, our "cosmic homelessness," that is, the inability to answer questions about our own essence, drops a shadow of doubt with regard to the being of the universe itself. (Our "cosmic homelessness" can be qualified as *nonbeing*.)[13] Then it is from this perspective of our own finitude, mortality, non-attunement to, and incommensurability with the universe that one must have

12. Ibid.
13. It is worth recalling that Kierkegaard expressed in a dramatic form his anxiety about the impossibility of describing one's position in being: "One sticks his finger in the ground in order to judge where one is. I stick my finger in existence—it feels like nothing. Where am I? What is the 'world'? What does this word mean? Who has duped me into the whole thing, and now leaves me standing there? Who am I? How did I come into the world; why was I not asked, why was I not informed of the rules and regulations. . . . How did I come to be involved in this great enterprise called actuality? Why should I be involved in it? Am I not free to decide? Am I to be forced to be part of it? Where is the manager, I would like to make a complaint!" S. Kierkegaard, *Repetition* and *Philosophical Crumbs*, trans. M. G. Piety (Oxford: Oxford University Press, 2009), 60.

the courage to think of the universe in order to assert ourselves. However, this assertion of ourselves has a particular spiritual importance only for those who still value the humanity of humans, the naturalness of nature, justice of the police, and other perennial values that crown man in the center of the world, for whom this world is given to fulfill the "infinite" task of finding its destiny in union with the Creator of the universe and the giver of life. Thinking of the universe leads one to thinking of God, and it is in this that thinking follows a hidden theological commitment.

It is not difficult to see that thinking of the universe as if we think of thinking itself at present allows one to establish certain liturgical connotations as articulations of the overall temporal span of the universe, its past, present, and future, in conscious acts that fight oblivion, which pertains to the eternal flux of being. When articulated, the universe is being remembered not only as its realized past. The question of active remembrance of the universe is the question of such an understanding of human life in which past, present, and future are not considered anymore as signs of an all-annihilating *Kronos*, but as able to be integrated through remembrance in the image of humanity living in tension between a thanksgiving for existence and a hope for its eternal sense.[14]

To study the universe, though, does not mean to establish a simple vision of the world on the grounds of mundane curiosity or personal needs. It rather forms a vision of that "selfhood" of the universe (as the *makro-anthropos*) which is truly important for one's existence and which brings to unconcealment the truth of human existence. When we speak of the "self" of the universe, we do not presume that it has hypostatic features but, allegorically speaking, humanity by looking

14. As was suggested elsewhere, the universe as its past, even if human beings know their meaning only precariously, can be *respected*, as certain ancestors of our being, so that this respect can establish a sense of communion with the universe that overcomes loneliness and despair (Primack and Abrams, *The View from the Centre of the Universe*, 291).

at the "face" of the universe sees this "face" as looking at themselves, and it is this all-penetrating "glance" of the *makro-anthropos* that forms the image of humanity as its ability to see the infinite in the finite. In a certain sense human beings, as they are sustained by this last-mentioned glance, want to respond to it, thus asserting not only their longing for commensurability with the universe, but also their infinitely transcending lordship over the universe, resisting their cosmographic insignificance and fear of being crushed under the weight of astronomical facts. Pascal, for instance, compared humans to reeds, thinking reeds, in the universe, the weakest but thinking element in the chain of being, so that a drop of water can kill a person; the universe does not need to arm itself in order to crush anyone. "But even if the universe should crush him, man still would be more noble than that which kills him, since he knows he is mortal, and knows that the universe is more powerful than he is: but the universe itself knows nothing of it. All our dignity, then, consists in thought. It is through thought alone that we have to lift ourselves up, and not through space or time which we cannot fill."[15]

The freedom in thinking of the universe, however, has its delimiters; this freedom does not imply an arbitrary rule in thinking, first of all its spiritual arbitrariness. When we brought the reader to the thought that thinking of the universe is accompanied by thinking of God, we were conscious that there always was a danger of a "divinization" of the universe. This does not mean a naïve and outdated pagan perception of the cosmos as a living organism or the place where gods corresponding to different astronomical objects are abiding. It is a much more refined form of spirituality that is implied here, rooted in the sense of immanence of the universe, its infinity as

15. B. Pascal, *Pensées*, 199 (cf. 113), trans. Louis Lafuma (Paris: Éditions du Seuil, 1962), 103 (cf. 67). This ET: Pascal, *Pensées*. Selections, trans. and ed. Martin Jarret-Kerr (London: SCM, 1959), 78 (cf. 39).

an actually existent mystery that does not need any reference to the personal transworldly ground of the world. Cosmology in this case becomes a spiritual exercise since it is based in the life of the spirit; however, the demarcation between such a "spiritual cosmology" and theological commitment arises at that point when human beings make a distinction between the universe as a necessary condition of their existence (that is, an immanent medium of their inhabitation) on the one hand, and God, as an underlying transcendent *sufficient* ground of the very possibility of life and vision of the universe on the other hand.

Speaking theologically, there is here a difference of a *soteriological* order, so that to avoid arbitrariness in thinking of the universe means to follow a theological commitment referred to the salvific sense of the universe. A difference of a soteriological kind was pointed out by V. Lossky when he commented on the place of cosmology in the writings of the Fathers of the Church and, in particular, in the case where cosmology loses the sense of the centrality of humanity in the perspective of salvation, for example: ". . . Copernican cosmology, from a psychological or rather spiritual point of view, corresponds to a state of religious dispersion or off-centredness, a relaxation of the soteriological attitude, such as found in the gnostics or the occult religions."[16] An example of such a dispersion and relaxation of the soteriological attitude can be found in modern "spirituality without God," according to which the immanence with the world goes together with no belief at all (for there is nothing to believe in since everything is already here and now) and despair (as there is no hope for anything since everything has already happened), which correspond to the idea that a human being is already there, in that reality which theology names "the age to come."[17] Thus if the state

16. V. N. Lossky, *The Mystical Theology of the Eastern Church* (London: James Clarke, 1957), 195.

of affairs is such as it is just described, the question of salvation as a personal spiritual endeavor, as an intensive anthropological transformation (*metanoia*) may be abandoned as irrelevant. One has everything, which is given in its existential concreteness, and all this represents an unsolvable mystery with which we have to live and die. In a certain sense the immanent and infinite universe is treated as that realized "kingdom" of being in which everything is given and one does not need to enquire in the facticity of this givenness. It is at this point that the theological commitment, in contradistinction to the spirituality without God, aspires and breaks toward the transcendent, enquiring into the origins of being in the perspective of the human life and the sense of its coming into existence. Theological commitment reveals itself as a concern with the *sufficient* conditions of the human existence, implying that life is not only a gift of existence, but a gift of relationship and communion with the eternal.

Thus the delimiters in free-thinking of the universe proceed in the long run from the freedom of human beings made in the image of God: all thoughts and articulations of the universe always contain in themselves traces of the divine image. Even when cosmology proves the insignificance of humanity in the universe, the divine image remains, exactly because the human mind always resists all attempts to circumscribe its life in rubrics of the natural, finite, and transient. Human beings aspire to understand the underlying sense of beings and things, not according to their "nature" (which is unfolded in the sciences) but according to the final causes of these beings and things in relation to the place and goals of humanity in creation.[18]

17. See A. Comte-Sponville, *L'esprit de l'athéisme. Introduction à une spiritualité sans Dieu* (Paris: Albin Michel, 2006).
18. Humanity, first of all, is not satisfied by that vision of its own place in the universe which positions it in the same way as "marble is in the bag" or a "cat is in the house," or "a teacher is in the classroom"; ". . . it is at this point that a kind of rebellion takes place: the full reality

This understanding cannot be explicated only through physics and biology. It is based in views on humanity as the crown of creation made in the image of God. And this is the reason why in a godlike fashion humanity wants to recognize all sorts of beings (either simple physical objects or living organisms) not according to their nature, that is, according to their compelling givenness, but as results of humanity's *free will*.[19] The image of eternity is retained in any cosmological theory created through the free-willing even if this theory predicts the finitude of all actual forms of existence and life. Free-thinking of the universe is thinking of freedom of the incarnate human person, brought into being in the Divine image by the will of the Holy Spirit.[20]

One can briefly summarize the objective of this book as the unfolding of theological motives in humanity's perception of existence in the universe, which, on the one hand, outlines human beings as its slaves, constantly "crushed by the ever-increasing mass of the astronomical facts," and which, on the other hand, manifests the sense of human life in the universe by elevating it beyond the world order through a belief, hope, and love in the perspective of eternity.

of the individual is surely not exhausted by statistics, and the identity of the person demands an appreciation of his situation in the world distinct from one's situation in the world." M. Natanson, "Being-in-Reality," *Philosophy and Phenomenological Research* 20, no. 2 (1959): 233.

19. The analogy comes from St. Maximus the Confessor's discussion on whether God knows created things according to their nature. His answer is negative: God knows things according to his will: ". . . when Christians were asked by some outsiders puffed up with their learning, how they can claim God knows existent things . . . and that he knows intellectual being intellectually and sensible things sensibly, they replied that he neither knows sensible things sensibly nor intellectual things intellectually. For it is out of the question that the one who is beyond existent things should know things in the manner proper to beings. But we say that God knows existent things as the products of his own acts of *will* . . ." *Ambigua*, 7, PG vol. 91, 1085B. ET: P. M. Blowers and R. L. Wilken, *On the Cosmic Mystery of Jesus Christ: Selected Writings from St. Maximus the Confessor* (Crestwood, NY: St. Vladimir's Seminary Press, 2003), 61–62. Emphasis added.

20. "Authentic theology consists not in the conjectures of man's reason or the results of critical research but in a statement of the life into which man has been introduced by the action of the Holy Spirit." Archimandrite Soprony (Sakharov), *St. Silouan the Athonite* (Moscow: St. Trinity and Sergius Lavra, 1999), 171 (in Russian).

This dichotomy between the infinitely small, finite physical existence and the feeling of the light of eternal life was felt by the Fathers of the Church and the great mystical philosophers, as their personal vision of the darkness of hell and the light of the Spirit—to both of which human beings are constantly turned and in the presence of which they must not only continue their life, fighting cosmic homelessness and despair, but also fight to find the sense of themselves and all creation. Theological commitment in cosmology is thus a characteristic expression of the visible and invisible universe as it appears to man in the perspective of communion, that is, through the eyes and senses enlightened by the Divine presence. Numerous books on cosmology discuss the role and place of humanity *in* the universe. This book brings the universe inside humanity, making the universe that mirror of its soul which humanity desperately wants to find.

"Theological Commitment" as a Different Form of the Dialogue between Theology and Science

There is an element of socio-historical reality that sheds light on the reasons behind the proposed enquiry into the theological commitment in cosmology. First of all, cosmology always (in particular before its twentieth-century developments) was a part of theism. Cosmological arguments for the finitude or infinitude of the world in space and time were employed as different arguments for the existence or nonexistence of God. Theistic inferences are still alive and very popular among some philosophers and cosmologists who attempt to use cosmology in both apologetic and atheistic conclusions.[21] However, this dimension of the debate is not our

21. The literature on this topic is vast. See a concise and eloquent review of recent discussions in H. Halvorson and H. Kragh, "Theism and Physical Cosmology," in *The Routledge Companion to Theism*, ed. S. Goetz et al. (London: Routledge, 2012), 241–55.

primary concern, because the alternative of existence or nonexistence of God is not an option for this research, which takes an explicitly theistic stance (that is, theological commitment) on the grounds related to the facticity of human persons who are the subjects of cosmological knowledge. Correspondingly, we do not analyze cosmology from the perspective of an explicitly theistic stance based on some dogmatic propositions of God's existence; rather we proceed cautiously from what we call theological commitment as an existential, experiential mode of communion with God.

Secondly, the topic of research is related to the dialogue between theology and science in general that became a matter of scholarly discussions in the last twenty to thirty years. The question is: Has this dialogue, in the form it has been conducted, succeeded so far, that is, has it achieved any results that have had impact on both science and theology? The author believes that a negative answer is provided by the unceasing scientific and technological advance (in particular in the exact natural sciences), which continues with no recourse to the dialogue between theology and science whatsoever. All discussions on whether science and theology are in conflict, or in "peaceful coexistence" with each other, do not have existential implications: the problem remains, and its ongoing presence points to something that is basic and unavoidable in the very human condition. This net result indicates that the method of conducting this dialogue at present is unsatisfactory in the sense that it does not address the major question as to what is the underlying foundation in the very *distinction, difference, and division* between science and religion as those modes of activity and knowledge that flourish from one and the same human subjectivity. But this type of questioning makes any scientific insight irrelevant simply because science is not capable of dealing with the question of its own facticity, that is, the facticity of that consciousness which is the "pillar and ground" of science.

Theology can respond to this question from within the explicitly belief-based ground, namely faith, in that the knowledge of the world represents natural revelation accessible to humanity because of the God-given faculties. Knowledge is possible only by human persons whose basic qualities are freedom and capacity to retain transcendence with respect to all they assimilate through life and knowledge. In this sense the universe as articulated reality has existence and sense only in a mode of personhood, which is a divine gift. Since science does not account for the very possibility of knowledge, that is, personhood, it is automatically prevented from participation in the dialogue with theology on equal footing. It is logical then to express doubt on the meaning and value of such an existing "dialogue" with science at all. If one insists on this "dialogue," it becomes obvious that science and theology cannot enter this dialogue as symmetric terms. And if there is no impact of this "dialogue" on the logic and development of science, what remains for theology is to exercise an introspection upon science, to conduct a certain *critique* of science from a position that is, by definition, above and beyond not only scientific thinking, but secular thinking in general related to particular socio-historical and economic realities. Thus symmetry between theology and science, theology and cosmology in particular, is broken from the very inception. It is this asymmetry that constitutes the approach to science-religion discussions that we describe in terms of *theological commitment*. Theological commitment is such a stance on human being that always positions it above and beyond those realities disclosed by science alone. It appeals to those meanings of existence that do not compel the recognition of the science in the manner that natural phenomena do. These meanings originate in an innate quality of human beings to long for immortality, that is, communion with the unconditional personal ground of the whole world, which humanity names God. And it is through this longing that the universe acquires a certain sense as that constituent of God's creation which makes it possible for human persons to fulfill God's promise for eternal life and communion. Theological commitment is thus existential commitment.[22]

Thirdly, theological commitment is the reaction to modern atheism. Indeed, in its goals and tasks the dialogue between Christianity and science is to oppose atheism. However, if one carefully looks at how this dialogue has been conducted so far, one easily realizes that the existing forms of this dialogue are adapted to that which is imposed by atheism. Such a dialogue turns out to be no more than a reaction to atheism, sometimes attempting to unconvincingly justify the very fact of this reaction. Contemporary atheism manifests itself not only as freedom from historical authorities and tradition (that is, liberation from freedom in a Christian sense) and not only as the unprincipled following of the proclamation "enjoy life, for there is no God," that is, not only as the worst form of the unenlightened slavery of Plato's cave in which signs of the Divine presence are not recognized and the very ability to see them in the world is reduced to nothing. Atheism promotes a cult of immanence, the actually existent infinity of the given,[23] appealing *de facto* to deprivation of the senses and the vision of the transcendent (and hence to the relaxation of a soteriological moment). Since modern science, and technology in particular, encourage individuals to be transcendent-blind, creating the immanent images of the transcendent, the advocates of atheism appeal to science. By so doing atheism adjusts to the demands and moods of modern time. It is much easier not to deny the presence of the Divine in the world, but to

22. On an Orthodox Christian appropriation of existentialism see L. Puhalo, *Freedom to Believe: Personhood and Freedom on Orthodox Christian Ontology* (Dewdney, BC: Synaxis, 2001), 48–59.
23. See a more elaborate formulation of a mysticism of immanence, for example, in Comte-Sponville, *L'esprit de l'athéisme*, 145–212.

claim that all spheres of human activity are self-sufficient and do not need any reference to God. Since from a philosophical point of view the question of God's existence or nonexistence cannot be decided at all (the philosophical mind remains in the "negative certitude" with respect to this question), then why should one try to answer it. Would it not be easier to recognize that science, art, literature, and so on are just given in rubrics of that which is *unconcealed* to humanity? Here atheism reveals itself as secularism, as a kind of trans-ideological *läicité*, as a servility to nobody's interests, and as a servility to the alleged ideal of humanity understood only empirically, as that humanity which is alive here and now.[24] (It is supposed that this ideal of humanity has in itself a universal criterion of its own definition.) To define this humanity in simple categories that overcome racial, national, and class differences one needs a universal language. It is science which pretends to be such a language; to be more precise, it is that scientific form of thinking which reduces the phenomenon of humanity in all its various manifestations to the physical and biological. It is clear from here that modern atheism as a certain form of "immanent humanism" is no more than a scientific atheism. However, this atheism positions itself as more aggressive[25] and sinister, more advanced philosophically and anti-theologically[26] than was the case in the Soviet Russia. The reason for this is that modern atheism is ultimately motivated by the logic of material

24. As was argued by G. Goutner, the alleged ideal of humanity, understood for example as its unity, simply does not exist. One can think of it only in a modality of hope that has a religious nature. See G. Goutner, "The Unity of Humanity in an Eschatological Perspective," in *Theology of Creation*, ed. A. Bodrov and M. Tolstoluzhenko (Moscow: St. Andrew's Biblical and Theological Institute, 2013), 230-36 (in Russian); as well as his "The Idea of Humanity. Epistemological and Ethical Aspects," in *Methodology of Science and Anthropology* (Moscow: Institute of Philosophy of the Russian Academy of Science, 2012), 170-92 (in Russian).

25. See examples of this in, for example, R. Dawkins, *The God Delusion* (London: Black Swan, 2007), and V. J. Stenger, *God the Failed Hypothesis: How Science Shows That God Does Not Exist* (New York: Prometheus, 2008).

26. See, for example, Comte-Sponville, *L'esprit de l'athéisme.*

production and human resources, that is, by the needs of the developing economies and not an abstract ideology.[27]

The freedom from traditional and philosophical authorities as well as historical values inverts in modern atheism toward slavery to the scientifically articulated and verified. It is paradoxical, and fundamentally different from the Soviet model of atheism, that a slogan that "knowledge is power" is not appreciated in the economically advanced societies, for all-encompassing knowledge, that is, knowing too much, is potentially socially dangerous. This entails in turn that knowledge and science both function in society in a reduced and popular form that does not allow one to judge of its certitude, quality, and completeness. Scientific knowledge becomes a world-outlook, an ideology and a filter of social loyalty and adequacy. As a result, the abuse of science becomes a norm that creates an illusion of its efficiency and truth in all spheres of life. The scientific method is treated as self-sufficient and not being in need of any justification and evaluation. Science proclaims the truth of the world from its own rationality, which functions in the disincarnate collective consciousness. Supported through a system of grants from economically powerful groups, it is allegedly done for the sake of human good. However, by functioning in society science forgets the simple truth that science is a human creation and its initial meaning was to guard the interests of people and not to make them slaves and hostages of the scientific method.

The situation with the dominance of the scientific approach to all aspects of life becomes even more paradoxical when one realizes that human beings do not become happier and freer from the aspects of material existence. They cannot escape social injustice, the hardship

27. This point was emphatically defended by C. Yannaras in his article "The Church in Post-Communist Europe," in *The Meaning of Reality: Essays on Existence and Communion, Eros and History* (Los Angeles: Sebastian Press & Indiktos, 2011), 123–43.

of mundane life, diseases, and moral losses. This happens because science as an ideology does not spell out what is most important, namely that it *does not know the goals and ways of its future development.* In its grandeur science has to intentionally disregard those aspects of reality that are not described by it or that behave sporadically and unpredictably with respect to scientific prognosis. Economic growth and welfare of developed nations that are used to living in comfortable conditions, the cult of consumption and greed, demand more technological development related to the exploitation of natural resources. Every new discovery in physics is employed for the optimization of the production of goods and energy, so that one can speak about merciless exploitation of physical reality in general. It is very seldom that the question of the legitimacy and justification of such an exploitation, or, as some say, "the rape of nature,"[28] is even thought of. By making nature an object of manipulation, scientific consciousness forgets its humanitarian duties in respect to nature: nature must be "respected" simply because we live in it and because there is the light of that all-embracing reason (Logos) which we, human beings, carry in ourselves as little *logoi*.[29] The objects of nature are inseparable from their creator, and the oblivion of this fact leads to the loss of love of them in the same sense as the loss of love for other people. A careful insight of a philosopher or a theologian will unmistakenly identify the root of the problem, namely that the atomization and disassemblement of physical reality in the course of its exploitation has its origin in the ethical individualism of those who know this reality, that is, the loss of love for nature in the scientific

28. This was the title of P. Sherrard's book *The Rape of Man and Nature: An Enquiry into the Origins and Consequences of Modern Science* (Ipswich, UK: Golgonooza, 1991), where he aggressively criticized modern science for exaggerating the sphere of applicability of its methods and the resulting dehumanization of humanity and desanctification of nature.
29. See an antology of papers on the ecological approach within Eastern Christianity: J. Chryssavgis and B. Foltz, eds. *Toward an Ecology of Transfiguration: Orthodox Christian Perspectives on Environment, Nature, and Creation.* (New York: Fordham University Press, 2013).

community. Individualism consists in the fact that the exploration and acquisition of physical reality becomes an affair of that human spirit which is divided in its narrow professional and corporative interests, in which the element of catholicity with nature through the divine-given existence is forgotten and love does not rule anymore for the interest of knowledge and longing for the perpetual good.

The ambitions of immanent secular reason, supported by scientific achievements, seem even stranger if one realizes that modern science, in spite of its successes, manifests symptoms of a deep crisis related to the uncertainty of its goals. Scientific activity is purposive to the extent that accompanies any human activity. Any particular research has a concrete objective either to satisfy a practical interest or simply curiosity. However, when we speak of the uncertainty of goals of science in general, we mean something different: the scientific quest is spontaneous and unrelated to the spiritual, infinite tasks of humanity. The practical purposiveness of scientific research thus unfolds only a particular sector of nature, so that there remains a gap between that which has been known through scientific phenomenalization and that which cannot be known by science at all. This fact manifests that nature has a propensity to remain *concealed* and react with respect to human experiments unpredictably. As an example, one can point to nuclear physics, which by acquiring the mysteries of the microworld risks creating a state of matter that can threaten human existence on this planet.[30] There is a danger in nuclear experiments of trespassing the boundary of the unconcealed, related to human existence, when constructed devices and artificial states of matter may behave in a nonhuman way, contradicting the

30. For futurological accounts based on the threats originating in modern science, see J. Leslie, *The End of the World: The Science and Ethics of Human Extinction* (New York: Routledge, 1996), and M. Rees, *Our Final Century, a Scientist's Warning: How Terror, Error, and Environmental Disaster Threaten Humankind's Future in This Century—On Earth and Beyond* (London: Heinemann, 2003).

initial objectives of experiments and turning science against humanity. A simple example from philosophical discussions of the 1950s is the atomic bomb, which brought humanity to a new situation in which the conditions of its existence are not controlled anymore only by the natural processes, but depend on the good will of people making decisions to use or not to use nuclear weapons, thus influencing global natural processes.[31] Another example is the ecological crisis. The melting polar icecap of Greenland, the extinction of some animal species, and the forthcoming migration of peoples living in the Arctic region show that technological applications of science apart from *moral reason* lead to problems of the social and political order. Science through technology is not neutral anymore to economics and politics and, on the contrary, becomes their result and prophet. The process of exploration and knowledge of the surrounding world, and thus its "transformation," becomes involved in the sphere of interests of the world's powers and classes so that its ethical significance is determined by its belonging to this or that social-economic demand. That which has been said entails that scientific knowledge and the very idea that society can and must develop only on the basis of scientific progress becomes an ideological dogma, the following and defending of which in turn becomes a matter of social loyalty. However, without understanding its logic and definite goals, scientific progress, being *de facto* unavoidable and irreversible, carries within itself a potential danger because of the unpredictable nature of it applications. Human beings want to live better and longer; however, this natural desire does not supply a clear understanding of the goals of science, whereas

31. N. A. Berdyaev prophetically argued in the 1930s that humanity is entering a new era in which the stability of the world will depend on moral decisions of humanity as to how to use the technology available through scientific advance. See his paper "Man and Machine," *Issues in Philosophy* (*Voprosy Filosofii*) 8 (1991): 147-62 (in Russian).

humanity is becoming more and more dependent on its achievements and applications.

The fact that scientific advance leaves huge realms of being unexplored and unknown becomes even more evident in theoretical sciences, in particular in cosmology. On the one hand cosmology provides us with a comprehensive theory of the universe supported by observations. On the other hand it has to admit that those forms of matter in the universe which are physically understood constitute only 4 percent of its material content (the remaining 96 percent associated with the so-called dark mass and dark energy remain as of yet beyond the reach of experiments; their existence is a matter of theoretical conviction). The more that cosmology refines its scenario of the universe's evolution, the more it realizes the abyss of the physically unknown. Speaking philosophically, cosmology makes clearly seen the boundaries of the *unconcealed* that is related to humanity: it is only 4 percent of matter in the universe that can be said to be consubstantial to human physical and biological form. Amazingly, however, in spite of all evidence for the limited nature of our knowledge of the universe, cosmologists sometimes position themselves as "prophets and priests" of the universe, preaching of it as if they know the absolute truth of the world. As we demonstrate in this book, such a conviction with respect to knowledge of the universe originates in a naïve representation of the universe as a whole as an "object" whose phenomenality can be exhausted through the logic of scientific signifiers.

One of the major attributes of modern science that makes it powerful is its radical mathematization of nature. Physics and cosmology, through mathematical models and theories, predicate realities inaccessible in direct experiments. There is a paradoxical shift of representations of reality here: unobservable intelligible entities are treated as more fundamental and responsible for the contingent display of visible nature. As we argued elsewhere, mathematization of nature is accompanied by the diminution of humanity, in particular the personal dimension of existence.[32] Person disappears from scientific discourse in spite of the fact that all articulated facts are made by persons. Science is being effected in the name of human persons, but this same person turns out to be outside of scientific description. Persons are needed for the anonymous objectives of science to disclose reality, but they do not exist for science as agencies of other nonscientific truths and individual lives. Science as a social process needs scientific workers but not persons as unique and unrepeatable events of disclosure of the universe. The same is true with respect to society, which needs not persons but masses of individuals that are much more easily adapted to the norms of materialistic thinking and behaviorist stereotypes based on the consumeristic results of technological progress. Modern atheism exploits this aspect of modern science by insisting on the effective nonexistence of personhood as a philosophical and theological notion. The oblivion of the person is treated by Christian theology as an encroachment on the absolute priority of the human world and those communal links in human societies that have formed the spirit of Christian civilization and the integrity of its historical paths through communion with God. The oblivion of the person is the encroachment on the significance of its history impressed in the architectural image of European cities, masterpieces of art and literature, in the very way of European thinking and its values. The oblivion of the person constitutes an attack on all traditional forms of societies and life, which by the logic of the economical must cease to exist or become unobservable.

32. See A. Nesteruk, *The Universe as Communion* (London: T. & T. Clark, 2008), 188-205.

To defend the person and to reinstate it to its central status in the dialogue between theology and cosmology becomes a leading motive of theological commitment. To reinstate the person means to understand that the problem of theology and science manifests the basic distinction and division of two attitudes to life in one and the same human person. The dialogue between theology and science becomes the explication of the split between intentionalities that the human spirit attempts to reconcile. This, by using the language of Husserl, forms one of the infinite tasks of the human spirit to understand the meaning of existence. The very fact that this dialogue exists attests that human beings transcend the conditions of their physical-biological existence, the self-realization of a special place in the universe in which the function of the Divine image in man is realized.[33] Thus the fact of the dialogue attests also to the fact that it contains the elements of transcendence and asymmetry between theology and science related to the human condition which is called personhood. It is this asymmetry, articulated in reflection, that we call theological commitment, by confirming once again that this is an existential commitment. Correspondingly, it seems doubtful that the dialogue between Christian theology and cosmology is possible without faith that both theology and cosmology represent modalities of the relationship between humanity and the Divine. Thus the dialogue ultimately contributes to growth of faith in God, to that infinite task which aims to restore the salvific Divine image in man.

Theological Commitment as Knowledge in Love

The defense of person in scientific discourse, particularly in cosmology, implies that the thinking of the universe must have

33. N. Berdyaev, *Slavery and Freedom* (London: Centenary, 1944), 94.

exclusively human features. In other words, the assertion of persons through theological commitment must be ontological and not abstract academic *ad extra*. Theology knows only one way of such an assertion, namely through the events of communion. What makes communion uniquely human is the acquisition and advancement of humanity through love. And it is this love that changes the ways of approaching knowledge, making this knowledge the expression of love to God who created that universe which can be known. This proclamation proceeds from the church because the church, in contradistinction to any scientific ideology and metaphysical convictions, asserts the gospel's message of Christ's "trampling death by death" as the possibility of transforming the process of our physical dying into the event of communion of our person with the person of God. By this, humanity is provided with the opportunity to be transfigured from the state of biological existence to life in freedom from any delimiters to love, including love in knowledge. Christ's coming into the world confirmed that the Logos is indeed truly existent not as some given and inarticulate necessity that governs the universe, but as love, according to whose *logos* the world was created. Christ's coming into the world is the unconcealment of that modality of the Divine which is called *love*. But this love is the search of understanding of that whom it loves. Here one sees the link between love and truth, for love to Christ and to a neighbor in Christ can withstand the trial by the world only if this is love of truth. Then one can propose that the dialogue between Christianity and science has sense only if a theological truth based in love of God and the world, created by him, enters a dialogue with love of truth in science, which assumes love of the world and of humanity which knows it.

As a historical reference one can point toward St. Augustine, who formulated a thesis that love (understood as charity, that is, *agape* [Gr.] or *charitas* [Lat.]) is the condition that what is being known is true: "The only way to truth is by love."[34] In other words, truth is rooted in love as its epistemological condition, not because truth cannot be fully disclosed without love, but because it is love itself that is the ultimate and only foundation for the possibility of seeing and grasping truth. One must love truth in order to conceive it. In his *Confessions*, with a reference to Christ (John 8:4), Augustine gives an example of the contrary, that is, that the encounter with truth can lead to hatred, for truth, as a "saturated phenomenon," discloses and shows ourselves to ourselves. Truth forms us differently depending on whether we love it or not. And it is when there is a lack of spiritual power in us in order to overcome hatred with respect to that which this truth discloses as negative in us, we receive retribution: we do not want to be disclosed by truth; however, it will manifest us despite our will, but it will not manifest itself to us.[35] This is why, only in love, when the truth of all "being wrong" in us is manifested and is accepted with love so that hatred is being overcome, will it be possible for this truth to be manifest to us. Love of truth or wisdom assumes that the desire for experience of transient things has been overcome so that knowledge becomes edifying and not arrogant.[36] Pascal expressed differently a similar thought: "With respect to human things it is said that it is necessary to know them in order to love, . . . the saints, on the contrary, say, of divine things, that they must be loved in order to be known and that truth (*vérité*) is manifested only through love (*charité*)."[37] Love, however, is not that which is simply commensurable with experience of mundane reality. It demands that one overcomes the sense of despair and futility of

34. St. Augustine, *Contra Faustum*, Bk. 32, 18. ET: NPNF, Ser. 1, vol. 4 (Grand Rapids: Eerdmans, 1994), 581.
35. St. Augustine, *Confessions*, Bk. 10, XXIII, 34.
36. St. Augustine, *On the Trinity*, Bk. 12, XI, 16.
37. B. Pascal, "De l'art de persuader," en *L'esprit géométrique* et *De l'art de persuader* (Paris: Bordas, 2002), 29.

human existence, which was articulated by Pascal in his *Pensées*, and acquires love of God: "what a great distance between knowledge of God and love of him."[38]

The presence of love in knowledge of the universe, which is the subject matter of this book, means a very simple thing: even if cosmology persuasively convinces us of our own insignificance and the transience of our position in the universe, up to such an extent that one can start hating life on the grounds of its absurdity and ontological homelessness, the love for the truth of life itself provides us with a powerful capacity to see in humanity the truth of its origin proceeding from the truth of God himself.

Love as a phenomenon that is unique would be inaccessible without Christ: "God is Love" (1 John 4:16; cf. Luke 10:27). Thus those who are living in God are those who love him, that is, those who love each other (1 John 4:20). It is only this love that can guarantee an access to the "great reason," for it is love given to us in the revelation of the Word, the Logos, which reveals itself as *logos*, that is, as rationality, which makes it possible to approach the phenomena that are closer and more intrinsic to us, those that are experienced by the spirit-bearing human flesh and exceed the capacity of discursive thinking. Christ's revelation has shown that love has a reason that sees and makes articulate those things that the collective and anonymous reason leaves outside of its scope. Christ has demonstrated the *logos* of love by his passion and through his resurrection; he has disclosed it and accomplished it to "the end" (John 13:1). In Christ, truth reveals itself as the ultimate and initial truth, the truth according to which all other truths are possible and that will consummate all truth at the end of times (John 14:6). In a cosmological context, where one aspires to knowledge of the

38. B. Pascal, *Pensées*, 377 (trans. Lafuma, 161).

universe as a whole, which is physically incommensurable with human beings, one needs archetypically to have the mind of Christ-Logos, by whom and through whom the universe was created and for whom, after the incarnation in flesh, the whole universe was intelligibly given in its fullness as an instant of the Divine love, and who remains the Lord of the worlds: "In his right hand he held seven stars; . . . his face shone like the sun in full strength" (Rev. 1:16).

Christian love assigns to knowledge available through reason the status of true knowledge. For ". . . there is nothing love cannot face; there is no limit to its face, its hope, its endurance" (1 Cor. 13:7). Love loves unconditionally, in particular without a condition of mutuality: love does not need anything in response to its gift. In this sense the very creation of the universe manifests a perfect love as the unconditional primacy of love with respect to being: God created the universe out of his love and he does not expect any acknowledgment of this from creatures, for God is above and beyond any sentiments originating from that which he has created. If one refuses to accept the creation of life and the whole universe as a gift of love by declining the invitation to participation in this love, this very love does not suffer from this, remaining still a perfect love. Correspondingly, studying the universe as the created represents a hidden knowing of the divine love, and not only *that which* has been created. To know the universe as a whole means to know God's love of all creation, including oneself.

Christian love justifies the very possibility of true knowledge, for to the one who loves, that is, believes in God, everything is possible, including knowledge of truth. The loving human being has the same privilege that God has (Mark 10:27): "Everything is possible to one who believes!" (Mark 9:23). Christian love justifies the very possibility of knowledge, for it is this love that allows human beings to become aware of their centrality in creation as bearers of the divine image, the image of that one to whom everything is possible. However, as was said by St. Paul, true knowledge is not knowledge achieved through one's own effort (1 Cor. 8:2). To know truly one should be known by God, and for this, one must love God (1 Cor. 8:3). To know oneself truly and to know the universe truly means to have knowledge not by my own thought, but by the thought of the one who discloses himself only to those who love him. Knowledge of the universe as self-knowledge in order to be true requires the acquisition, as its archetype, of the "mind of God" through love for him. But this is not what is given to humanity in its natural propensity. To know through love and to know through God's "mind and eyes" requires one to exercise one's will. Christian love, as contributing to the theological commitment, thus means the ability to approach things and to see in them the divine presence contrary to their compelling empirical evidence, that is, to see that presence which can only be manifested to humanity as a carrier of the free will and love. To know things through love and will means not their passive contemplation, but active participation in them, a sort of communion with them which allows one to get access to an otherwise inconceivable transcendence of other creatures and essences. Here love takes responsibility for that which is affirmed in philosophy and science because it Christologically reinstates the definition of philosophy, and hence the sciences, to their proper sense as "love of wisdom" or "love of truth." Without love, human reason is limited in interpreting the world, by transforming it into objects of possession and hence dealing only with their deficient phenomenality, thus abandoning an ontological question on the facticity of their existence. Christian love confesses a hope that the time will come when the "great reason of love" will respect the objectivized world as the other, which is worthy of love, that is, it will see the world by the eyes of the Logos through whom and by whom all is. To respect and love the world means to love and accept every personal life as that potential center of disclosure and manifestation of the world through the gift of the divine image. Correspondingly, no scientific theory will have any existential relevance if it does not take into account the uniqueness and absolute value of human life. The rationality of secular consciousness here must be complemented and overpassed by the *logos* of Christian love.

Theological Commitment as Conciliar and Ecclesial Knowledge

Some aspects of modern science, on being critically assessed, reveal the presence of contradictions arising from its pretension to independence (autonomy) and freedom of research. The idea of freedom to explore turns against humanity because it becomes the freedom for futility to be imprisoned in the cycle of materiality, or as one ancient writer put the comparison, humans dealing with the enclosures of the material world act as "animals turning the mill."[39] The issue is not whether scientific progress brings its fruits for simplifying and varying human life; the question is about the intrinsic *telos*, which seems to be absent from introspection by science. When Heidegger qualified science as not thinking, he partially meant that science does not understand its own goal. It is functioning in the world as an autonomous human activity, but its ultimate sense is obscured. Humanity, exercising its freedom of exploiting and subjugating nature to its utilitarian needs, thus forgetting about the sacramental and thanksgiving attitude to nature, becomes a tragic hostage of this freedom to "explore." The alleged freedom of exploration of nature as being devoid of theological ground and reflection leads to its own self-negation by reducing humanity to

39. Gregory of Nyssa, *Funeral Oration on Placilla*, PG 46, 888D–889A.

mere hermeneutics of biological survival and depriving it of its dignity and theo-centeredness. The freedom of persons from spiritual authority eliminates personhood as the issue, reducing humanity to no more than a futile consubstantiality of the material and thus disintegrating community into dispersed biological applications endowed with the function of indefinite consumption. Contrary to this, Christianity affirms freedom by placing human beings in the center of all questioning and articulations of the world, by implying that human dignity does not simply follow the impersonal scientific view of humanity as a thing among other things, but refers to that invisible origin by whose will humanity strives to fulfill its destiny.

The theological commitment in the dialogue with science sees the split between faith and reason, or the split of intentionalities in one and the same human subjectivity, as the loss of perception of centrality of human person in the dialogue: therefore the dialogue with science has to naturally follow theology, which "has to do with existential needs of the human person."[40] While reasserting personhood as the uniting mode of articulation and exploration of the world and God, one must not be naïve in order to understand how different this personal knowledge of the world and God can be in physical or social reality. Personhood, or personal mode of existence, implies not only radical *otherness* with respect to the substance of the world, but also *communion* with this substance, which, in particular,

40. J. Zizioulas, "The Contribution of Western Orthodox Theology," *THE MESSENGER, Journal of the Episcopal Vicariate of Great Britain and Ireland* 6 (May 2008): 42–43. In accord with this, Fr. D. Staniloae writes: "Man today is not content to be just a consumer of the products and distractions provided by technology; he demands to be a man of ever closer relationship with his fellow men, and consequently a man who, in a manner much more acute now than before, lives out his obligation to find those ways and means which will assure that these relationships do not become painful and inimical, but instead remain friendly and responsible. These ways and means can only be discovered however by experience, by coming into contact with those higher realities that man thirsts for in order to escape the deadly monotony of purely material distractions." *Theology and the Church* (Crestwood, NY: St. Vladimir's Seminary Press, 1980), 217.

is realized in human communities as the living chain of interpersonal relationships. Communion here incorporates much more than a simple sharing of living space and exercising love, empathy, and compassion with respect to fellow-neighbors. It implies an element of corporate responsibility for persons as community in a particular environment that is conditioned not only by simple social norms, economic interests of a particular group, or purely by the instinct of survival. Communion of persons assumes a certain reference to the source of their existential otherness in the Divine.[41] It is through this reference that the collective responsibility is linked to the concept of *wisdom*, as distinct from that of knowledge. It is often implied that scientific advance takes place in a stream of wisdom: science makes human life better and longer; it gives knowledge and conquers ignorance. But this scientific wisdom does not address the issue of existence: in its success in answering the question "What is the universe?" it does not answer the question "Why is the universe?" Scientific wisdom operates in the limits of the pre-given, which is already accessible to the discursive mind. In a way, scientific wisdom is tacitly embedded in a wisdom of another kind, that is, the wisdom of being, which is affirmed through the very fact of our existence in the universe; and it is the understanding of this ultimate existential wisdom that became a major preoccupation of Christian theology since its early patristic period.

In early patristic times wisdom meant that knowledge was inconceivable without reference to the source of its facticity in God. Knowledge can rather be connoted with created wisdom which in turn is paralleled with philosophy, or with reason. In this case the wisdom of the church, that is, divine wisdom, contrary to created wisdom, represents a saving knowledge to which all mundane

41. See on the theological dialectics of communion and othereness J. Zizioulas, *Communion and Otherness: Further Studies in Personhood and the Church* (London, T&T Clark, 2006).

activities such as science and philosophy contribute and with which they cooperate. But by cooperating in attainment of truth, the wisdom of the sciences and philosophy never exhausts this truth because they are contingent upon this truth existentially: they contain a glimpse of the divine wisdom through the sheer fact that science and philosophy exist. The sciences and philosophies aspire to this wisdom, but in themselves can never attain it within their own boundaries. Reason was sanctified by the spiritual intellect as that link between man and God that made wisdom accessible to human grasp.[42] It is important to stress, however, that the presence of wisdom in human communities was secured not through particular visionaries and spiritually advanced persons, but on the level of corporate participation in it which originated in the charismatic nature of the church and its Councils, which were gatherings of bishops, not of academics, so that the Councils were liturgical events through which the church affirmed its truth. This wisdom is described in Eastern theology by using such words as "catholicity" (universality or conciliarity). One of the features of catholicity is to establish decisions about what is good and right for men not on the grounds of what is good for this or that individual but what is good for all humanity in its entirety, including past, present, and future generations: thus catholicity in its essence refers to the fullness of humanity understood eschatologically.[43]

42. St. Augustine articulated wisdom as the link-piece between creation and God by making a distinction between uncreated and created wisdom, as it appears to the human spirit (*On the Trinity* XII, 14.22). Augustine insists that in spite of the fact that both the word of wisdom and knowledge are given by the same Spirit (1 Cor. 12:8), they are distinct (*On the Trinity* XII, 15.25). And wisdom in this context as being created is dependent upon something else, which originates beyond creation.

43. The intuition of fullness encompasses all possible generations of human beings who will ever live is the idea of fulfillment of the *pleroma* of humanity, that is, of the fullness of the "body" of humanity in Christ. St. Gregory of Nyssa argues that when the Holy Scripture says "God created man according to His image and likeness," it does mean ". . . the entire plenitude of humanity was included by God of all, by His power of foreknowledge, as it were in *one body*. . . . The whole race was spoken of as one man. . . . Our whole nature, then, extending

Conciliarity, in its depth, appeals to such a morality whose subject is not a single person, or a particular political group, but all humanity. The gift of finding the ultimate common background for human intelligence in the transcendental sphere was granted to many visionaries and deep thinkers. However, it did not prevent the atomizing tendencies in human communities, remaining thus no more than a philosophical pointer toward some common truth of humanity, but not reaching the truth of its unity in full. What was missing in all such findings is the *charismatic* and *eucharistic* dimension of this truth as present and manifested in the church. This gives another dimension to the notion of conciliarity: it is only through being in church, that is, being in council with all people, and being under the veil of the Holy Spirit, that it is possible to know truly. The reality of the church, its tradition as the continuity of the historical revelation of God in the World, as well as the constant presence of the Holy Spirit in the church's liturgy, forms the setting for the ultimate *transcendental and multihypostatic* "subject" to show its own presence in the conditions of its empirical absence. It is through the wisdom of this "subject" that all outward articulations of the world possess truth, understood in an ecclesial and hence eucharistic sense, as truth of life.[44] If humanity is brought into existence in order to realize its ecclesial function by building the picture of the universe together with the universal church, its destiny is to take care of the universe by bringing creation back to union with God. The whole history of the universe, seen previously only through secular eyes and displayed

from the first to the last, is, so to say, one image of Him Who is." Gregory of Nyssa, *On the Making of Man* 17, NPNF, vol. 5, 406 (emphasis added).

44. The assertion of human existence as ecclesial existence received a symbolic interpretation in St. Maximus the Confessor who interpreted the entire universe as the universal Church (*Mystagogy* 2). If one takes into account another parable of Maximus, that is of similarity between man and the universe (*Mystagogy* 7), then one can in the same way infer that there is analogy between man and Church, so that in some sense the Divine image in human beings is essentially the image of the universal Church. Maximus develops this theme in *Mystagogy* 5.

as a natural process, will transform consequently toward its ecclesial mode, that is, through the prism of the history of salvation.[45]

By relating humanity to Christ, whose hypostasis, after Pentecost, was transmitted to the church, theology implicitly affirms that the Christ-event, as central for our comprehension of the possibility of knowledge of the entire universe, has some cosmological significance. Then one can conjecture that the development of the universe has, theologically speaking, a drastically different meaning before the Incarnation of the Logos on earth, and after it. It was necessary for the universe to be in a state of constructive development in order to sustain life on earth and to allow God to condescend to us and to assume human flesh in order to initiate the new stage of salvation history. After the Incarnation and Pentecost, realizing humanity's ecclesial standing, humanity becomes fully responsible for the fate of the universe in the sense that cosmic history becomes a part of the history of salvation.[46] Humanity then must only be understood in the context of the promise of God for its salvation as constituting the locus point of the meeting of God and his creation, as the mediating agency, which is supposed to bring the whole universe through its genuine knowledge to new creation. The wisdom of what we have just discussed is formed by what the church is left with after the resurrection and ascension of Christ, the wisdom we know through church tradition and its ever-experienced liturgical epiclesis.

45. St. Maximus the Confessor gave a metaphorical expression of this transformation in *Mystagogy* 7. In analogy with Maximus, for Gregory of Nyssa the fulfillment of the *pleroma* of humanity will be accompanied by cessation of that time which we experience as temporal flow of physical events and by cessation of procreation, that is effectively by cessation of the biological function of human beings as we understand it today (Gregory of Nyssa, *On the Making of Man*, 22). But this will imply that human nature will experience a change that will lead to a change of the embodiment constitution which determines intentional consciousness at present.

46. The "anthropic principle" that relates the fact of the human embodied existence with the physical parameters of the universe, that is with the possibility of the Incarnation of God, becomes now rather related to the Church as the Body of Christ that comprises not only the living but also the dead and saints.

In the same way that Christians experience an eschatological presence of Christ through liturgy, ecclesial wisdom in the knowledge of the universe through science discloses to men the presence of the hypostasis of Christ. This wisdom reinstates to their eucharistic unity the existing split between the ecclesial and scientific intentionality in studying the universe, that is, unity in communion with God, revealing thus the work of scientists as a *para-eucharistic work.*[47]

Here the wisdom of the Christian church makes itself clearly distinct from philosophical and scientific knowledge as being natural predispositions of human reason since ancient times. The ancient Hellenistic world, as well as all philosophies and sciences which followed its intellectual pattern, did not feel the modes of gratitude and thanksgiving as a beginning of thought. If for the ancient thought there was nobody who had to be thanked, for the modern thought it has always been a fight against the transcendent who might be thanked. The absence of eucharistic intentionality in philosophical and scientific visions of the world results in a desire for unlimited and unconstrained possession of knowledge of things in order to use them for some particular utilitarian goals. Because the possession of things, even in their abstract knowledge, destroys a loving relationship to them, the intentionality of thanksgiving ceases to function as the gratitude for the very fact of existence of those things in creation that are supposed to be loved. To reacquire that eucharistic intentionality in knowledge, one is required to exercise *metanoia* when abstract knowledge and ideas become a manifestation of that *image* which discloses the One who stands in communion with the human spirit and who makes it possible to see behind scientific proofs a certain witness of the One. This *metanoia* represents a mode of ecclesial reality; "thus, it is the church as eucharistic mystery which

47. Cf. J. Zizioulas, *Being as Communion* (Crestwood, NY: St. Vladimir's Seminary Press, 1997), 120; Nesteruk, *Light from the East*, 2; *The Universe as Communion*, 217.

gives us knowledge of a universe which was created to become a eucharist."[48] *The universe acquires the sense of sacrament and is thus a correlate of the eucharistic intentionality of humanity*. The Christian church as carrying and sustaining this intentionality reveals itself as that ultimate multihypostatic subject which unfolds the universe in the state of communion and loving relationship.[49]

Theological Commitment and Critique of Secular Thought

One may now discuss the importance of the ecclesial dimension in knowledge from a different direction, namely as to why theological commitment implies a certain critical function with respect to science. Contemporary science is historically rooted in so-called modernity (sometimes historically associated with the fall of Constantinople in 1453), which has been responsible for dualisms such as the opposition between faith and reason, the dualisms that formed the grounds for excluding the divine and transcendent. It is modernity that is responsible for the claim that truth is based on universal reason, which tells us what reality is like. In this historical setting theology (as a way of life) was forced in the dialogue with science following the rules of modernity but not its own intrinsic logic of communion with God. These rules effectively dictated that theology entered the dialogue between faith and reason along the lines of adopted secular standards of scientific truth or normative rationality, assuming a particular notion of the knowing subject (as impersonal and disembodied collective subjectivity) that is sharply opposed to the theological way of asserting truth through events of

48. Ignatius IV (Patriarch of Antioch), "Three Sermons," *Sourozh: A Journal of Orthodox Life and Thought* 38 (1989): 2.
49. This, as we will see below, is similar to the criterion of coherence of epistemic justification in cosmology, which requires mutual agreement among cosmologists in their community.

incarnate hypostatic lives. Seen theologically, the secular standards of affirming truth are themselves based in certain myths and beliefs (for example in the existence of a universal human reason) so that they can affirm objective values only precariously. One of the dimensions of the modern dialogue between science and theology is exactly to put modernity's claim for the universality of truth under question (at least in what concerns the human sciences, including philosophy and theology), asserting that modernity's ways of appropriating truth were in a certain deviation from the unified vision of the world that was based in the characteristic alliance between faith and knowledge, both originating in communion with God.

Modernity can be characterized as a change in the very way of questioning God. According to the Christian biblical thinking the question never was formulated as "Whether I must believe in God?" and hence "Does God exist?" For a Christian the main question was a biblical one: "Who is that God whom I must trust?" There is a fundamental difference between belief in existence of God and trust in God. The God of the Bible requires from one much more than a recognition on the level of fact or theory of its existence. It demands from us an "existential commitment," an entrusting of our lives in God's care. This does not mean a lack of a rational element in such understanding of faith and trust. In early Christianity one can meet rational arguments in favor of the existence of God. It is enough to mention St. Athanasius of Alexandria, who pointed out that one can deduce the existence of God from observing the order and harmony in the world.[50] One must remember, however, that these arguments were aimed at Christians, that is, at those who believed in God in order to reveal a rational element within Christian faith.[51] Rationality in faith aids one in elucidating the sense of this faith as faith in God,

50. St. Athanasius of Alexandria, *Contra Gentes*, 35:4. ET: NPNF, series 2, vol. 4.
51. Clement of Alexandria, *The Stromata, or Miscellanies*, Book I, Ch. 5. ET: ANF, vol. 2.

but not as belief in happiness and the pleasures of life, for example. Correspondingly, knowledge as such does not lead us immediately to knowledge of the Divine, for the Divine participation in us is not an object of direct "observation" or intuition, but is revealed only through rational reflection.[52]

It is probably only starting from Descartes in the seventeenth century that arguments for the existence of God came to be used in order to convince skeptics that God indeed exists. The intellectuals in Western Europe started to follow a view that religious beliefs are based first of all in rational convictions. This view corresponded to the rise of knowledge rooted in empirical justification and "scientific method." The arguments for the existence of God, beginning from the times of modernity, based on the logic of scientific demonstration, became the only legitimate foundation for faith in God at all. Divine revelation and personal experience were no longer considered a responsible judgment. The witness of the Scriptures was accepted only after the rationality of such an evidence was established through other independent methods. One can notice here how an increasing wave of secular thinking detached from faith began to impose its standards in the realm of faith. Secularism meant not simply that in arguments related to faith one must use reason, but that this reason is independent of faith and immanent to the logic and laws of this world, being thus the only measure of all human activity, so that faith in God and religious experience acquired any sense only from within this reason.

By criticizing faith for its unconvincing arguments and accusing it of being subjective and related to the overcoming of existential

52. The importance of rational faith, as the faith reflected and elucidated by reason, was accentuated by Thomas Aquinas (*Summa Theologia*, 1a,2,1) in his polemics with St. John of Damascus (*Orthodox Faith*, 1,3), according to whom faith in God is an innate quality. See discussion of this issue in F. Copleston, *A History of Philosophy*, Vol. II, *Medieval Philosophy: Augustine to Scotus* (London: Burns, Oates & Washbourne, 1950), 336–37.

anxieties, "modernity" with its scientific method falls into another extreme: by assigning to reason a disembodied universality, the discourse of humanity is thus being thrown away from the sphere of reason. Science becomes a world-outlook that deals not with concrete human beings (with their immediate concerns, history and life), but as an indefinite and abstract, anonymous and nonempirical structure in being which "controls" every particular embodied creature. Nobody argues against the efficiency and importance of science in the modern world, but it must be remembered that the adoption of the scientific method, in fact, manifests a certain belief in the efficiency and existential importance of the anonymous and disincarnate way of knowledge in opposition to other forms of experience and views of the sense of the human. By decentering the cosmos, that is, removing it from the sphere of the personal standing "in front of" (as an aesthetical category) and making it an extensional arena of blind physical forces (i.e., making it an object), human beings make a preference for a different mythology whose existential meaning is reduced to a simple desire to doom itself in the meaningless and contingent non-purposiveness in the universe, that is, to be dissolved in the "cosmic homelessness." It is this decentering of the cosmos that is disclosed from within the theological commitment as based in a *philosophical belief* in the very possibility of such a decentering. Since this philosophical belief is not demonstrable from the ways of science itself, its very application remains no more than an empirical and contingent fact that points toward the contingent facticity of that world-picture which is built upon it. Correspondingly, all judgments of the scientific method on the possibility or impossibility of making inferences regarding the transworldly foundation of the universe remain precarious.

Modernity's stance on knowledge is seen as a certain deviation from the unified and spiritually universal approach to knowing reality that existed in late antiquity and the Middle Ages. Science, or more precisely knowledge (*episteme*), received its interpretation by theology, which elucidated the sense of knowledge and the foundations of its contingent facticity. Knowledge understood by modernity excluded communion, that is, the living participation and ontological relation with that which is being known. (This relation, by using the language of Heidegger, one could interpret, on the one hand, as letting human essence be controlled by the circle of the *unconcealed*, and, on the other hand, for human beings to be able to remain *concealed* in relation to this circle, retaining its hypostatic, irreducible to nature qualities.) Correspondingly, the truth of such knowledge was limited to individual comprehension and to the correspondence of thought to its object (*veritas est adaequatio rei et intellectus*). *Ratio*, understood as a reduced and transformed version of *logos* of Greek Patristics, corresponded to the transition from the epistemic priority of communion to the priority of the individualized rational concept. The subject of logic and knowledge of modernity acquired a new qualitative feature of being able to wander at large over reality without being aware of its own fundamental otherness with respect to it, that otherness which, nevertheless, allows one to enter relation with reality and interpret it as words imbued with energy (*logoi*). The distortion of the sense of *logos* lied in the foundation of all standards of thought originated in the scholastics (and later were present in Descartes) was a considerable deviation from Christian theology of late antiquity.[53] The change of view on

53. The problem of changing views on the role of subject and the sense of knowledge in history of philosophy (regardless to the Christian context) was carefully traced by Heidegger in his book on Nietzsche (M. Heidegger, *Nietzsche*. Volume IV: *Nihilism* [San Francisco: Harper, 1997]). Contraposing modernity to medieval scholastics (which he links to knowledge "associated with the order of salvation"), Heidegger points out that "man, independently and by his own effort, contrives to become certain and sure of his human being in the midst of beings as a whole" (89). According to Heidegger, the major task is search for the ways of such assurance, a method that inevitably led to the Cartesian formulation "Cogito ergo sum." However, the proclamation

knowledge as originating not in a hypostasis but in an individual was a certain distortion of anthropology in favor of psychological individualism and also the evaluation of a human subject on the basis of the juridical criteria.[54] As a result, the understanding of objectivity, corresponding to this change, as being devoid of the living communion and based on the realities of a law-like order, led to the formation of the scientific and technical civilization and methods of knowledge rooted in utilitarian principles related to social rights and goals.

Whereas Western theology had to adapt to the demands of modernity and hence accept secular norms in its arguments on the presence of the Divine in the world and its interaction with the sciences, the theology of the Eastern Church retained the "premodern" experience of seeing sciences and knowledge without adapting to the secularism of modernity. It is because of this that Christian theology, in particular in its Eastern Orthodox form, being faithful to the tradition of life in communion with God dogmatically and liturgically, thus transcending all historical divisions, feels empowered to question the foundational premises of modern science and the ways its dialogue with theology is organized.[55] Since modernity is seen by the Eastern Orthodox as a certain deviation

of this thesis does not liberate the theologically committed *ego* from asking a question about the foundation of the very contingent facticity of *cogito*. This, as the Fathers of the Church asserted, was not a question of knowledge as such, but a question about the *logos* of this knowledge, of its very possibility, that is, a spiritual question, inspired by the Holy Spirit.

54. Nowadays such a distortion of anthropology leads to the formation and cultivation of a type of person whose ability to function in conditions of dynamical communion is in a state of decline. This concerns first of all the limited freedom of speech, independent thinking and judgment, in conditions where the abilities of the imagination and even more, volition, are essentially suppressed.

55. This is a particular dimension of a possible response to the question posed by A. Walker more than twenty years ago: "Given that modernism by definition wants to scrutinise and criticise all traditional ways of thinking and expression—and modernism is no respecter of confessions, for all historic and traditional commitments are grist to its critical mill—is there any way we can critically evaluate modernist thought from the perspective of historic Christianity?" *Different Gospels* (London: Hodder & Stoughton, 1988), 4.

from the view that any knowledge contains a deposit of faith (let it be simple existential faith), Christian theology is given the right to use the language and critical methods developed within modern and postmodern philosophy, and other human sciences, in order to explicate those "faith-like commitments" that underlie modern science with its claims for truth and hence the imposed form of its dialogue with theology. If scientific claims for truth will be seen as endowed with the certainty of belief, the dialogue between theology and science will rise to a different level, namely that the distinction and difference between theology and science will be seen as the differentiation of intentionalities and constituents of one and the same human subjectivity. *In this, the dialogue between theology and science will acquire features of a phenomenological project* where phenomenology is employed as a particular method in exercising a critical function of theological commitment. In different words, the discourse seen through the theological commitment cannot avoid phenomenology as a method of explication of this commitment.

The appearance of phenomenology in theological discourse shows once again that theology in the mode of operation by reason is a form of critical thinking, because phenomenology is itself a mode of operation of critical thinking.[56] The sphere of operation of theological critical thinking is in all realms where the church (ecclesial humanity) meets historical and cultural reality. Theology creatively and critically thinks of any emerging historical problem or scientific theme while remaining in the immutable state of the church's spiritual life, because this life is experience of God, that is, of eternity.[57] Thus theology always functions from above mass-religious consciousness, as well as "secular" scientific consciousness

56. "Phenomenology helps the partial sciences and the natural attitude by clarifying their partiality, by bringing what is absent to them, and showing that what they identify can be seen from perspectives they do not enjoy." R. Sokolowski, *Introduction to Phenomenology* (Cambridge: Cambridge University Press, 2000), 198–209.

which claims its freedom from any faith commitments; theology's unceasing task is to provide a constant and constructive critique of these modes of consciousness by referring them to the original divine image in humanity.

In doing the above critique theology asserts itself as a *meta-discourse*, that is, as that form of critical thinking about different modalities of social activity, including a scientific one, which expresses the Divine presence and action, and which is not being bounded or exploited by some other particular human activities as their "prophetic" voice, be it the socio-historical sciences or a kind of all-encompassing transcendental philosophy. The critical function of theology with respect to other discourses never allows theology to slip into such a position that its scope and place will be determined by other discourses, for example by the science-religion dialogue as such. In this sense theology can never be defined and positioned by secular reason and thus it does not accept the idea of a complete autonomy of that sphere of the worldly reality which is asserted through rational, scientific understanding.[58]

One must take into account that by promoting theology to be a critical thinking we imply that this theology is in ecclesial setting, that is, its inseparability from experience of God through historical tradition, liturgy, and other forms of communion. This entails that by being critical with respect to various forms of thought, theology

57. In words of D. Staniloae: "The very existence of the Church is an effect, continually renewed of the action of the Holy Spirit in creating communion" (*Theology and the Church*, 218). "The door of the infinite riches of the personal or interpersonal divine being has opened up before the reflections of Orthodox theology, and with it the prospects of an endless progress of the human spirit within the divine" (ibid.). "The paradox of the Church mission in 'this world' is just in that the power of the ecclesial influence of the world directly depends on the ability of the Church to be 'bigger than the word,' to transcend the world and to see it through the 'Divine vision.'" Metropolitan Filaret of Minsk and Slutsk, *The Way of the Life-Asserting Love* (Kiev: Duh I Litera, 2004), 53 (in Russian).

58. As it is emphatically advocated by J.-L. Marion, theology deals with the saturated phenomena, whose phenomenality cannot be embraced by means of scientific analysis.

represents the voice of the church as that "place" in the universe where God meets humanity. This implies in turn that in order to remain critical and encompassing with respect to other discourses, the church must remain *independent* in its voice and not to be easily adapted to the requirements of secular reason, and in particular, to the demands of the dialogue with science.[59] If this were not to be the case and secular reason uncritically claimed its right for neutrality, objectivity, and independence from any faith commitments, faith and reason would remain parallel and nonintersecting in this age.[60] But the separation of faith and reason is the consequence of many other divisions in one and the same subjectivity, and this separation in turn divides, in this subjectivity, the Divine and the created. It is this very division (Gr. *diairesis*), which St. Maximus the Confessor described as the moral tension between the Creator and the created, and whose alleviation is the ultimate goal of the human accomplishment of the Divine likeness.[61] If the tension between faith and reason is to be overcome, it is clear that it can be done only within a strong *faith-commitment*, and secular reason alone is incapable of attempting this mediation in a nontotalitarian and nonreductive way. However,

59. This makes Christian theology flexible to any scientific developments without being assimilated by them. According to V. Lossky, Christian theology . . . is able to accommodate itself very easily to any scientific theory of the universe, provided that this does not attempt to go beyond its own boundaries and begins impertinently to deny things which are outside its own field of vision" (Lossky, *Mystical Theology of the Eastern Church*, 106). This accommodation means to remain critical of all scientific claims for monopoly of truth, that is, to remain "meta-discourse."

60. This thought was discussed by archbishop John of San Francisco (Shakhovskoi) in the context of the views of the physicist Max Planck on science and religion. M. Planck, in "Religion und Naturwissenschaft," *Vortrag gehalten im Baltikum (Mai 1937) von Dr. Max Planck*, 2te unveränd (Leipzig: Auflage Joh. Ambrosius Barth, 1938), compared the growth of scientific knowledge and of religious experience with two parallel lines. They have a common point of intersection, infinitely distant from ourselves, that is, distant from the present age and being in the age to come. See Archbishop John of San Francisco (Shakhovskoi), *On the Mystery of Human Life* (Moscow: Lodiya, 2003), 15 (in Russian).

61. See, in this respect, L. Thunberg, *Microcosm and Mediator: The Theological Anthropology of Maximus the Confessor* (Chicago: Open Court, 1995); J.-C. Larchet, *La Divinisation de l'homme selon Saint Maxime le Confesseur* (Paris: Éditions du Cerf, 1996).

if theology submits itself to the logic of the secular (for example, assuming a scientific form) it would become one particular, although very special, mode of activity separated from other modalities of human reality that do not fit the rubrics of secular demands.[62] As an example, following the logic of the secular, theology has to deal with the issue of biological evolution and origin of humanity which, in the perspective of science, accentuates the physical and biological, that is, the corporeal and hence collective. But no theory of evolution can literally say anything about the origin of hypostatic human subjectivity, that is, of persons who articulate this same evolution as such. It is persons who have empathy and love, who can rejoice and suffer, but whose account is impossible in science. Theology, in contradistinction to science, is an existential enterprise never abstracting from the concreteness of the human person and its desire to attain immortality. Theology is experience of communion with the invisible origin of all life, so that it encompasses all reality in which humanity is present, not only physically (through the senses and discursive cognitive faculties) but through insight (Kant would say *judgment*); thus it is intrinsically present in all disclosures and manifestations of reality by human beings, so that all reality's articulations are referred to and *judged* by the theological modality of life.

The theological commitment in the dialogue with science means the radical stance on science following from the objective requirement that ecclesial theology must draw a clear borderline between the dispassionate contemplation of what happens in modern science and its involvement in it.[63] The criteria of delimiters can be

62. Such a "secular" theology, for example, would not be able to take into account liturgical rites, communities, and communion as an indispensable component of experience of the Divine.

63. In a general context this implies the possibility of the critical evaluation of modernist thought from the perspective of historic Christianity. In the words of A. Walker, this means to "demonstrate our commitment to go beyond rational critiques of our culture and modern

set in words of Jesus Christ: "What does anyone gain by winning the whole world at the cost of destroying himself?" (Luke 9:25; cf. Matt. 16:26). In modern terms it would sound like this: "What does humankind gain by exploring and subjecting the world to its curiosity and utilitarian needs at the cost of losing the sense of integrity of existence and the vision of humanity's infinite tasks and spiritual goals as linked to the transcendent?" Theology must not, it has no right to be involved in, those movements of a new nihilistic spirit originating, in fact, from the modern apology for atheism that dares address to modern men the questions like this: What is the point of the humanity of humans, the naturalness of nature, the justice of the polis, and the truth of knowledge? Why not rather their opposites, the dehumanization of humans to improve humanity, the systematic raping of nature to develop the economy, injustice to keep society more efficient, the vast ocean of distracting and existentially irrelevant information to escape the commitment to truth? Since these counter-possibilities are no longer just a hypothetical speculation but nearly the sole program of the ideologies that have dominated in history since the beginning of the twentieth century, the church and all those for whom the humanity of humans, the naturalness of nature, the justice of the polis, and the truth of knowledge remain absolute values, must have a decisive and radical response to it by conducting the systematic critique of those forms of secular (and scientifically based) thinking which encourage mass-consciousness to the "winning," autonomous scientific part of the world at the cost of destroying communion with the whole. However, its theological radical critique of the scientifically asserted world does not preclude this same theology from being radically

thought forms . . . and confront modernity (our advanced societies) with the gospel" (Walker, *Different Gospels*, 4). "In the light of the gospel, we have sometimes been called to stand firm against prevailing philosophies and intellectual movements that are not only against the Christian Church but also against humanity" (ibid., 1).

positive with respect to science and the world. What Orthodox theology judges is the alleged autonomy and independence of the scientific view of the world from the very intricate inherence in the human and hence in the Divine.[64] The positive judgment of science and the world originates from the sanctification that existential ecclesial theology undertakes by bringing all fruits of human labor, including science and its picture of the world, to their correct operation in the wholeness of communion.

Here not only a dispassionate critique of a scientific secular mode of thinking is implied.[65] The Christian imperative calls into question the ethical value of pure secular science with its pretense of objectivity and neutrality and its claim for the truth of being, as if it is devoid of any faith assumptions and possibility of transcendence. It calls into question some *gnostic* ambitions of modern science to be the power that helps people solve problems of physical survival. In these pretensions modern science denies not only theology's right to predicate reality, but it denies also philosophy (as love of wisdom) for the uncertainty of its judgments (*certitudes négatives*). The scientific secular mind aspires not to philosophy, but to *gnosis*, that is, precise and demonstrable knowledge. Its aim is to justify the thesis that one

64. Cf. P. Nellas, *Deification in Christ: Orthodox Perspectives on the Nature of the Human Person* (Crestwood, NY: St. Vladimir's Seminary Press, 1997), 93–104.

65. In fact, not only scientific but also Christian thinking. In the words of T. Torrance: "[I]f you detach Christianity from Christ, then it becomes attached to society; then it is immediately engulfed in the whole socio-political world, so that then you've got a radical secularisation of Christianity" (Walker, *Different Gospels*, 52). This is the reason that the approach of Christianity to science must take place within the basic doctrinal issues such as Creation, the Incarnation, Resurrection, etc. Any loose adaptation of Christianity to the impostures and demands of modern scientific culture threatens to dissolve Christianity and, as was said above, to make it one particular modality of the activity that excludes all that is human which does not fit in the vision of this particularity. D. Staniloae expresses a similar thought: ". . . we do not think it is necessary to give up our stress upon the spiritual content of dogma when we are faced with the argument that such content can say nothing to the man of today, and therefore we must insist only on the conformity of dogmas with the results of the natural sciences. We believe that dogmas can only be preserved by emphasizing the spiritual meanings they contain" (Staniloae, *Theology and the Church*, 216–17).

must keep silent about that which cannot be spoken in terms of the rubrics of reason. Theological commitment advocates the opposite in a sophisticated, apophatic, sense: one has no right to keep silent about things of which we cannot speak (using pure reason), for in this case we pass over in silence the essence of our existence. Yet one can talk about ultimate existential things only through metaphors and aberrations in being clear that the fullness of essential questions cannot be exhausted by the faculties of reason. When the precision of judgment becomes an absolute value so that all questions beyond this "precise gnosis" are abandoned, the human being feels lost and deprived of its own existential anxieties. In this sense, faith implied in the theological commitment never threatens science and philosophy: on the contrary it protects them from the all-pervasive pretensions of *gnosis* (and, as a result, from atheism and soul-corroding nihilism).[66] Theology based in faith, being all inclusive, needs both philosophy and science because faith operates in the conditions of an incarnate humanity that seeks and asks for truth. Faith has its duty with respect

66. Cf. J. Ratzinger, *Wesen und Auftrag der Theologie* (Freiburg: Johannes Verlag Einsiedeln, 1993), s. 40.

to reason,[67] but it still remains critical with respect to both philosophy and science.

There must be made, however, a comment on the sense of our usage of the word "critique" if it is applied by theology with respect to secular thinking, including science. In fact, there must be made a distinction between our sense of "critique" and atheistic criticism of theology based on the grounds of the so-called "critical rationalism." Briefly, the essence of this philosophical trend descending from K. Popper, and being first of all represented by H. Albert,[68] consists in appeal to constant understanding and revaluation of the achieved landmarks of knowledge, revising and amending all empirical results as well as intellectual constructs. This methodology of research and seeking for truth is treated as being genuinely rational. Correspondingly, the attitude of the atheistically oriented representatives of this "critical rationalism" is to criticize religion, and Christianity in particular, for being dogmatic and using the "strategy

67. One implies here the duty of faith with respect to philosophy. Christian teaching on the Incarnation, in order to reveal God in its humanity, appeals to a new and superior reason that pertains to human reason. Christians do not have choice in possessing the "kind of reason," that is *logos*, because they bear name of that one who is the Logos himself. This is the reason why Christians had to acquire the achievements of Greek philosophy and sciences (one may recall Clement of Alexandria who argued in favor of this). St. Augustine asserted that Christianity cannot be compared with ancient religions (*theologia civilis* and *theologia fabulosa*—political theology and mythological theology), with the only exception *theologia naturalis* (natural theology), that is, with an attempt at rational knowledge of God through studying celestial movements. Augustine insists that the term *theologia*, for Christian faith, must be understood only as true knowledge of the Divine. Since the notion of truth is employed, ". . . comparison must be made with philosophy"; thus faith becomes, first of all, the subject of philosophy because as Augustine affirms, "the true philosopher is the lover of God" (Augustine, *City of God*, Bk. 8:1. ET: trans. H. Bettenson [New York: Penguin, 1980], 298). In spite of the fact that philosophy is not identified with knowledge of God, it is obliged to Christian theology in what concerns its rationality. It is because of this obligation that one could develop *theo-logia*, that is, a knowledge of God whose foundation comes from God himself. In this sense faith has its duty with respect to reason because it has duty with respect to itself. See J.-L. Marion, "La foi et la raison," *Le croire pour le voir. Réflexions diverses sur la rationalité de la révélation et l'irrationalité de quelques croyants* (Paris: Parole et Silence, 2010), 17-29.

68. See H. Albert, *Treatise on Critical Reason* (Princeton: Princeton University Press, 1985), who develops an idea of critical methodology of knowledge following from their criticism with respect to theology.

of immunization" from any criticism with respect to basic dogmas of faith. This concerns first of all the central theological conviction of the existence of God. Since no rational demonstration of such an existence is possible, so that any possible statements about God can be doubted and hence criticized, theology immunizes this aspect of faith from any rational critique and thus it falls into the fallacy of dogmatism, that is, an arbitrary chosen premise for any further deductions. In view of these accusations, our main concern here is the following: If the followers of "critical rationalism" charge religion and theology with dogmatism and irrationalism, how can theology, according to our view, be a universal tool for criticizing secular thinking and science? How is it possible that behind the alleged irrationality of theology there is something that makes it "rational" in the sense of being able to oversee and justify all discourse based on the rationality of reason. In other words, could theology respond to "critical rationalism" through its critique? The answer to this question is already present in the history of thought, and its further explication could form a separate volume on science and religion.[69] However, this is not an objective of this book and we would like to provide a very brief response, related to our claim that theology is destined to exercise a critique of secular thinking.

Any philosophical strategy with respect to science, including that of "critical rationalism," leaves one basic question unanswered despite a continuing criticism and revision of scientific views of the world. This question is about the very facticity, that is, the very possibility of scientific advance that allegedly goes along the lines of "critical rationality." As we mentioned above, this advance and the "critical rationality" acting upon it does not understand its sense and its goal.

69. See, for example, H. Küng's response to Albert's critical assessment of theology in its pretense for rationality in H. Küng, *Does God Exist? An Answer for Today* (London: SCM, 1978), 324–39, 439–51.

Its *telos*, if it is somehow envisaged, must, from the point of view of critical rationalists be corrigible and amendable, that is, contingent and historically adjustable. But such a critical approach to the ways of human knowledge as part of life leaves humanity in the state of despair, cosmic homelessness, and non-attunement to the universe that has been mentioned before. Theology in this sense provides the human search for truth with teleology in the sense of orientation, recourse to what is solidly existent and to the sense of what human life is about in the perspective of eternity. It breathes a meaning into the disenchanted world so that human beings no longer suffer from being mere spectators staring at the universe's emptiness. A critical function of theology is to override any form of mundane criticism based on scientific rationality and to extract out of this criticism an ultimate positive core of human existence in the perspective of the promise of salvation. To overcome the critique of theology from the side of critical realists, this theology must establish critique of the critique. Thus it must transcend in the name of restoring the home place for humanity. Theology and theologians cannot permit anyone to prevent them from advocating and defending the humanity of humans, the naturalness of nature, the justice of the police. Any cosmological theory with its advance of corrigible findings and mind-boggling discoveries must be subjected to an existential and hence theological critique: the divine image of humanity must be preserved even if this cosmology "crushes man under the weight of the astronomical facts."

Theological Commitment in a Phenomenological Modality: The Centrality of Person

In its intrinsic critical function upon all social undertakings theology manifests itself in a phenomenological modality, that modality which studies, analyzes, and qualifies states of human consciousness by referring them to their ultimate source in persons who inhere their image in the Divine. Theology deals with phenomena, in human consciousness, of the presence of God, so that in this sense theology is the domain of phenomenology. However, unlike classical philosophical phenomenology, which fights against transcendence, ecclesial theology assumes the possibility of transcendence in its stance on humanity which, belonging to this world and articulating it, does not allow this world to swallow and reduce humanity to nothing. Here transcendence means the ability of human persons to preserve their otherness with respect to the universe even when this universe is effectively humanized through human articulation. However, this otherness, understood theologically, if it is realized and preserved, and if it is developed and extended through the articulation of the universe, contributes to the growth of religious faith. Thus theological commitment in cosmology, as the movement "beyond secular reason" implies, through bringing personhood to its God-given centrality in being, the deepening and acquiring new experience of God which is manifested in new forms of thought and philosophical language. If cosmology considers itself as self-generating knowledge in which the God-given centrality of humanity is lost or distorted, human thinking of the universe and of its own position in it becomes imbued with the existentially irrational, the sense of homelessness, not being attuned with and incommensurable to the universe, leading to death which strips all sense and value from human life. Seen in this perspective any cosmology, if it loses commitment to existential faith (not to mention its Christian sense), that is, if it does not promote human life on this planet, can become a spiritually damaging practice, where the boundary between the human and inhuman in the universe can be trespassed and the sense of life is lost.[70] All this means that the explication of theological commitment in cosmology implies the work of spiritually enlightened reason with the aim of explicating persons and their communal affinity to the Divine.

By using phenomenology as a methodological tool in explicating theological commitment, we assume its extension beyond its "classical" sphere associated with Husserl and his followers.[71] From a philosophical point of view the extension of phenomenology toward theology (a so-called "theological turn") is not unproblematic[72] since it deviates from the initial objectives of phenomenology, which was fighting against transcendence and made a methodological doubt of God through a phenomenological reduction.[73] The question is:

70. S. Horujy argues in a similar vein that since in the Christian vision, it is impossible to speak about creation without speaking about humanity, it follows that "any discourse, all the contents of which are restricted to natural phenomena only, is devoid of religious content and has no connection with theology. This conclusion can be used as a useful criterion or test in discussions of theological problems of modern natural sciences." If a naïve methodology of such discussions is not related to anthropological reality, it is not related to religious reality either. "In such cases what we see are illusory problems and pseudo-religious discourse in religious disguise." See Horujy, "How Exactly Is Spirit Present in Creation? The Hesychast Reception of Natural Theology and Its Modern Implications," in *The Spirit in Creation and New Creation: Science and Theology in Western and Orthodox Realms*, ed. M. Welker (Grand Rapids and Cambridge: Eerdmans, 2012), 105.

71. The scholars of Husserl pointed out that Husserl never talked about religion, God and mysticism explicitly in his published works. Mall lists three books concerned with religious matters: *The Crisis of the European Sciences*, *Erste Philosophie*, *Zur Phänomenologie der Intersubjektivität*. Husserl, nevertheless, discusses religious issues in his unpublished manuscript. See details in R. Mall, "The God of Phenomenology in Comparative Contrast to That of Philosophy and Theology," *Husserl Studies* 8 (1991): 1. Assessing Husserl's tension between his attitude to the problem of God as being the founder of phenomenology, on the one hand, and being a Christian believer, on the other, Mall states that "the chasm between the God of phenomenology and that of theology remains unbridged till it is bridged either by a fulfilment of intended meaning of the concept of God or the reality of God makes its entrance unto human consciousness via the routes of a mystic experience, revelation, faith or grace. The path phenomenology has legitimately to traverse is only the former one and not the latter. Husserl might have reconciled the two in his own person. But that's a different story, then . . ." (13). See also E. Housset, *Personne et sujet selon Husserl* (Paris: Presses Universitaires de France, 1997), 265-90, where the reader will be able to find a comprehensive bibliography on Husserl's involvement in religious issues; as well as A. Bello, *The Divine in Husserl and Other Explorations*, Analecta Husserliana XCVIII (Dordrecht: Springer, 2009).

72. See D. Janicaud, *Le tournant théologique de la phénoménologie française* (Combas: Éditions de l'Éclat, 1990).

73. E. Husserl, *Ideas Pertaining to a Pure Phenomenology. First Book*, § 58.

Do phenomena associated with the presence of the Divine retain in them something that does not exhaust them and does not allow their complete acquisition by consciousness?[74] Such phenomena change the classical philosophical stance on the a priori character of the cognitive faculties and the ability to constitute phenomena as phenomena of consciousness. Theology benefits from such a philosophical discussion because theology, having had duties with respect to reason, can be theoretically advanced for the dialogue with science to become more articulated in modern philosophical, linguistic, and semantic formulae.[75] This, in a way, constitutes a patristic ideal; correspondingly, being in the same tradition, contemporary theology should learn from the early theologians how

74. In order to clarify the sense of what is meant by this, it is worth quoting T. Torrance where he refers to the question posed by K. Barth: "How do we come to think, by means of our thinking, that which we cannot think at all by this means? How do we come to say, by means of our language, that which we cannot say at all by this means?" There always remains incongruence between God as the known and man as the knower. However, if the knowledge of God is to take place it must rest upon reality and grace of the object known. In this case the reality of things reveals itself to us and acts in us even in that case when the link between our knowledge and language is irreducible to the intrinsic relations between thought and speech. "We are, therefore, restricted to the sharp alternatives: either to be entirety silent, that is, not even to venture the sceptical question . . . as in regard to the rationality of nature or the laws of thought . . . ; or to ask questions only within the circle of the knowing relationship in order to test the nature and possibility of the rational structures within it." T. Torrance, *Space, Time and Incarnation* (Edinburgh: T. & T. Clark, 1997), 54-55. Rephrasing this in terms of postmodern philosophical theology, how one can speak of the transcendent? Or, in other words, how can one speak of that which is incongruent with language and orders of conceptual thinking? How can one conceptualize that which is intrinsically nonconceptual, preconceptual, or pretheoretical? Will not any speaking about phenomena whose phenomenality does not allow them to be conceptualized be a sort of violence and distortion with respect to these phenomena, reducing their phenomenality to the circle of immanent consciousness and thus depriving them of their otherness, that is, if that which is retained in them is beyond their phenomenalization by consciousness? See, for example, Marion, *In Excess*; J. K. Smith, *Speech and Theology: Language and the Logic of Incarnation* (London: Routledge, 2002).

75. Metropolitan Filaret writes in this respect: "Theologians of Christian antiquity were in a constant dialogue with philosophy of their age. While appealing to the Fathers we should learn from them how to conduct such a dialogue. This is necessary for contemporary theologians to enter a similar dialogue with contemporary philosophical thought. Perhaps one should develop a new theological language and this, certainly, does not mean to become unfaithful to Church's dogmatic teaching; on the contrary this new language will facilitate to such an expression of this teaching, which allow for this dialogue to take place" (*The Way of the Life-Asserting Love*, 44-45).

to conduct a dialogue with science and philosophy by employing contemporary philosophical thought. In reference to the practices of the Fathers, this constitutes a neo-patristic dimension of the dialogue of theology with science as another dimension of theological commitment. Correspondingly, the explication of the theological commitment in cosmology becomes a contribution to this synthesis as the extension of old forms of thought and existential meanings toward our age.

The theological commitment in cosmology implies a certain stance on anthropology, or personhood. It sees the split between faith and reason, or the split of intentionalities in one and the same human subjectivity, as the loss of centrality of the human person in the dialogue. Correspondingly, it is because of the advance of technology and science, which diminishes a personal dimension of existence, that there is a growing concern about the respect of human dignity and freedom in our time. Humanity, understood theologically as events of hypostatic existence, manifests the living transcendence (through communion) and thus the possibility of reaching out to the transcendent, as that personal rationality through communion with which the world receives its meaning as the means of the dialogue between humanity and God. The dialogue between Christian theology and science becomes a radical form of intellectual, cultural, and spiritual *mediation* among all splits and disintegrations in human life. It demonstrates that there is a common ground for all humanity in the otherness of God, as well as in consubstantiality with the universe. This position confirms an old patristic view that theology is a mode of life and the essence of the human condition is communion with the divine. Within this perspective the dialogue between theology and science rejects either the dominance of pure faith or pure reason, for it considers both of them within the integrity of hypostatic humanity as two modes of participation in the divine.[76]

At the same time it does not want to consider theology as a dialect enclosed in itself and unrelated to other spheres of human reason, although it never adapts to any unquestioned norms of secular reason.[77]

In a phenomenological perspective the problem of mediation between theology and science can be formulated as the reconciliation of the two types of experience in one and the same human subjectivity. On the one hand, in science, this experience is empirical and theoretical, delivering to human subjectivity knowledge of things "present in their presence." This is achieved by the fact that all phenomena related to the outside world are constituted within the immanence of the *ego*. Regardless as to whether one means empirical observations and controllable measurements, or mathematical statements, what is evident is that in all these cases the "reality of the outer world" is affirmed through the structures of the constituting subjectivity. On the one hand consciousness poses scientific phenomena outside itself making them objects, as if they exist separately and independently of the subject; on the other hand the form of the content of these phenomena is generated by the human subjectivity so that the form of these phenomena is immanent to

76. This implies, according to Fr. D. Staniloae, that "any progress in understanding dogma depends in part on the progressive understanding that science has of the world." However, and here he accentuates the theological commitment, "theological thinking cannot be separated from spirituality," and this is the reason why Orthodox theology "takes scientific progress into account only in so far as science makes a contribution to the progress of the human spirit, and only in so far as it deepens in man the experience of his own spiritual reality and of the supreme spiritual reality . . ." (*Theology and the Church*, 216).

77. Here our position is reminiscent of some ideas from the so-called "Radical Orthodoxy" theological movement. However, one must insist on a cardinal difference between our Eastern Orthodox Christian advocacy of theological commitment within ecclesial setting from a vague reference to the church in "Radical Orthodoxy." For a survey of "Radical Orthodoxy's" ideas see, for example, J. K. Smith, *Radical Orthodoxy: Mapping a Postsecular Theology* (Grand Rapids: Baker Academic, 2004). It will also be useful to take into account a volume on the dialogue between Eastern Orthodoxy and "Radical Orthodoxy": A. Pabst and C. Schneider, eds., *Encounter Between Eastern Orthodoxy and Radical Orthodoxy: Transfiguring the World Through the Word* (Farnham, UK: Ashgate, 2009).

human consciousness. Thus the transcending tendencies in scientific knowledge are present in the very foundation of its natural attitude, whereas phenomenology attempts to clarify the sense of this transcendence by referring it to the subjective pole of knowledge. Scientific phenomena, articulated as certain transcendences of the sphere of subjectivity, can be represented discursively as objects that manifest themselves as being poor in terms of their intuitive content, that content which forms an invisible and silent context of that objective reality which appears through the procedures of science. In other words, by making phenomena objective in a scientific sense, what is left behind is the intuitive content of the life of consciousness, which cannot be phenomenalized at all; thus human subjectivity cannot be reduced to that which is phenomenalized; hence there is something in this subjectivity—its personal, hypostatic center—that passes over any scientific presentation of life.[78] Thus human persons transcend the scientifically organized universe in a very sophisticated sense: even if cosmology asserts human existence as insignificant, it cannot remove the intuitive content of the transworldly dimension, which pertains to human persons as divine-made agencies. Thus when the scientific mind poses physical reality as objective and independent of human insight, it is not as if human history has been "cosmosized," that is, placed in the cosmic context, being reduced to the necessities of substances and the laws of the universe. It is completely opposite: the universe is being humanized, becoming the content and structure of human subjectivity as part of the unfolding human history. The universe becomes immanent to humanity, whereas humanity retains its transcendence to it. In spite of the fact that this transcendence is always in place and is the motivating force of any scientific enquiry, which never stops because science

78. See on unkowability of human beings for themselves J.-L. Marion, "*Mihi magna quaestio factus sum*: The Priviledge of Unknowing," *The Journal of Religion* 85, No 1 (2005), pp. 1-24.

never abolishes the freedom of humanity to progress beyond the already achieved, it is this very science that cannot give an account for the ground of this transcendence because it does not account for persons. The dissatisfaction of science by its inability to account for the contingent facticity of personhood leads, in a paradoxical way, to its fight against transcendence as the retaining of those foundational intuitive existential contexts that make possible any scientific articulation.

Theology exhibits a clear difference with science: it claims that it is possible to accept the phenomena of the divine as absolute, unconditioned by thought or speech, that is, to retain as "present in absence" that which is beyond the expression of what is given or revealed, that is, beyond that which can be phenomenalized. On the one hand the phenomena of the divine are immanent because they belong to human experience; on the other hand they are transcendent because they cannot be phenomenalized within the rubrics of pure thought and language, that is they cannot be exhausted by means of signifiers of that which they suppose to signify. One can say that theology retains transcendence in immanence. Here classical phenomenology, with its philosophical respect for immanence, enters an irresolvable conflict with theology. According to classical phenomenology the phenomenality of God, as well as the underlying facticity of science, would be forbidden, insofar as they reestablish transcendence as opposed to the reduction that attempts to neutralize it. Science is not subjected to this problem to the same extent because the scientific discourse does not attempt to see the "other" side of physical phenomena, that side which is responsible for their contingent facticity. In science human subjectivity operates in the natural attitude by affirming objects of the universe as existing outside and independently of this subjectivity. However, the facticity of that givenness of objects of physics, that is, their articulated phenomenality, lies in that same subjectivity which attempts to be abstracted from them. If in theology the problem of the phenomenalization of the divine coincides with the problem of theology's facticity, in science the obvious phenomenalization of the finite things and events does not naturally bring human subjectivity to an enquiry about the facticity of the science that explicates these phenomena. Science can effectively function within the sphere of immanence of that subjectivity which generates it, remaining merely an efficient tool, the very possibility of which remains obscure.[79]

Then a reasonable question arises with respect to the dialogue between theology and science: What is really meant by the dialogue between theology, which implies transcendence in its very definition, and science, whose monism, as immanentism, is implanted in scientific methodology so that transcendence is precluded? The situation is aggravated by the fact that transcendence is not self-evident even in theology if it is taken in a purely philosophical mode. That philosophical theology which considers God in terms of existence and real transcendence, causality and substance, is subject to a phenomenological critique: God is disqualified from being a phenomenon. In contradistinction to this, the theology of experience is based on facts and manifestations linked to the Scriptures and eucharistic communion, and here we deal with such *phenomena* that render a sort of concealment, not being fully disclosed through those aspects of intuition that cascade toward expression. In other words, theology understood as experience deals with phenomena that are pretheoretical. Hence there is a general problem of how to express, theoretically, pretheoretical experience; for example, how to employ thought and speech in order to express that which, by intuition, cannot be thought and spoken of, that is, that which exceeds the

79. Cf. A. Gurwitsch, *Studies in Phenomenology and Psychology* (Evanston, IL: Northwestern University Press, 1966), 399–400.

limits of the constituting ego. In other words, how is it possible to retain the transcendence of God while speaking and thinking of him within the immanence of human subjectivity? Evidently, a similar question must be formulated with respect to the universe: how to retain the transcendence of the universe as a whole while speaking and thinking of its observable and nonobservable parts within the immanent subjectivity.

The Greek patristic response to such questions would be that knowledge of God cannot be exhausted by reason and its linguistic means; theology operates with metaphors and allegories that, however, reflect existential, precategorical, and pretheoretical truth. The challenge to philosophical theology, which appropriates existential truth of God within the limits of reason, is to overcome the phenomenalization of the transcendent and thus to preserve transcendence in immanence. In different words, theology has to deal with the intrinsic ambivalence of the givenness of the divine, that is, with its "presence but in absence." If theology, being scrutinized by philosophical thought, is in need of justification of its own ability to retain transcendence within the sphere of phenomenality of consciousness, cosmology, if it intends to engage with religion, needs a similar sort of justification but to a much wider extent. This implies that the problem of mediation between theology and cosmology requires one to deal with a generic issue of the possibility of transcendence in cosmology. In more specific words: In what sense does intrinsic immanence of scientific assertions about reality retain in itself the elements of transcendence; or how does the theoretical speech of scientific discourse retain the signs of that otherworldly ground of overall facticity, which is fundamentally pretheoretical? One must not be surprised that this question is formulated by persons. In other words, the retaining of transcendence means here the transcendence of persons as not reduced to that which is phenomenalized by the sciences. Seen in this way the difference between science (cosmology) and theology can be described in terms of the difference in expressing experience of transcendence.

The explication of this last-mentioned difference and the outline of the ways of reintegration of intentionalities employed in science and theology can be made on the level of those borderline situations where the excess of intuition of a phenomenon effectively blocks its discursive exhaustion and renders in it something that has not been intended and conditioned by experience. Here cosmology provides us with at least two issues relevant to our concern: cosmology of the origination of the universe as a single and unrepeatable *event*, and the issue of position of humanity in the universe in the perspective of an unrepeatable and incommunicable *event* of embodiment (incarnation) of every human person. The universe appears to humanity as given in its contingent facticity, but its sense and origin are not comprehended by humanity. Humanity, on the contrary, is comprehended on the basis of the event of communion with the universe in the very measure that this event is not comprehended. Similarly the event of birth as contingent hypostatic incarnation (the event of hypostasis in Levinas's terminology[80]) is not comprehended by the personal subjectivity (this event does not show itself to subjectivity), but this person is comprehended on the basis of this event in the very measure that the person itself does not comprehend the event.[81] It is the inability to comprehend the sense of embodiment

80. See E. Levinas, *Time and the Other* (Pittsburgh: Duquesne University Press, 1987), 42-43.

81. A hypostatic human being appears to itself without controlling the conditions of its contingent appearance but attempting to phenomenalize it through the flow of life as directed to the future. In different words, subjectivity is extended here toward a nonintentional immanence, or reversed intentionality where the *ego* finds itself subject *to*, but not subject *of*, a givenness. The *I* no longer precedes the phenomena that it constitutes but is instead called into being as the one who receives this intentionality. The sense of the *I* is driven not by preconceived forms of subjectivity but by events whose sense is not immediately accessible to subjectivity but unfolds in time: the more we progress in time in seeing the universe, the more we comprehend the

in its hypostatic facticity that makes the problem of origin of the universe as well as the problem of origin of personal existence one and the same unsolvable metaphysical mystery.[82] Correspondingly, one can speak about incommensurability of the universe as well as the facticity of personal existence to all forms of conceptual thinking. It is the inability to comprehend the pretheoretical and preconceptual in the givenness of the universe to the human person as well as the givenness of the person to itself, that indicates that the universe and person show themselves in rubrics of immanent consciousness while remaining incomprehensible and retaining inexhaustibility in terms of conceptual thinking.

From the Image of the Universe in the Garments of Skin to the Image of the Universe in the Divine Image

In this concluding part of the introduction, we intend to discuss perhaps a most difficult aspect of the theological commitments in science and in cosmology in particular, namely that aspect of the Judeo-Christian faith which is related to the Fall. If, before this point, the theological commitment could be thought of as a kind of philosophical extension of the vision of the world, which complemented some theology-related ideas on creation and the possibility of knowledge of the world, the appeal to that theologically understood Fall dispels an illusion of the neutrality of science with respect to theology. The stance of faith originating in Scripture and tradition as those elements that reveal and manifest the Divine providence with respect to humanity and the world place scientific activity in the context of the history of salvation and transfiguration

sense of its past; the more we grow in our life, the more sense we constitute out of the fact of our coming into being.

82. Cf. G. Marcel, *Being and Having* (London: Collins, 1965), 24.

of the universe. In this perspective cosmic history, contrary to the common opinion that human history is included in the cosmic evolution, becomes a part of the history of humanity.[83] From the point of view of the physical cosmology, such a vision of cosmic history can be seen as subjective and nonscientific. However, if it is seen from within a phenomenological stream of thought, this view of cosmic history as part of human history can be justified by referring to the fact that the constitution and phenomenalization of the universe is being done by human subjects. Here an existentialist motive is present, namely the primacy of the fact of life and consciousness, which is not subjected to any doubt and interrogation in terms of that which allegedly preceded it temporally or logically. Christianity complements such an existentialism in that it relates the primacy of human life and consciousness to their createdness, that is, their origin in God, in the gift of life in his image.

The assertion that cosmic history is included in the history of salvation is a very strong aspect of the theological commitment, for here one assumes that humanity is not only the microcosm, that is, human beings are connatural with the visible universe, but that, *de facto*, humanity at this stage of the history of salvation determines the fate and future of the universe (even through its knowledge). This conviction entails not only geocentrism, related to the planet Earth as that place where history is being created, but also a spiritual anthropocentrism implying the vision of humanity not only in terms of nature subordinated to the necessities of the physical and biological order, but as *persons*, that is, hypostatic existence, from within which one can only talk about the existence of the universe as an articulated

83. As it was asserted by J. Ortega y Gasset in a different context: "Because all other reality is included in human life, that life is the basic reality; and when a reality is the reality, the only one properly to be considered as such, it is obviously transcendent. This is why history . . . is the superior science, the science of fundamental reality—history and not physics." J. Ortega y Gasset, *Man and Crisis* (London: Allen & Unwin, 1959), 122.

reality. When theologians affirm humanity as hypostasis of the universe,[84] they implicitly point toward Christ as the archetype of the human person, Christ as the incarnate Logos, in whose hypostasis the universe as well as man exists. In this sense the geocentrism and anthropocentrism of cosmology inherent in the theological commitment mean Christocentrism, for it is here, on the planet Earth, that the meeting of the Divine and human, uncreated and created took place, and it is because of this that earth is spiritually central as that place from which the disclosure and manifestation of the sense of the created universe takes place. One must also take into account that the Christocentric view of cosmology provides a certain justification of the very possibility of knowledge of the universe as a whole. Indeed, it is only the archetype of Christ (who became incarnate in human flesh in one particular point of space and who did not leave his place at the right hand of the Father, and thus was hypostatically present everywhere in the universe created by him) that gives us a historical example of such a vision of the universe in which its separated and distanced objects at the same time constitute the unity with respect to that by whom and through whom they have been created. One can speak of a "theogenic" homogeneity of the universe, as that which "is situated" at the equal "separation" (*diastema*) from the Logos who creates the universe.[85]

Theology treats humanity made in the Divine image as microcosm and mediator, whose task was to bring the universe to the union with

84. Clément, *Le Christ, Terre des vivants*, 91.

85. See an explication of this idea in A. Nesteruk, "The Cosmos of the World and the Cosmos of the Church: St. Maximus the Confessor, Modern Cosmology and the Sense of the Universe," in *Knowing the Purpose of Creation Through the Resurrection*, ed. Bishop Maxim (Vasiljevic) (Alhambra, CA: Sebastian, 2013), 297-333. (A shortened version of this paper can be found in *The Messenger: Journal of the Deanery of Great Britain and Ireland*, Part 1: N 22 (March 2013): 13-36; Part 2: N 23 (July 2013): 16-32). See also A. V. Nesteruk, "Man and the Universe in Patristic Thought: the Teaching of Maximus the Confessor and Modern Cosmology," *Journal of Siberian Federal University. Humanities & Social Sciences* 7, no. 6 (2014), 959-991.

God through overcoming the divisions (*diairesis*) in created being.[86] In application to the modern view of the universe this would mean the overcoming of its split into causally disconnected regions, into lifeless eons and lethal forms of matter. Relating this to what has been said before, the transfiguration of the universe would imply the overcoming of its perception through the prism of "cosmic homelessness," non-attunement with it and incommensurability with it. Human beings, brought into the universe without their consent and doomed to the "cosmic mediocrity" of its spatial location asserted by the cosmological principle, can transfigure the universe, that is, to think of it and participate in it through the creative love of God, in which the cosmic "abyss" of its indwelling will become a symbol that calls the human being for another infinite "abyss" of living in God. The predicament of humanity is that its cosmic duty, and its initial image originating in the Creator, underwent a catastrophic transformation in that metahistorical event which theology calls the Fall.

Leaving aside the details of the theology of the Fall, it is important for us to understand what is the meaning, in the context of knowledge of the universe, of a theological conviction that we see the universe through the eyes and mind distorted in comparison with that which was implanted in the lost Divine image. Where does theology see that boundary in knowledge which separates our vision of the universe from that one which was granted to the incarnate Christ as its creator and carer? Theology defines the human condition after the Fall in terms of the so-called "garments of skin."[87] Sometimes the "garments of skin" are associated with corporeity. However, for us it is important to concentrate on, so to speak, the epistemological

86. See more details on these divisions—for example, in Thunberg, *Microcosm and Mediator*, ch. 6.
87. The details of theology of "garments of skin" can be found, for example, in Nellas, *Deification in Christ*, 23-104.

consequences of the "garments of skin." Since all aspects of human activity, including exploring and learning of the universe, are related to the postlapsarian condition, our sense of reality of the universe and of our place in it is affected by that obscurity which was imposed after the Fall on the initial human faculties. According to the teaching of the church, before the Fall it was the unity between man and the universe through which the universe was to follow man to its "end," analytically described by Maximus the Confessor in *Ambigua* 41 as the overcoming of divisions (*diairesis*) in creation. Man's transgression set nature off course, making it develop in enclosure with itself, isolated and blind, devoid of any *telos* and doomed to futility. Matter was deprived of its development toward the spirit; it stopped being humanized and being subjected to transfiguration. Humanity did not change its place in creation, but it did change its *relation* with creation and hence its perception and understanding of the created universe, its sense and meaning as related to the task of mediation, which was handed to man and which he did not fulfill. What is characteristic for the present condition is not only related to deceptive desires and passions, but that the very process of learning of the external world is the direct consequence of this condition as the necessity of adaptation and biological survival as well as the still archetypically present, but obscured, desire to grasp the sense of humanity in the universe. Maximus the Confessor refers to that initial "wisdom" granted to man whose loss led to the demands of scholarship and learning. And the learning itself is the result that there appeared between man and God some obstacle, some division which, in order to be overcome, must be studied. The very process of learning of the world is treated by Maximus the Confessor as the loss of superiority, the lordship of man over creation.[88] This leads humanity to see the universe in the

88. Maximus the Confessor, *Ambigua 45*, PG 91, 1353C–1356A.

image of its own moral decline, so that man builds the world in its own image, that is, the image of the Fall.[89] As was expressed by O. Clément, "creation is distorted in man—'dust you are and to the dust you will return' (Gen. 3:19),—by replacing the mystery of the unity in plurality by knots of appearance, the individual or collective phantasms. The imaginary worlds of individuals and groups intersect without ever coinciding in the hereditary illusion which distorts the world. Man ceases to see the real world, that which God created in his glory, *because God is not revealed in creation.* He sees the universe in the image of his decline and makes the world in his image."[90]

Theologically speaking, the learning activities that are pursued by the sciences represent the content of what is meant by the garments of skin. This entails an important conclusion: the vision of the world through human consciousness dressed in the garments of skin is limited by this very fact. There is a natural limit in such a knowledge, which means that our vision of the universe is limited by that "moment" in the past of human history beyond which our consciousness cannot make any insight because it is itself functioning in the conditions of the garments of skin which have some originary connotations. In other words, while exploring the world human consciousness finds its own limit, its boundary and its hidden foundation which it cannot explicate, that is, answer the question of its own facticity. This consciousness projects this limit onto the world through different sciences. Citing once again O. Clément, "the discoveries of geology and paleontology inevitably stop at the gates of Paradise because it represents a different modality of being." The same can be said about cosmology: there is an absolute limit in all speculations of the universe in what concerns its remote past. The idea of the Big Bang manifests itself not only as a methodological

89. Clément, *Le Christ, Terre des vivants*, 102–3.
90. Ibid.

limit in scientific knowledge of the universe by man; it represents a limit in understanding of the origin of the present state of human consciousness as such, that is, the mystery of this consciousness. The cosmology of the Big Bang leads consciousness to the "gates of Paradise" in the sense that this same consciousness understands that it is rooted in something that it cannot explicate, but whose archetypical memory is retained.[91] The meaning of what we have called above the "positive incertitude" of science can be reexpressed by saying that science is functioning in the conditions that theology calls the Fall and it "cannot get above the Fall, because it is included in such a cosmic modality which is provoked by it and because it is inseparable from the temporal, spatial and material conditions which have appeared as a result of destruction of the state of paradise."[92] The presence of material references that are transient and spatially varied creates the "incertitude" of scientific knowledge as related to this age, that is, to the age that retains only glimpses and shadows of the "age to come" lying in the foundation of the present display of the universe. That which cosmology calls evolution of the universe, represents, in a spiritual sense, the process of objectification, a *diastatic* split in "all in all," the alienation of the first Adam as the universal Man, including humanity in its totality and the universe. The Big Bang, in this vision, becomes a cosmological symbol of the Fall, of the inception point of that cycle of physico-biological existence in which humanity became involved as a result of its transgression. The trace of the Divine image in humanity, however, is retained at least in that

91. One can find similar ideas in Bishop B. Rodzyanko, *Theory of the Universe's Decay and Faith of the Fathers: Cappadocian Theology—the Key to Apologetics of Our Time* (Moscow: Palomnik, 2003) (in Russian), as well as in Fr. S. Sokolov, *The Other World and the Time of the Universe: Time and Eternity* (Moscow: Kovcheg, 2008) (in Russian). See my commentary on Rodzyanko's ideas in A. Nesteruk, "The Problem of Faith and Scientific Knowledge in Russian Religious Thought of the 19th-20th Centuries," in *Interpreting Nature and Scripture*, ed. J. M. van der Meer and S. Mandelbrote (Leiden: Brill Academic, 2008), 395-99.

92. Clément, *Le Christ, Terre des vivants*, 106.

this same humanity can critically assess and explicate the meaning of the Big Bang theories.

All that we have described as the present human condition and its effect on knowledge of the universe, nevertheless, does not diminish the positive aspects of the garments of skin, for they were granted to humanity after the Fall with the purpose not only of physical survival, but of the recreation and renewal of those obscured aspects of being created "in the image of God," which were not destroyed and did not perish entirely. God did not strip human beings of their reason, as a manifestation of dominion over creation, and it is through the empirical and theoretical acquisition of the outer reality, that is, through knowledge and scientific practices, that the world was shaped in a coherent image of the cosmos. However, the most important and constructive positive usage of the garments of skin comes from the inherent possibility to search through the world in flux and mutability for the permanent good and the foundation of the world, that is, as we said before, to preserve an essential dimension of the human condition to transcend the world, that is, to resist being spiritually suppressed by the immensities of the universe through which cosmology portrays man as its speck of dust, to retain in humanity its difference from the world, its centrality to creation through archetypical memory of the initial communion with God.

In light of our previous discussion the theological commitment makes it possible to interpret cosmological theories, related to physical reality involved in flux and decay as those elements of instability and disorder, causing anxiety and despair in human hearts dressed in the "garments of skin," which advance them back to the archetypical state, that is, toward that which is beyond it, to that which, in a paradoxically temporal sense, belongs to the age to come. It is only through the reversal of the "path of Adam" through spiritual insight into the sense of creation, as the process directed to the future, that the task of relating the universe to its creator can be fulfilled. The theological commitment makes it definite that the sense of cosmology can only be unfolded if its study is going on along the lines of the "positive" use of the garments of skin, that is, the recreation of that hidden impetus which is left latent in humanity since its creation "in the image." This constitutes an unavoidable existential and theological commitment to any dialogue between science and Christianity. Correspondingly, the objective of this book is the explication of the Divine image in humanity through studying cosmological theories; its methodology is the positive use of human consciousness of the universe dressed in the garments of skin in order to reveal and restore that archetypical memory of "all in all," that is, the vision and communion with the universe through communion and union with God.

Concluding Remarks

As a result of a long discussion of the philosophical and theological objectives of this book, one can state its *credo* in an encapsulated form. The existential and phenomenological explication of the theological commitment in modern cosmology as the unfolding of the sense of the universe from within communion events entails that cosmology, in a way, turns out to be "subordinated" to anthropology. Philosophically this means that the interpretation of cosmological ideas is based on the epistemological centrality of humanity as such a kind of being from within which that which is called "the universe" is disclosed and constituted. Theologically, this means that the sense of the universe is established from within the relations between God and man, that is, from within a concrete earthly history being an arena for these relations. As was expressed by C. Yannaras, if "the entire fact of the world to be constituted as an existential fact, then every reality is recapitulated in the relationship of humanity with an active reason (*logos*) as an *invitation-to-relationship*, which is directed towards humanity alone."[93] In both philosophical and theological aspects of such an approach to the universe one can find a phenomenological reversal of the anthropological problem: humanity is not inserted in the allegedly preexisting cosmic history but, on the contrary, cosmological evolution has its origin in the history of the human as that primary and inherent existential beginning of any possible articulation of the world. This beginning expresses that which G. Marcel called the initial and unresolvable mystery of human existence.[94] It is that mystery which is associated with the fact of humanity's creaturehood, its mystical coming into being through the act of the Divine love. It is the stance on the spiritual centrality of humanity in the universe and the presence of the Divine image in articulations of the universe that constitutes the essence of the theological commitment in studying the universe. A phenomenological method of treating the content of cosmological theories as the content of human experience, so to speak, their interiorization by the ego, explicates a simple eidetic truth that cosmology manifests the spiritual condition of humanity, that condition which is subjected to a trial of free-thinking of the universe. Physical cosmology mercilessly dooms human beings to homelessness in the universe, their mediocrity and effective nonexistence in the divided, and sometimes non-consubstantial layers of physical reality. By so doing it subjects the human spirit to a severe test of resisting despair and oblivion and encourages transcendence, that is, the sense of its commensurability with the eternal as the God-given ability to contemplate all temporal and spatial extensions

93. C. Yannaras, *Postmodern Metaphysics* (Brookline, MA: Holy Cross Orthodox Press, 2004), 137.
94. Marcel, *Being and Having*, 24.

(*diastema*) as having origin in the same otherness of their creation that consciousness has.

If one places such a vision of the universe in a cosmographic context, the resulting "spiritual cosmology" turns out to be geocentric, because it is anthropocentric. However, this egocentricity has a theological foundation, for the meeting of God with humanity took place on earth and it is in this event through which the basic divisions in creation have been archetypically overcome and brought to the unity. Then that geocentrism which pertains to the fact that the universe is disclosed from a specific and contingent location in the universe becomes an expression of the Christocentric essence of cosmology, for the very possibility of the integral knowledge of the physically disjoint world has its origin in the archetype of the Divine image in man, that is, of Christ, understood not only as a carrier of the human nature, but also as the Logos-Word of God who did not cease to be at the right-hand side of the Father and who continuously sustains creation and its economy at all scales and all remote corners of the universe. It is this archetype, when Christ is treated as the Lord of the world (Rev. 1:16), which is gifted to humanity in order that it could know the universe at scales that incommensurably exceed in depth as well as at large the physical and biological parameters of human existence. One can say that the very possibility of knowing the universe becomes in a certain way the experiencing of the event of the Incarnation of the Lord of the worlds from within which the universe manifests itself as an event of human history.

If, for a moment, one disregards a theological stance on human existence and approaches it on the grounds of the physical and biological, as well as cognitive facticity, including the faculty of rational comprehension of the world, the universe will appear to us from within the transcendental delimiters that pertain to the human condition. The universe is constituted from within these delimiters so that the picture of the universe comprises not only that which can be phenomenalized, that is, represented as objects, but the very conditions of the possibility of such a constitution. The possibility of elucidating these conditions in the strict discourse of the natural sciences seems to be doubtful. If one assumes that the cognitive faculties as well as human reason have foundation in something physical and biological, one loses here the problem of hypostatic, that is, personal existence, for personhood is that aspect of the individually unrepeatable, isomorphic to the world's existence with respect to which science can only think in terms of riddles. It is because of this that theology enters a cosmological discourse as a pointer to that from which the transcendental delimiters in the constitution of the universe can originate, namely to the Divine image in man. In this sense the explication of those epistemic procedures that are employed in cosmology, in its essence, will be the explication of content of the idea of the divine image in man, or, to be more precise, of that impetus which is still acting in humanity's postlapsarian condition and which attempts to restore the distorted image. Correspondingly, the method of such an explication, based in transcendental philosophy and phenomenology, becomes intrinsically manifest in that theological commitment which is implicitly present in cosmology. Our desire to reflect upon knowledge of the universe from within the experience of life corresponds to that endeavor of the modern philosophy of religion which overcomes that which Heidegger named "ontotheology." In view of the objectives and tasks of the present research, this would mean the overcoming of "ontocosmology" as that abstract science of the universe as a whole which, ultimately, in analogy with ontotheology must lead to the "death of the universe," certainly not in a physical, but moral sense, as that kind of being which is devoid of the value and beauty by which the cosmos of ancient Greeks was filled in.

It is the reader who must judge as to what extent this project will have succeeded. Since, in contrast to the existing vast collection of volumes on philosophical theology with phenomenological interventions, there are practically no books on philosophy of cosmology written through the prism of theological commitment, the author is conscious of the fact that his project is novel and original. He takes responsibility for all its faults and is ready for a critical collaboration with all.

There is a note on bibliography. Unlike the books on Christian theological cosmology, the list of references on physical cosmology amounts to an astronomical figure, increasing on a daily basis through journal publications and Internet archives. This is the reason why we quote only recent major monographs that summarize cosmological research up to a certain date, reducing references to journal publications and electronic archives to a minimum.

1

The Universe and Humanity

Cosmology and Anthropology as Two Parts of the Book of Being

> *As the only being in the world conscious of itself, we are, at the same time, the consciousness of the world. . . . We cannot be aware of ourselves without being conscious of the world and of the things in it. The better we know the world, or the more aware we are of it, the more conscious we are of ourselves.*
> *–Staniloae, Experience of God, p. 3*

Contemporary physical cosmology is a well-established and vast enterprise that includes astronomical observations, space programs, research institutions, and funding strategies. Cosmology develops fast: numerous conferences, workshops, and public lectures are held constantly, resulting in further publications of collective volumes, and numerous new studies, academic and popular, appear daily on the Internet and in bookstores. Apart from physical scientists, cosmology attracts historians and philosophers of science, as well as millions of those who adore science and trust its final word on the nature of things. This is a dynamic set of enquiries about the world around us that constitutes an integral part of contemporary scientific and intellectual culture. The popularity of cosmology in mass media and among ordinary people demonstrates how deeply it affects the collective consciousness of people (including their ideology) and that it has existential and ethical implications. Cosmology is involved intimately in dialogue with religion too: it becomes an arena of theistic inferences and justifications of otherworldly transcendence when the results of its theories are brought into correlation with theological convictions. Contemporary cosmologists are often seen as exercising a certain priestly role in modern society,[1] signaling how cosmological ideas can have immediate existential and social impact that catch and fascinate public opinion.[2]

1. This happens in spite of appeals of some critics of cosmology that "it would be sad however to abandon the whole fascinating area [that is cosmology] to the priesthood." M. J. Disney, "The Case Against Cosmology," *General Relativity and Gravitation* 32 (2000): 1125-34. Disney implies here the ecclesial priesthood, not that modern-age scientific "priesthood" which definitely reveals itself through the culture of popular scientific books and exercises a certain ideological function in societies. The "priesthood" of scientists, who implicitly or openly oppose any idea of the Divine standing behind or lying in the foundation of the universe, creates a new and even more effective "scientific apologetics" for things that are not visible but are invented and believed. One can wonder then, what really sustains their credibility and how do these beliefs constitute the body of modern science in conditions where the criteria for truth are not sustained through the same traditions and practices that advocate access to truth in ecclesial context.
2. Commenting in March 2004 on Stephen Hawking's renunciation of his hope for the near end of theoretical physics through creating a theory of everything, J. Cornwell points out that the success of Hawking's theories in mass-consciousness was connected with the fact that "his fans came to regard him with an enthusiasm tantamount to religious fervour," which originated in Hawking's "rash philosophical peregrinations that appeared to enthral the students." As Cornwell writes, "[H]e [Hawking] was fast earning a reputation as the world's leading scientific guru." J. Cornwell, "Hawking's Quest: A Search without End," The Tablet (27 March 2004): 4. The religious "fervour" of Hawking's claims becomes even more obvious if one realizes that all his mathematical work incarnate in imagination about the universe did not bring any predictable and observable novelty. His invocation of intelligible realities (like the no-boundary condition or Euclidian structure of space time, etc.), whatever mathematical ingenuity is embedded in them, cannot be subjected to direct testing and indeed makes the whole enterprise very similar to those religious practices of the past and present whose aim is to proclaim access to truth through a direct appeal to basic anxieties and fears of humanity's sense of existence by bypassing the route of their empirical validation or reference to events of life. The lack of

In spite of all this, some claims of cosmology for possession of complete knowledge of the universe or, in a weaker form, that cosmology is approaching to being able to explain the essence and contingent facticity of all in the universe, are far from justified. Some cosmologists raise general doubts whether cosmology can pretend to follow rigorously the scientific method,[3] usually understood to be that all knowledge (including mathematical theories) is ultimately related to experimental verification. It is a fact of cosmology's sociology that there are extrapolations and conjectures in its claims to truth that go beyond scientific justification,[4] and this involves the whole field in an interdisciplinary discourse in which the criteria of validity and truth are much more vague than those in the natural sciences.

As asserted previously, the objective of this book is to elucidate the nature of cosmology's claims for the value and truth of its theories in a wider context: it deals *not* with physical cosmology *per se*, but with its standing in the cross-disciplinary field of links between humanity, its history, and the universe in its entirety.[5]

Physical Cosmology and an Input of Philosophy

Cosmology, understood as part of theoretical physics, forms a subject matter that by its very nature tests the boundaries and the very

existential meaning of cosmological symbols used for incantation of the universe points toward the fact that the "priestly character" of cosmology exhibits itself in the worst idolatrous sense unless it is subjected to a proper philosophical and theological critique.

3. A careful analysis of methodological weaknesses in cosmology has been done by George Ellis, "Issues in the Philosophy of Cosmology," in *Handbook of the Philosophy of Science, Philosophy of Physics*, Part B, ed. J. Butterfield and J. Earman (Amsterdam: Elsevier, 2007), 1183–1283.
4. Scientific justification is problematic *per se*, but in cosmology we have a situation where explanation exceeds the level of physical causality and appeals to the "facts" of intentional consciousness, which are themselves not subjected to any scientific methodology.
5. This is in the spirit of W. Stoeger, who argued that astronomy and cosmology play an important integrative role in phases of cultural evolution along different ways. See details in W. Stoeger, "Astronomy's Integrating Impact on Culture: A Ladrièrean Hypothesis," *Leonardo* 29 (1996): 153.

possibility of scientific explanation. Indeed, cosmology describes itself as a science that deals with "the universe as a whole," the universe as the all-encompassing, singular, and unique "object" of cosmology. However, the usage of the word "object" applied to the universe as a whole is problematic simply because the mainstream philosophical thought of reality does not allow the concept of the universe to fit in it. Indeed, according to this thought the universe consists of independent individual things (objects) that are embedded in space-time. These things as objects are *individuals*, because they have a spatio-temporal location (through their separation in the astronomical display, for example), they are a subject of predication of properties, and they are distinguishable from each other through some properties. The universe as a whole cannot be thought as an object (or as an individual) because as a whole it is not embedded in space-time (it is rather a *totality* of space and time, the totality to which such features of space and time as extension cannot be applied).[6] The universe is unique but not in the sense of an individual: one cannot relate it to anything outside it or to anything of the same qualitative order of existence; in other words, its uniqueness cannot be affirmed through a belonging to a set that would entail a sort of relationality. The universe cannot even be distinguishable from

6. The word "object" cannot be legitimately applied to the universe because the universe as a singular and self-contained whole cannot be detached from human insight and thus positioned as something outside and devoid of the human presence. There is a fundamental inseparability between the universe and the knowing subject who is always a part of the universe. Another problem emerges from the universe's uniqueness, which cannot be set among other universes. The modern view of the universe as developed from a singular state (Big Bang) that *de facto* encodes the universe in its totality invokes a counterintuitive sense of the universe as a singular, that is, unrepeatable *event* (not an object!) with respect to which the natural sciences experience a sense of fear and desire to explain it away. Edgar Morin expressed the idea that the cosmos reveals itself as the universe and event. On the one hand the physical universe is constituted through regular repetitive features; on the other it is a singular event as phenomenon, a phenomenon that evolves for more than ten billion years. The temporal unfolding of the universe that appears to human contemplation lies indissolubly in the advent-event (avènement-événement) of the world. E. Morin, *Science avec conscience* (Paris: Fayard, 1982), 120.

anything carrying particular properties inside the universe because, by definition, it comprises everything. The universe's individuality could be imagined if it were to be a hypostatic being who could affirm its existence from itself in a Divine manner of "I am who I am." However, it is exactly here that cosmology differs from theology: cosmology predicates the universe as totality devoid of any personal features; in theology, on the contrary, the universe rather appears as being disclosed and manifested by humanity as such a uniqueness and hence "individuality" that follows from human nature.

The predication of the universe as a whole in terms of properties is problematic because the universe does not attain original givenness in the manner characteristic of particular individual things.[7] The constitution of an individual thing as an object, that is, as a thing subjected to thematization and objectification, assumes as a condition the release from "environmental confinement"[8] or the context in which a thing is looked at. The universe as a whole cannot be released from such a confinement because it is itself, by definition, the ultimate environment and context for everything.[9] Thus the standard meaning of the phrase "object of explanation," as if its identity had been defined, cannot be applied to the universe with ontological

7. It is because of the inseparability between the human observer and the universe that the conditions of the universe's observability and mathematical expressibility are constitutive of the very concept of the universe. In this sense the "physical objectivity" of the universe cannot bear an independent reality in a classical sense. Indeed, unlike in classical physics, the basic conditions of the constitution of the universe as a whole have not been permanently available and thus have to be questioned. Cf. M. Bitbol, P. Kerszberg, and J. Petitot, eds., *Constituting Objectivity: Transcendental Perspectives on Modern Physics* (Dordrecht: Springer, 2009), 4, 18.

8. Terminology from M. Heidegger, *Being and Time* (Oxford: Blackwell, 1998), 413.

9. Here one sees an original sign that cosmology in a way has some features of the human sciences, because, it is known, that the releasing from environmental confinement is not necessary for thematization and objectification in the human sciences where a perspective on reality is crucially dependent on the researcher's intentionality originating in the existential and socio-historical condition (an thus cannot be environmentally free). Applied to cosmology, this would mean that if one implies the in-itself of the universe (as its identity) to be studied, it must preserve this identity as free from any change through the release from environmental confinement, that is, from the inherent subjectivity of a knower of the universe.

clarity.[10] But we do indeed understand and use this expression, "the universe," and therefore there must be a manner in which the universe is given, a consciousness of the universe that bestows sense on such language. Therefore, before any philosophical deliberation or scientific thematization of the universe, there must be experience of the universe as the recognition that there is something permanent and persistent in the background of the change or the variable. There is the sense of identity of the universe as an intentional correlate of subjectivity, but this identity remains an unfulfilled ideal. A possible scientifically reductive approach to the identity of the universe as an inherent and nonrelational aspect of an entity or a logical subject does not clarify the ontological status of this identity. Thus the universe of cosmology, being thematized, naturally represents the ultimate noematic limit in the process of scientific exploration and explanation. Nothing further is empirically or theoretically accessible to which recourse can be made in order to explain the most general properties of the universe as a whole and the facticity of its own existence.[11] In one way or another, natural scientific explanation stops right there.

The very existence of the universe turns out to be the precondition for physical science: the latter describes and explains phenomena that take place in the universe as something already given. This is the reason why the universe (as the totality of being) is not itself subject to a physical explanation. The phenomena with which physics deals have to be present. Physics simply takes the existence of its objects for

10. Cf. Theses A1 and A2 in Ellis, *Issues in the Philosophy of Cosmology*, 1216.

11. This was always realized by cosmologists themselves. As an example one can refer to Dennis Sciama's interview of 1978 where he underlined the existence of a borderline between the ultimate questions about the universe's facticity and the exploration of its properties: "None of us can understand why there is a Universe at all, why anything should exist; that's the ultimate question. But while we cannot answer this question, we can at least make progress with the next simpler one, of what the Universe as a whole is like." Quoted in H. Kragh, *Cosmology and Controversy* (Princeton: Princeton University Press, 1996), xi.

granted. The laws of physics are laws that hold *within* this universe; they are not supposed to be laws that hold across "universes" (these laws would be universal for many universes with different contingent properties), whatever that would mean.[12]

Physics is not able to enquire into the underlying facticity of the phenomena within the universe (in this sense physics is not a descriptive history). If this facticity is associated with the contingent appearance of phenomena (as contingent outcomes of physical laws[13]) as if these phenomena manifest the radical coming into being of that which has not been before, physics cannot link the being of these phenomena with that "something" (in a vague parlance "nonbeing") from which they come. In other words, physics can deal with the manifestations of being but not with the alleged ground of these manifestations in the otherness of being. It deals with something that obeys laws already in being. In technical philosophical language the same idea can be expressed differently: since physical cosmology is capable of apprehending the interior of the universe, the universe exhibits itself as intelligible; but because of the contingent nature of this intelligibility (it is not self-explanatory in the way of necessary

12. One must emphasize that we imply *physical laws* manifested through facts as individual phenomena particular to this actual universe. If, in a style of Leibniz for example, one defines laws as universal conditions that determine the structure of facts in a class of conceptually possible universes, namely that class of universes subject to those laws, then it would be better to drop the adjective "physical" in application to these laws. From this wide perspective, if a presumed physical law is valid for this actual universe only, it does not qualify as a law at all, according to the very definition of "law," but can at most be seen as an arbitrary generalization.

13. However, the very contingent appearance of things in the universe points toward the laws whose outcomes supply these appearances: there must be these laws in order to have these particular things. It is difficult to separate in the universe as a whole between its factual (material) and nomic (law-like) features. In this sense one can talk about the facticity of physical laws themselves as linked to the boundary or initial conditions in the universe. See, for example, in this respect Y. Balashov, "On the Evolution of Natural Laws," *The British Journal for the Philosophy of Science* 43, no. 3 (1992): 354-56; "Two Theories of the Universe: Essay Review," *Studies in History and Philosophy of Modern Physics* 29, no. 1 (1998): 147; "A Cognizable Universe: Transcendental Arguments in Physical Cosmology," in Bitbol et al., *Constituting Objectivity*, 269-77.

being, otherwise it would not be contingent) the universe embodies a semantic reference beyond its manifestations. Cosmologists cannot themselves retain the pointers to the otherworldly (if they are present at all) and conduct a proper philosophical work. Thus the clarification of the sense and ground of cosmology's own facticity in a way necessitates a transcendence of the scope of the natural sciences and appeals to the methods of the human sciences functioning in the conditions of irreducible historical contingency. Some physicists, in an attempt to address the foundational questions in cosmology, make manifest a "philosophical" mode, not because they adhere to a realm of "philosophy" but because they do not follow the normal ways of theory-assessment in the natural sciences. This was the original motivation, for example, for inflationary cosmologies, which aspired to explain away the problem of the special initial conditions of the universe responsible for the contingent display of the astronomical universe. A similar motivation lies in ideas of a multiverse. These models, having a developed mathematical basis and being successfully employed for problem solving, raise philosophical problems and need appraisal through methods of philosophy and of the human sciences.

One can generalize by saying that on the one hand, physical cosmology avoids touching upon ultimate questions; on the other hand, because of the special status of its subject matter, that is, the universe as a whole, as well as the fundamental inseparability of human subjectivity from the universe, cosmology is imbued with these questions and in order to attend to them one has to invoke a philosophical attitude to cosmology.[14] By conducting a philosophical

14. It is this mentioned inseparability that makes the cosmological idea (that is, the idea of totality of the world) fundamentally different among other ideas of reason, such as the idea of the soul or the idea of God. Kant wrote that neither psychological nor theological ideas entail contradiction and contain antinomies. I. Kant, *Critique of Pure Reason*, A673/B701, trans. N. K. Smith (London: Macmillan, 1933), 552. Practically, this means that one can easily deny the existence of a soul (let us say, on materialistic grounds) or deny the existence of God (on atheistic grounds). However, it is impossible to deny the existence of the universe, for it

analysis of cosmology one can on the one hand articulate the *qualities* of cosmological theory that make it scientific, and identify the naturalistic *limits* within cosmological methodology. On the other hand by transcending these limits through an enquiry into cosmology's facticity one inevitably brings cosmology beyond the scope of the natural sciences, since, de facto, here humanity enquires into the facticity of its own historically contingent subjectivity. Philosophy here manifests itself as a method of enquiry into the sense-forming activities of human subjectivity in the subject area of the universe as a whole.[15] However, since philosophers do not have a supply of knowledge about nature in advance, on which they can draw or to which they should refer, it would be wrong to take philosophy of cosmology as dealing with issues independently of the research going on in physics and mathematics. But in spite of the fact that the origin of scientifically motivated facts lies within cosmologists' thinking, which is obvious, the sense of cosmological ideas and their significance for the constitution of human subjectivity

would deny the empirical world of sense, which is part of the universe and which contains the foundation of all knowledge about universe. The antinomian nature of reasoning about the universe originates exactly here: by being in the sensible world one cannot disentangle from the universe; at the same time the universe as totality is never fully materialized in the world of the senses.

15. The fact that the encounter with the problem of the universe as a whole represents more of an epistemological issue than anything that can be associated with the natural sciences, was long ago understood by such thinkers as Nicholas of Cusa and Kant. The very concept of "learned ignorance," which amounts in modern terms to the apophaticism of knowledge in general, and which had been drawn from astronomical-cosmological considerations, had most of all an epistemological meaning pointing toward the limits of reason and the puzzles it has to encounter while dealing with such a limiting concept as the universe. See, for example, A. Koyré, *From the Closed World to the Infinite Universe* (New York: Harper & Brothers, 1958), 5-19. A similar sense was attached by Kant to his famous cosmological antinomies, which were indications of the fundamental paradoxical structures of reason rather than any constructive theories of the universe. Here is a characteristic quote from a contemporary study on Kant: "Because reason examines itself in order to extract laws from within itself, instead of simply greeting these laws, the cosmological antinomy is the place where the innermost depths of our humanity manifest themselves. In the antinomy, nature speaks to our inquiring minds in the most direct possible way, precisely because, as a complete whole, it is exposed to the danger of being lost in obstinacy or despair." P. Kerszberg, *Critique and Totality* (Albany: State University of New York Press, 1997), 101.

escapes the scope of the natural sciences and requires a philosophical approach not restricted in its scope to the causality of nature.

The Special Status of Cosmology as a Natural Science: From Substance to Manifestation

The special status of cosmology among other natural sciences is determined by two decisive factors: its subject matter is unique and cannot be posed as an outside object. In other words, there is a fundamental inseparability of the enquiring intellect and the universe as a whole. Said philosophically, the universe enters all forms of human cognition as the ultimate horizon of contexts.[16] Here we are confronted with a question about the status of cosmology as a natural science. In an attempt to study some aspects of these contexts, cosmology exhibits some features of the human sciences in the sense that the humanly made choice and emphasis of topics of investigation through their naming, methods, and goals have a genetic *historical* priority over the physicalist claims of cosmology about the universe, as if it is independent of the human-centered insight. The seeming epistemic priority of the human sciences' element in cosmology is linked to the fact that the human world (or the "premise-world") associated with the conditions of embodiment has object-noematic priority over all "other worlds." In other words, cosmology admits not only a bottom-up explanation (that is, based on an ascending series of physical *causation* from the macroscopic empirical

16. Here, in analogy in the Husserl's definition of the "world-horizon," the universe as such is never given in a manner pertaining to ordinary objects. The universe as a horizon of all contexts in the physical and mathematical enquiry into the structure of the world cannot be an object and is distinct from any object given in the background of contexts. The universe is co-perceived as the necessary horizon of all individual entities (astronomical or terrestrial) that are immediately experienced. E. Husserl, *Phenomenological Psychology* (The Hague: Nijhoff, 1977), 70-73. See also A. Steinbock, *Home and Beyond: Generative Phenomenology after Husserl* (Evanston, IL: Northwestern University Press, 1995), 104.

phenomena to the additive totality), but also a top-down *inference* based on *intentionality* of human subjectivity.[17] This intentionality includes, for example, the very idea of the universe as an overall totality.[18] From the point of view of the empirical physics the invocation of this idea is optional. Correspondingly, the idea of the origin of the universe does not proceed from earthly physics: it enters

17. This distinction can be elucidated by a long quote from a paper of C. Harvey, "Natural Science Is Human Science. Human Science Is Natural Science: Never the Twain Shall Meet," in *Continental and Postmodern Perspectives in the Philosophy of Science,* ed. B. E. Babich, D. B. Bergoffen, and S. V. Glynn (Aldershot, UK: Avebury, 1995), 121-36: "It is common parlance to say that whereas the natural scientists seek to explain, the human scientists seek to understand. This distinction between understanding and explanation, however, is itself predicated upon the deeper distinction between *intentionality* and *causality*. If the natural sciences rely upon physicalistic causality as the human sciences rely upon intentionalistic motivation, and the intentionalistic motivation is shown to be prior to causal rationality, then natural science will be shown to be posterior to, because ultimately explainable in terms of, human scientific motifs" (125, emphasis added).
18. The meaning of that which we call "intentionality" here demands explanation. Intentionality is employed as an indication of an action of human subjectivity that is associated with the freedom and potential inexhaustibility of the process of conscious acquisition of existence. Unlike physical causality, it is difficult to reduce intentionality to a scheme of operations in which even abstract terms are related according to a certain rule external to them. In this sense intentionality is free from logical constraints that follow from the physical universe as well as from constraints of any particular, for example discursive, mode of thinking. It works by association and free-willing acts based in existential certainties. One can say that intentionality is a relation of transcendence, directed either to an object of intention (the universe as a whole, for example), or to the other hypostatic human being. But intentionality pertains to a concrete human being, remaining as consciousness with itself, thus working within the given self-identity of consciousness and its potential indestructability. In going beyond the limits of consciousness, intentionality envisages the world as an infinite system of this consciousness's own possibilities. Intentionality envisages itself as the truth of all its partial experiences and the universe (seen, for example, through physical causality) to which it has access. In intentionality, consciousness and the universe belong reciprocally together: in other words, communion is effected. Intentionality (as manifestation of communion) is not operation and hence cannot be understood as such by abstract effectuation: it is impossible, for example, to re-effect in outward terms the movement of consciousness, which attempts to make human life commensurable to the whole universe. Yet, intentionality is accessible for detection and explication through the reflexive acts of consciousness: while being in communion with the universe one can invoke introspection upon this inseparable being, thus formulating the very initial quest for the position of this embodied consciousness in the universe. The disclosure of the functioning intentionality related to communion with the universe and which underlies the visible products of human life in apprehending the sense of the universe (for example in cosmology) must then be self-referential in the sense of the ultimate horizon of meanings that originate either in the universe, or in human subjectivity as such. In short, intentionality is thought of itself, that is, a fundamental movement of disclosure, the coming to light of signification in existence itself.

discourse through an intentional interrogation into the ground of all things, an interrogation that is not part of the physically causal enquiry. Ultimately intentionality means an appeal to intelligible, invisible entities (for example, the multiverse) in order to explain, or to be more precise, to interpret the phenomenal world. Cosmology in this respect provides an endless chain of illustrations. When we speak about the language of intentionality we assert that physical cosmology, in spite of being a science of the abstract and detached from human reality, yet is placed in the context of human affairs, thus exhibiting in a characteristic way the interplay between the nonhuman and human in man as an enquiring subject. This intrinsic ambivalence in cosmology originates in the paradoxical human condition of being in the world and constituting the world. In response to this paradox it is plausible to conjecture that the *content* of cosmological knowledge (that is, astronomical facts and theories of the universe as a whole, including its alleged origins) must be considered not as contraposed to human subjectivity, but as transcendentally constituted. In other words, cosmology itself must be seen as part of the transcendental discourse, that is, the discourse of the conditions that allow the universe to manifest itself (in particular, through mathematical expressibility). Correspondingly, one should make a subtle distinction between the principles that coordinate knowledge of the universe and those connecting principles (expressed mathematically) that state the relation between the properties of objects already constituted. It is this transcendental constitution through the stabilization of patterns of thought that has fundamental human origin in the very act of expression of interest and participation in the universe which gives itself for being constituted.

Seen in such a way, the intended "subject matter" of cosmology (the universe in its totality) exceeds the scope of the physical sciences, for it refers not only to the content of what has already been manifested, but to the conditions of this manifestation, which are not part of the physical description *per se*. Seen in this perspective only, the phenomenal universe is a sort of static image in the ongoing process of manifestation. By its constitution, physical cosmology provides us with a particular, logically and physically accessible pattern in the interpretation of the universe which, however, does not exhaust the whole sense of human presence in the universe as the condition for its manifestation.[19] The transcendental sense of cosmological discourse arrives from the recognition that the universe is not that which is manifest, but that it is *the manifestation* related to humanity. In this sense the universe is always our universe. By its sense the discourse of the universe as *the manifestation* has to comprise not only the current scope of observations and theories about the universe, but the whole history of formation of views on the cosmos as well as all philosophical and theological issues on the conditions of knowledge of the universe, the *telos* of this knowledge and its value. The universe as manifestation implies a constant participation or communion with it, which is tantamount to saying that the universe as manifestation means life.

19. This concerns first of all the dimension of personal (hypostatic) embodiment. Indeed the discursive or linguistic expression of experience of the universe does not rule out the immediate corporeal presence of the universe on the level of sheer consubstantiality between human beings and the universe. Correspondingly, if this dimension is overlooked, then the perceived inability of cosmology to make results personally meaningful can be alienating and frustrating for nonspecialists: for example, the sheer insignificance of humanity in cosmic scales can create a sense of anxiety and despair related to the meaning of human life. However, cosmic physics does not exhaust the sense of the human experience of space, or of astronomical objects. Our experience of the universe as that mysterious environment with beautiful night skies and warming presence of the life-giving sun exceeds and is much richer than just knowledge of astronomy or solar physics. The problem is that the formalized and mathematized science sometimes makes an effect of de-legitimizing and de-appreciating other ways of communion with the wonders of space. See, for example, A. Nieman, "Welcome to the Neighbourhood: Belonging to the Universe," *Leonardo* 38, no. 5 (2005): 383-88.

The conditions of manifestation of the universe are always implicitly present behind all its empirical appearances and theoretical representations, yet escape an explicit constitution. They reveal themselves through an excess of intuition over logical simplicity and mathematical thoroughness, which delivers the paradoxical sense of *presence* of the universe, the sense that is never disclosed in discursive terms, thus leaving one with an immanent awareness of the universe's *absence*. The incompleteness of the physical description of the universe brings us to that stance in knowledge which is called "apophaticism," that is, such a mode of experience in which what is intended to be signified through discursive description is never exhausted through its signifiers. One can only mention that the "apophatic" conviction applied to some limiting situations in cognition is well known in the history of philosophical and theological thought.[20] The ambiguity of "presence in absence" of

20. Yannaras describes as "apophatic" that linguistic semantics and attitude to cognition which refuses to exhaust the content of knowledge in the logic of signifiers. C. Yannaras, *Postmodern Metaphysics* (Brookline, MA: Holy Cross Orthodox Press, 2004), 84. In philosophy, for example, it originates from an epistemological argument pertaining to a sort of linguistic reformulation of the Kantian transcendentalism (which is typical for poststructuralism) that language conditions the accessibility and intelligibility of reality. In this approach the very phrase "there is" points to a referent that the language cannot capture because the referent is not constituted by language and by definition is not the same as its linguistic effect. According to this view, there is no access to the referent outside the linguistic effect, but the linguistic effect is not the same as that referent it attempts but fails to capture. This situation entails, in analogy with theology, a variety of ways of making such a reference. A phenomenological philosopher, J. Ladrière, without using the notion of apophaticism, points toward the same feature of any knowledge, more precisely to the apophaticism of the kind in which the human existent approaches the encounter with the world. An object is never a pure reference to itself, but is also a revelation of the fashion of its comprehension. "This revelation is at once partial and total: partial because it cannot of itself exhaust that which essentially presents itself as inexhaustible, total because the project [that is, the fashion of comprehension] is indivisible and completely present in each of its manifestations." J. Ladrière, "Mathematics in a Philosophy of the Sciences," in *Phenomenology and the Natural Sciences*, ed. T. J. Kiesel and J. Kockelmans (Evanston, IL: Northwestern University Press, 1970), 448; see also 450. The whole range of cognitive situations that fall under the scope of apophaticism can be found in works of J.-L. Marion under the name of "saturated phenomenon." See his *In Excess: Studies of Saturated Phenomena* (New York: Fordham University Press, 2002). Theology on its side affirms that any knowledge of God if it pretends to be real and true cannot escape the conditions of incongruence between human knowing and the reality of God. This incongruence is the immanent feature of any

the universe deprives a genuine cosmological project of any flavor of *foundationalism* understood as an epistemological correlate of the notion of an ontological ground, be it the constituting subjectivity of the self, or the outer universe as underlying substance. Cosmology has to function in the conditions of the classical paradox of human subjectivity in the world, which arises in this context and points to the fundamental difficulty in attempting to formulate the sense of being of the universe in terms of ground-grounded relationship.[21] The universe as manifestation thus escapes any accomplished definitions and descriptions, and it is because of this that human subjectivity itself is being constituted through its openness to the universe to the extent that this very subjectivity cannot comprehend the universe. One sees thus that the cosmological discourse (as a

serious theologizing: one cannot raise questions on the reality of God from some position that excludes a particular feature of the Divine manifestation included in the very irreducibility of any speech about God. See, for example, T. Torrance, *Space, Time and Incarnation* (Edinburgh: T. & T. Clark, 1997), 52-55.

21. This paradox is a perennial problem of philosophy and was anticipated by ancient Greek philosophers and Christian thinkers. This paradox is discussed in detail in the next chapter. Anticipating our discussion below, one can briefly summarize the paradox as follows: on the one hand human beings in the facticity of their embodied condition form the center of disclosure and manifestation of the universe as a whole, modeling it as overall space and time which exceeds the limits of the attuned space related to humanity's comportment on the planet Earth (the home place). On the other hand the depicted universe as a vast continuum of space and time positions humanity in an insignificant place in the whole totality, making its existence not only contingent (in physical terms) but full of nonsense from the point of view of an actually infinite universe. Said bluntly, the actual infinity of the universe is attempted to be articulated from an infinitely small part of its formation. Cosmology as the discourse of the universe as a whole brings one face to face with a general philosophical objective of avoiding any sort of foundationalism in knowledge of the universe that insists on the ground-grounded relation between humanity and the universe leading either to an idealistic reduction (subjectivity as the ground of the world) or to a materialistic, mathematically deterministic diminution of consciousness to illusion. In either mode of reduction the reality of the ground absorbs the grounded and the grounded is reduced to the categories of the ground. See in this respect J. Mensch, *Postfoundational Phenomenology: Husserlian Reflections on Presence and Embodiment* (University Park: Pennsylvania State University Press, 2001), 9-11. To avoid these reductions, the embodiment, as a premise of the person's grasp of the world, must be rather considered as that "over here," where a particular and immediate indwelling of life and the universe comes to *presence*. It is this coming to presence that determines that "place" which constitutes person as a center of disclosure and manifestation of the universe.

mode of the natural sciences) cannot pretend to be complete without recourse to the essence of the agency disclosing the sense of the universe, that is, of human beings.[22] An interplay between the dimension of the human and natural sciences in cosmology manifests itself as its intrinsic and unavoidable feature.

The Nature of Manifestation and Ontological Commitment

In some cases cosmology claims the existence of things on the grounds of theoretical consistency and a fit with other plausible constructs, but for which we can have no observational evidence (that is, the principle of direct correspondence with empirical reality is not applicable).[23] Such a situation, for example, happens in the extreme case of the construct of the multiverse,[24] where no direct

22. Cf. "A philosophy of nature and a philosophy of man are mutually complementary; . . . neither can be completed unless it shows itself as the counterpart of the other." G. De Laguna, *On Existence and the Human World* (New Haven and London: Yale University Press, 1966), 81-82; as well as the statement "Anthropology and ontology are indeed two sides of the same coin," in H. Köhler, "The Relation between Man and the World," In *Astronomy and Civilisation in the New Enlightenment: Passions of the Skies*, ed. A.-T. Tymieniecka, *Analecta Husserliana* CVII (2011): 37. In the same vein J. Ladrière enquires: "Is there not, when we read it sufficiently profoundly, an analogy between the deep structure of nature and the structure of human existence as openness, creativity, possibility of accord with the event? The problematic of nature can thus be linked with the problematic of human existence." J. Ladrière, *Language and Belief* (Dublin: Gill & Macmillan, 1972), 186. However, one must not absolutize this analogy: cosmology provides only one particular dimension in the sense of disclosure of the human origin. It must not be treated literally. Cosmology as such does not provide any insight on the sufficient conditions in the facticity of human persons made in the divine image.
23. This, for example, can be related to the cosmological principle, which postulates uniformity of the universe beyond observational limits. Another example is the famous "inflaton" field, which drives the exponential expansion of the early universe.
24. Multiverse proposals in cosmology refer effectively to the old idea of the plurality of worlds understood either in a physical sense as an ensemble of worlds with all possible physical conditions, or a variety of mathematical structures that have or do not have their incarnation in the physical. In this case the existence of our universe in its contingent facticity is explained away through the inference that it simply belongs (in a generic sense) to an ensemble of universes that through its totality contains whatever is possible. The literature on the multiverse is vast; as an example, see M. Tegmark, "Parallel Universes," in *Science and Ultimate Reality: From Quantum to Cosmos*, ed. J. D. Barrow, P. C. W. Davies, and C. Harper (Cambridge: Cambridge University Press, 2003), 459-91, or B. J. Carr, ed., *Universe or Multiverse*

observational or experimental tests of the hypothesis are possible, and the assumed underlying physics is probably untestable in principle. These possibilities do not by themselves prove correct epistemic justification; even less do they point to the truth-content of what the theories claim. It is seen that here a sort of philosophical, that is, a trans-scientific insight is invoked.

In the case when cosmology predicates things beyond their verification through the principle of correspondence[25] it appeals first of all to the method of extrapolation (understood in a wide sense), which itself must be evaluated as tacitly committed to a sort of realism grounded in belief in the efficacy of extrapolation. Philosophically and scientifically the problem of extrapolation arises from those limits of scientific explanation that are set by the observational constraints inherent in our earthbound home-place. All that is in principle directly accessible to observation is positioned on the surface of the past light-cone with its apex on the planet Earth.[26] Outside that cone one has the uncertainties of extrapolation.[27] Thus the extension of a cosmologist's insight into the universe from earth in the attempt to encompass the universe in a single vision (including its absolute origin), requires an inference from what is already known to what is as yet only conjectured. For a form of knowledge that rests its claim

(Cambridge: Cambridge University Press, 2007), with a variety of papers on different aspects of the multiverse debate. In all multiverse proposals the question of existence, that is, of the contingent facticity of this universe, is thus quite illegitimately transferred to the question of selection, whereas the issue of the existence of the multiverse itself cannot be addressed at all for obvious philosophical reasons.

25. The sense of correspondence with empirical facts implied here must be distinguished from the principle of correspondence in quantum mechanics demanding that it should be constructed in correspondence with classical mechanics.

26. There is a tiny piece of the human observer's world line that relates to the immediate cosmic environment like the earth, planets in the solar system, and stars in our galaxy, which in terms of cosmic time and thus space are "close" to us, so that their separation from us is in a way "commensurable" with humankind's life span. We assert the existence of such objects in terms similar to those of earthly objects.

27. Thesis B1 in Ellis's "Issues in the Philosophy of Cosmology," 1220.

on its empirical, observation-based access to the world (most of the natural sciences), these limits raise clear difficulties.

One could claim that "extrapolations" (inferences) toward the fundamentally nonobservable and untestable are simply physical hypotheses that are assessed along a variety of lines, including observational tests as only one of these. These hypotheses may rely on appeal to analogy, on consistency with other cosmological contexts, on logical fertility and explanatory force, or a mathematical consistency and elegance. Over time they may be woven into a more and more tightly connected set of beliefs and ideas, each element of which derives support from the set as a whole.[28] One can claim even further that *extrapolations* in cosmology itself (whatever this means, including a shift of "home places"[29] in the cosmological principle, or a free eidetic variation[30] of the parameters of the whole world, which happens in theories of a multiverse) imply an extended sense of "scientific justification," for example epistemic coherence, which does not necessarily refer to tests and observations. This, in turn, entails a different commitment to realism. Let us briefly comment on this.

For example, in the models of origin of the universe, the major presumption is that one can extend the laws of physics

28. See, for example, E. McMullin, "Long Ago and Far Away: Cosmology and Extrapolation," in *Bang: The Evolving Cosmos*, ed. R. Fuller (Saint Peter, MN: Gustavus Adolphus College, 1994), 119-20.
29. This is the terminology of E. Husserl; see his paper "Foundational Investigations of the Phenomenological Origin of the Spatiality of Nature," in *Husserl Shorter Works*, ed. P. McCormick and F. A. Elliston (Notre Dame: University of Notre Dame Press, 1981), 222-33.
30. An eidetic variation in phenomenology is meant to subject to mental amendment some aspects of the intended "object" in order to reveal its true sense as that one which survives these variations. It is through this variation that the sense of individuality and identity of the universe is attempted to be constituted. An eidetic intuition of the universe moves us first to the realm of Platonic forms, where the universe as a whole is positioned as an intelligible "object" opened to our gaze. It is at this background that one can speculate about other possible universes as alternatives of that one where we live in. However no matter how much we may enjoy "living" among the intelligible forms, we never abandon the appearances of the real universe through which we can come to higher intelligibilities. See on eidetic variation in phenomenology, for example, Sokolowski, *Introduction to Phenomenology* (Cambridge: Cambridge University Press, 2000), 177-84.

(comprehended by us through mathematical formulae) toward something that cannot be physically independent of its mathematical gestalt. In other words, such an extension presumes effectively a set of beliefs that it is possible to catch the sense of reality beyond the sensible (corporeal, in the sense of physical equipment as an extension of one's bodily function) as its efficacious identity (which could be either on the level of the alleged substance or on the level of ideal forms) through time in spite of the postfactum resistance of reality to this.[31] The validity of these beliefs can only be justified on the grounds of their *coherence* as well as, to a lesser extent, agreement with that borderline physics which through observation is linked to empirical validation. The situation when justification is linked to beliefs is dealt with by that part of contemporary epistemology called the coherence theory of epistemic justification, which holds that a belief is justified to the extent to which the belief-set of which it is a member is coherent.[32] What is at issue in a coherence theory is a proposition's relation to other propositions, and not its correspondence with material facts.

Now we see that it becomes a task for philosophy to discuss the various sorts of hypothetical extrapolation that cosmologists make as a regular part of their work and the implied philosophical beliefs that drive them. As a matter of illustration, let us refer to the basic assumption lying in the very possibility and foundation of modern cosmology, that is, to the principle of uniformity of space-time and matter (the cosmological principle), which is based in extrapolation (in the certainty of a belief in an indifferent location of humanity in the universe)—that the average isotropic picture of the large-scale

31. This is a longstanding point made by E. Meyerson in his *Identity and Reality* (London: Allen & Unwin, 1964).
32. See, for example, J. Dancy, *Introduction to Contemporary Epistemology* (Oxford: Basil Blackwell, 1989), 116. The detailed discussion of the idea of coherence of epistemic justification in cosmology will be undertaken in chapter 4.

distribution of matter in the universe as observed from the Earth can be transferred to all possible locations (thus implying spatial homogeneity).[33] This extrapolation invokes a certain *philosophical* and, maybe, even a *theological commitment*, which acts in the cosmologist's mind as a regulative and indemonstrable belief.[34] The implication of this belief in cosmology is a particular causal structure of the global space-time of the universe; that is, this belief as an act of intentionality cascades down toward physical causality.[35]

Another illustration comes from inflationary cosmology: it confesses a belief that there exists a field φ (inflaton),[36] which is

33. One must point out that there are some discussions in scientific literature that try models that deviate from the spatial homogeneity of the universe.
34. Discussing the cosmological principle in close connection with the so-called Copernican principle, E. McMullin points out that the Copernican principle has to be understood in terms of what it rejects, namely older teleological beliefs about the uniqueness of the human and the likelihood that humanity has a selected position in space, for example being a cosmic center. "Indifference Principle and Anthropic Principle in Cosmology," *Studies in History and Philosophy of Science* 24, no. 3 (1993): 373. However, the desire to abandon the teleological explanation is itself based in intentionality, rather than any scientifically demonstrable conviction. The indifference postulated by the cosmological principle is indemonstrable because it itself lies in the foundation of the very possibility of scientific demonstration applied to cosmology. Thus it is based in the belief in knowability of the universe, which has a different motivation in comparison with that of teleology (but related to the latter).
35. Here one must make a comment by pointing out that the so-called "cosmological principle" (the principle of indifference or mediocrity) is applied not only in cosmology. Indeed the scientific method employs a principle of indifference with respect to similar objects that obey physical laws. For example, all atoms are considered as similar with respect to laws of atomic physics. However, this indifference does not deal with the outcomes of physical laws. Atoms are indifferent in what concerns them as constituting units, but their particular combinations, which give rise to complex structures (such as life-forms), are not subjected to laws of atomic physics and represent rather their outcomes, which can be different and specific. The cosmological principle states the uniformity of the universe at the level of the outcomes of physical laws, which led to a particular uniform structure of the observable universe. In this sense the cosmological principle is not simply the proclamation of the overall consubstantiality and uniformity of the universe at all structural levels, but a claim for a particular manifestation of this consubstantiality at the level of clusters of galaxies. The uniformity of the universe implies very specific initial conditions and topology, which theory attempts to explain away by inventing laws whose outcomes apodictically lead to the uniformity in the universe (an inflationary scenario, for example).
36. In spite of the fact that the hypothesis of this field, its very existence, is very efficient in a qualitative and quantitative modeling of observable phenomena, the physical nature of this field, that is, its relation to a certain class of observed particles, remains obscure. This is one of the major points of skepticism with respect to inflationary theories, which has been raised, for

described through a corresponding theory and which drives the evolution of the universe during the very early inflationary period. This belief coheres (as justification) with another scientific conjecture (belief) that there was a period of evolution of the universe with an exponential growth in time, which, in turn, solves some problems of radiation-dominated cosmology[37] and hence makes the so-called standard cosmological model even more coherent. One must stress here that the construction of a quite sophisticated theory of the inflationary universe is driven by the hidden desire to explain away the contingency of the initial conditions of the universe allegedly responsible for the contingent specificity of its present display. The very concept of contingency (as eventuality and historicity) is alien to physics, so that one observes a certain "pseudo-theological" commitment to overcome the "latent horror of the unique event."[38] A similar situation occurs with the idea of the so-called multiverse. Since no correspondence with empirical reality is possible, all

example, in the above-mentioned paper of Ellis, "Issues in the Philosophy of Cosmology," 1210. See a similar point made in R. Penrose, *The Road to Reality* (London: Vintage, 2005), 751; also S. Weinberg, *Cosmology* (Oxford: Oxford University Press, 2008), 202, 217.

37. These are the famous horizon, monopole, and flatness problems. See Weinberg, *Cosmology*, 201–8. See also R. Penrose, *The Road to Reality*, in what concerns a certain critique of the inflationary hypothesis, 753–57.

38. Through the discovery of the singularity in cosmology, as Torrance writes, "We have disclosed in the depth of our science a unique event of a staggering kind. It is this singularity that causes the trouble, for it arouses in so many people once again a latent horror of the unique event. It was rejection of the unique event that lay behind notions of the cyclic universe, or modern notions of the oscillating universe, . . . for there seems to be a regulative belief in many people's minds which insists that everything must be *repetitive* and thus equivalent, and that the world goes on as it always has in the past and always will go in the future." T. Torrance, "Ultimate and Penultimate Beliefs in Science," in *Facets of Faith & Science*, vol. 1, *Historiography and Modes of Interaction*, ed. J. M. van der Meer (Lanham, MD: University Press of America, 1996), 166–67. Since the unexplainability of the contingent original singularity initiates unexplainability of nearly all presently displayed features of the universe, the fear of singularity is equivalent, in a different language, to the fear of the *erratic* features of the universe, which are related simply to the fact that the universe is being comprehended by human beings who by virtue of their freedom constantly experience life as a breakthrough of the erratic in the allegedly present world order. To avoid singularity is tantamount to avoiding the fear and anxiety of erratic facts of life.

speculations about the multiverse are based in the belief that there is an extended meta-reality that comprises our universe, so that any justification for a theory of such a multiverse can only be based on the grounds of epistemic coherence, which is related to convention at the level of the community of cosmologists. The fact that the idea of the multiverse is driven by a pseudo-theological commitment to justify this universe through the reference to the "transcendent" can easily be detected by pointing to by-no-means rare discussions on how the multiverse competes with the idea of creation of the universe *ex nihilo* by God.[39] In the case of the multiverse, in fact, no realistic reference is even required. We are dealing here with a situation where the mental states (of cosmologists) affect our sense of reality and even contribute to its theory.[40] The idea of the multiverse can be approached from a different point of view if considered within the scope of phenomenology as an eidetic variation of the parameters pertaining to the actual universe. This variation takes place within human subjectivity and aims to articulate some apodictic features of this actual universe. In this case the invocation of the idea of the multiverse is a legitimate phenomenological procedure in order to reaffirm with new force the inevitability of the given contingency of this actual universe. But certainly this shows the causation implied

39. See, for example, discussion of this issue in J. Leslie, *Universes* (London: Routledge, 1989); D. Temple, "The New Design Argument: What Does It Prove?," in *Science, Technology, and Religious Ideas*, ed. M. H. Shale and G. W. Shields (Lanham, MD: University Press of America, 1994), 127-39; W. Stoeger, "Are Anthropic Arguments, Involving Multiverses and Beyond, Legitimate?," in *Universe or Multiverse*, ed. Carr, 445-57; R. Collins, "The Multiverse Hypothesis: A Theistic Perspective," in *Universe or Multiverse*, ed. Carr, 459-80. See a serious critique of the idea of multiverse in G. Ellis, "Does the Multiverse Really Exist?," *Scientific American*, August 2011, 38-43.

40. This thought was anticipated by Henry Margenau, who believed that modern physics could provide an evidence that the nature of its reality is determined not only through causation in empirical reality, but also through intentional acts of thought. In his approach to the nature of physical reality he posed a question: "Is sensed nature the only field of departure or arrival in the process of scientific verification, or will inspection of the eidetic structures of consciousness function in a similar way as dator of scientific fact?" H. Margenau, "Phenomenology and Physics," *Philosophy and Phenomenological Research* 5, no. 2 (1944): 278.

by the model of multiverse to be of a rather mental kind, so that the analysis of conscious states becomes the datum of scientific facts. Thus by an unexpected twist of reasoning, certain theories of cosmology can be used to illustrate phenomenologically the working of human subjectivity.

We see thus that the effectuation of coherence of epistemic justification in cosmology (which implies a communal, or transcendental dimension in cosmology) leads to a different stance on commitment to realism (that is, ontological commitment) in modern cosmological discourse. Cosmology is now seen as an enquiry into the condition of appearance of the universe, attaining reality such that it gives itself to be apprehended by human beings and their communities, the very reality of the world in which all sensible entities, astronomical objects, physical bodies, including human beings themselves, find their place and their meaning. However, this discourse of the appearance does not deal much with a description of what appears at the level of observational astronomy and constructs of theoretical physics, but deals in a more profound sense with a characterization of the very conditions that govern the possibility of appearance, or manifestation, of the universe. In other words, it is not, properly speaking, a discourse of the phenomena as such (related to knowledge of facts about the universe), but a discourse of the process of *phenomenalization* of the universe. In the traditional mode of language, a discourse pertaining to the *conditions* in which the phenomenon constitutes itself as phenomenon is called transcendental. By becoming more and more conscious of its own constraints and its own possibilities (as related to the place of humanity and its communities in being), the discourse of contemporary cosmology becomes more and more a transcendental discourse. This reveals the discourse of cosmology to be not only the discourse of the universe, but a discourse of human beings.

By being engaged in the discourse of the universe as a whole, human beings themselves are involved in and subjected to the process of their phenomenalization: on the one hand they take as their task to control this process through advancing (astronomical) praxis dependent upon their theories; on the other hand the universe remains that overall context and horizon of all horizons which escapes constitution by the discursive reason so that it is rather human subjectivity that is constituted by the universe to the extent it cannot comprehend the universe.[41] In this sense cosmology represents not so much that which is *manifest*, that is, the universe as such, but *the manifestation*, the manifestation that involves the universe and conscious human beings in the endless constitution.[42] Cosmology reveals itself as a contributor to the phenomenological project, as realization of a transcendental discourse.

Phenomenological Insight in Cosmology as Explication of the Human

A phenomenological insight into cosmology makes a reversal of its meaning by shifting the center of its enquiry from the noematic content to its noetic pole, that is, the generating human subjectivity. When scientific reason attempts to enquire into the origin of the universe in an absolute sense, the strategy of extrapolation acquires some features of *philosophical* transcendence. But here transcendence is not through physical causation (this would be an impossible break beyond the immanent), but through intentions based in acts of indemonstrable beliefs, beliefs that by their very existence and function point to science's limits of establishing the sense of the

41. The universe as a whole reveals itself in similarity with the "sublime" of Kant's *Critique of Judgment*.
42. Cf. J. Ladrière, *Language and Belief*, 169, 173, 176.

universe. Transcendence implies here an excess of intuition of the donation of the universe in the act of life, an excess over the discursive ability to grasp the sense of the universe. Transcendence points toward a simple truth—that the reality of the human embodied existence is not exhausted by its physical aspects.[43] Speaking philosophically, humanity stands at the crossroads of reality, being on the one hand embodied agency and, on the other, a dative of manifestation and nominative of disclosure. Correspondingly cosmology, if it is narrowed to the physical and expressed mathematically, cannot account for the ultimate sense of the universe because it cannot account for the ultimate sense of the human.[44] Since no science can give such an account, the question here is about the boundaries of the human in science. The atomic bomb, for example, being a human creation, characteristically points toward the inhuman, that is, to the limits of humanity as such. Thus the atomic bomb as a scientific achievement defines in an apophatic (negative) way the sense of the human. Cosmology plays a similar role: it provides some hints and pointers as to where human comprehension and articulation of the universe becomes paradoxically inhuman (the Big Bang, for example). In this sense the cosmology of the Big Bang becomes a characteristic, although apophatic, explication of the sense of humanity as that formation in being which is looking for

43. In the context of the so-called anthropic inference this was pointed out by M. Bitbol, "From the Anthropic Principle to the Subject Principle," *The Anthropic Principle: Proceedings of the Second Venice Conference on Cosmology and Philosophy*, ed. F. Bertola and U. Curi (Cambridge: Cambridge University Press, 1993), 91-100. In a wider philosophical and theological context this excess of humanity beyond nature was discussed in A. Nesteruk, "Human Image: World Image (Patristic insight into cosmological anthropology)," *Studies in Science and Theology* 9 (2004): 3-26; "Theology of Human Co-Creation and Modern Physics," *Mémoire du XXIe Siécle, numéro 3-4. Cahiers transdisciplinaires. Création et transcréation* (Paris: Du Rocher, 2001), 163-75.

44. The cosmic environment provides the necessary conditions for human corporeal existence (and this is exactly detected in anthropic arguments), whereas the sufficient conditions do not belong to the sphere of physics and point toward human morality, ethics, and some eschatological commitments. See discussion in A. Nesteruk, *Light from the East: Theology, Science and the Eastern Orthodox Tradition* (Minneapolis: Fortress Press, 2003), 200-214.

its own origin and its own history.[45] A phenomenological insight into the sense of cosmology as explicating humanity's quest for itself thus compensates the incompleteness of cosmology and reinstates its human creator to its ontological centeredness in disclosing and manifesting the universe.[46] At the same time, the limits of physics and scientific philosophy tested through cosmology in fact test the limits of humanity to understand its own sense of existence. The incomprehensible universe invokes in the human scientific mind humility and discernment in order to realize the limits of its pretensions for knowledge of the universe, the universe that resists disclosure and exceeds the capacity of understanding.[47] By being great and above the universe, humanity yet realizes its smallness. It is typically a human feature to have an idea of infinity in order to realize its own finitude.

Since cosmology, assessed phenomenologically, retrieves the "natural" centering of all non-egocentric tendencies of its world-building narrative in human hypostatic subjectivity, this assessment indirectly calls into question the purported neutrality and objectivity

45. One can point to similarities between the phenomenology of birth and aspirations of cosmologists to disclose the sense of birth (origin) of the universe. See Nesteruk, *The Universe as Communion: Towards a Neo-Patristic Synthesis of Theology and Science* (London: T. & T. Clark, 2008), 247-66.

46. The idea that a research into the underlying sense of science leads to enlightenment of the ways and *telos* of the human spirit was clearly formulated by many phenomenological philosophers starting from Husserl. Here is a quote from J. Ladrière: "The detail of the life of science must . . . be investigated in order to know something of the nature of reason and of its becoming. . . . The destiny of reason is outlined . . . in the incessant comings and goings that define the life of science. It is in the patient advance of its history that its finality reveals itself." J. Ladrière, "Mathematics in a Philosophy of the Sciences," 455.

47. The phenomenological construct of "presence in absence" can be easily applied to cosmology. For example: we see the universe back in time along the so-called past light-cone, so that the inference about the universe outside this cone can be considered as an attempt to deal with the universe as a whole which was present in its empirical absence. A similar thing can be said if one remembers that according to present-day models the visible matter represents only 4 percent of the whole material content of the universe. The other 96 percent (dark energy and dark matter) is postulated in order to balance the model with observations. In other words, the universe is present to us through 4 percent of what is visibly manifested but in empirical absence of the other 96 percent.

of some of its claims with respect to realities that are beyond empirical verification. It could suggest instead that such "neutral" descriptions of the world operate on the basis of existential concerns formulated in a set of *beliefs* (or myths, which may or may not be related to the *faith* of theology).[48] In this sense the phenomenological stance rejects that view that cosmological knowledge describes the world in itself;[49] rather these descriptions are seen as interpretations that are governed by controlled beliefs in the sense that they are related to a particular *path* of science in human history.[50] For example, if one is to understand and explain the past of the universe as constituted through human history, one must conceive it in terms of past possibilities of *this history* rather than as a defined and finished product. In this case the cosmologist's own historical consciousness is involved and in analogy with historical science, *cosmological discourse reveals itself*

48. A basic and unavoidable structure of any cosmological myth, including its contemporary scientific arrangement, is the duality between the factual and empirical on the one hand, and the intelligible (as allegedly stable and underlying) on the other hand. Here is how J. Ladrière describes this general typology: "The cosmological myth is an account of cosmogenesis. It tells how the world was made, how the contemporary world that stands before our eyes developed from what went before, from the non-world, the formless. . . . The schema of the representation is a successive unfolding in which there is a movement from homogeneous unity to a qualitatively differentiated multiplicity, passing through all the intermediary stages which progressively ensure differentiation from the origin and establish the link between the primordial moments and the complex, moving and multiform structure in which our cosmos was produced" (Ladrière, *Language and Belief*, 153).

49. As it was eloquently expressed by A. Gurwitsch, "The goal of phenomenology is not an exhaustive description of an infinite variety of immanent data, but the investigation of those contexts of consciousness owing to which there is a perceptible world, the universe of physical constructs, etc." A. Gurwitsch, "Comments on Henry Margenau's 'Phenomenology and Physics,'" in *Phenomenology of Natural Science*, ed. L. Hardy and L. Embree (Dordrecht: Kluwer, 1992), 43–44.

50. It is the presence of this concrete path of science that confirms our previous stance on cosmology as the working of constitution, that is, a reenactment of the production of the world. To clarify this point one can quote another paper of J. Ladrière: "The theoretical apparatus is thus not a description in the ordinary sense, as presentation of an entity, supposedly given, and of its properties, it is the characterization of something which is not a thing, but a *structural path along which a thing comes, from the ultimate horizon of every givenness, to the actual presence in which it is effectively given to apprehension.*" J. Ladrière, "Physical Reality: A Phenomenological Approach," *Dialectica* 43, no. 1–2 (1989): 138, emphasis added.

as a form of consciousness that humanity (as community) has of itself (although most cosmologists do not have it).[51] By revealing the *telos* in the historical path of cosmological explanation (as related to the representation of the ultimate origin of the universe),[52] phenomenological analysis discloses the hidden "theological" commitment in cosmological research, meaning that the beginning and the end of the universe in human thought is just a mode of this same thought speaking of its own beginning and its own consummation and implying a transcendent reference. Seen from a slightly different angle, this theological commitment corresponds to an attempt to know and see the universe as "all in all" being that primary existential memory in the human constitution which drives cosmological research.

In spite of doubts regarding objectivity and neutrality, scientific cosmology remains an extremely important and useful instrument in demonstrating just how human subjectivity affirms itself through its non-egocentric attitude toward the external world. The universe that science depicts as something different from us and devoid of our influence and presence, represents in fact the product of articulation through words and thoughts by humanity.[53] By creating a cosmological narrative, we affirm ourselves in a very nontrivial,

51. Cf. R. Aron, *Introduction à la philosophie de l'histoire* (Paris: Gallimard, 1938), 80.

52. See Nesteruk, *The Universe as Communion*, 250–54; "From the Unknowability of the Universe to the Teleology of Reason: A Phenomenological Insight into Apophatic Cosmology," in *Knowing the Unknowable: Science and Religion on God and the Universe*, ed. J. Bowker (London: I. B. Tauris, 2009), 78–81.

53. This is the reason why humanity, in a theological context, is called "hypostasis" of the universe. Cf. O. Clément, *Le Christ, Terre des vivants. Essais théologiques.* Spiritualité Orientale, n. 17 (Bégrolles-en-Mauges: Abbaye de Bellefontaine, 1976), 102; also D. Staniloae, *Experience of God* (Brookline, MA: Holy Cross Orthodox Press, 1998), 3. According to T. Torrance, "The fact that the universe has expanded in such a way that the emergence of conscious mind in it is an essential property of the universe, must surely mean that we cannot give an adequate account of the universe in its astonishing structure and harmony without taking into account, that is, without including conscious mind as an essential factor in our scientific equations . . ." *The Grammar of Theology* (Edinburgh: T. & T. Clark, 2001), 4, 6.

sometimes nonobvious sense.[54] Indeed, by creating a physico-mathematical narrative cosmologists lose control over the intentions that motivate them, since any deeper reflection upon what they have created is not in the focus of their reasoning. To understand the "data" lying behind this narrative one must consider them as manifestations of an expressive act, that is, to move from their *given meaning* to their *giving meaning*, from their pure phenomenality to the intentional life that generated them. By predicating the evolving universe and attempting to phenomenalize the mystery of its contingent origin, human subjectivity employs that intentionality which effectuates the *telos* of human subjectivity's ongoing incarnation as "coming to presence," assigning thus a dynamic character to personhood's manifestation.[55] As it was expressed by M. Munitz, "The goals of cosmology are goals of human beings." However, the universe as such benefits from these goals: "Through the measure in which they are reached, the universe becomes understood, perhaps for the first time anywhere throughout its vast stretches in space and time."[56] By reflecting the goals of humanity,

54. Cf. "By learning the ways of the universe and by reflecting upon them as they surface in the daily life of family and work and community, we take the first steps into a new form of human understanding and existence." B. Swimme, *The Hidden Heart of Cosmos* (New York: Orbis, 2005), 7. Cosmology allows one to know where one stands and to that extent has a sense of its own identity in relation to the world. Correspondingly, a possible lack of a cosmological sense deprives human beings and communities of their centeredness in the world as persons. According to F. Mathews, "A culture deprived of any symbolic representation of the universe and of its own relation to it will be a culture of nonplussed, unmotivated individuals, set down inescapably in a world which makes no sense to them, and which accordingly baffles their agency. . . . With no cosmological foundation for their identity, they invent precarious individual self-pictures, self-stories, ego-images, but their sense of who they are is tenuous." F. Mathews, *The Ecological Self* (London: Routledge, 1991), 5. As was differently expressed by Morris Kline, man's "ceaseless search for knowledge about himself, about the wonders of nature, the structure of the universe . . . gives point to lives which would otherwise spend themselves in orgies of meaninglessness." M. Kline, *Mathematics in Western Culture* (New York: Penguin, 1977), 423.

55. Cf. P. Heelan, "Nature and Its Transformations," *Theological Studies* 33 (1972): 493-502. See also in this context J. Compton, "Natural Science and the Experience of Nature," in *Phenomenology in America: Studies in the Philosophy of Experience*, ed. J. M. Edie (Chicago: Quadrangle, 1967), 82.

cosmology exhibits the traditional features of all mythologies, namely that the perceptible aspects of the universe are expressed in terms of human social, behaviorial, and existential concerns. In this sense the picturing of the universe as a historical process cannot avoid containing erratic facts associated with the human condition, to be more precise, with the intimacy of personal communion with the universe and the extent of not being attuned to it. Any imaginable attempt to disregard these facts and assess cosmology only on the basis of law-like ordered concepts would be incomplete and historically inadequate: in this case cosmology would provide us only with a fringe of the universe's phenomenality. The other "part" of the universe's phenomenality, which reflects the erratic fact of not being attuned to the universe, is rather reflected in poetic and artistic depictions relying on ecstatic acts of personal being in the universe as communion. This only confirms an already-formulated view that cosmology cannot dispense with anthropology—not only in a high philosophical sense, but in the mundane sense of human affairs.[57] The so-called mythological aspect of any cosmology thus naturally arises from the intention to interpret erratic features of the human universe through a reference to the astronomical order, and it is this aspect that brings with new force a "coherence" dimension in its epistemic justifications mentioned before: indeed mythologies never present gaps in their "explanations" and are invoked by the communities of

56. M. Munitz, "Kantian Dialectic and Modern Scientific Cosmology," *Journal of Philosophy* 48, no. 10 (1951): 338. As was differently expressed by D. Staniloae, through the reference to St. Maximus the Confessor, "man is called to become a world," that is, "macro-anthropos," through the eternal humanization of nature as eternal progress (Staniloae, *Experience of God*, 4).

57. As was provocatively conjectured by one author from the camp of the human sciences and arts, we need "a sort of mytho-scientific, neo-anthropomorphic" theory, one that would stay operational by combining the findings of mainstream science with conjectures based on mythological thought. This type of theory would map features of the universe through images taken from the domain of human social behaviour. . . . Although anthropomorphic theories might not be operational, they can lead to a better understanding of the universe." Y. Friedman, "Scientific Theory as One of the Fine Arts," *Leonardo* 26, no. 4 (1993): 361.

adherents on the ground of coherence of their claims. All this points to the simple truth that cosmology is not devoid of that dimension in its functioning which pertains to the human sciences.

2

Cosmology and Existential Phenomenology

Study of the Universe at the Crossroads of the Natural and Human Sciences

We know that man is closely allied with nature not only in the sense that he is part of it . . . but also, and even above all, in the sense that each impulse of his soul finds a profound and wholly natural substructure in the world, and in that way reveals to us a primordial quality of the structure of the universe.
–Minkowski, "Prose and Poetry," p. 244

Introduction:
Cosmology in Rubrics of Embodiment and Historicity

If cosmology, as a product of human activity, pretends to deal with the universe in its totality, assuming this totality in the natural attitude of mind as omni-spatiality and omni-temporality, it must exercise bravery in combination with a healthy skepticism in making pronouncements about the whole, by being only a tiny part of this whole. In spite of the fact that the philosophical mind, that is, a critical mind, accounts for its own incomprehensibility of this totality on the grounds of the finitude of humanity, this finitude is at the same time counterweighted by its alleged infinitude. For example, Kant, when discussing the analytic of the sublime, took it as the quintessence of human existence and asserted that as creatures dependent on the senses we are finite, while as creatures of reason we are capable of absolute freedom and independence. The universe can be paralleled with the sublime according to Kant's definition that "the sublime is that in comparison with which everything is small." Then, in his logic, nothing that can be an object of the senses can be considered as the sublime. However, and here Kant gives credit to humanity as capable of extending the realm of the senses: it is because "there is in our imagination a striving towards infinite progress, and in our reason a claim for absolute totality, regarded as the real idea, therefore this very inadequateness for that idea in our faculty for estimating the magnitude of things of the sense excites in us the feeling of the supersensible faculty."[1] It is this supersensible faculty that makes it possible to produce the idea of the universe and thus to pretend being commensurable with it. However, as was pointed out by J.-F. Lyotard, it is not in all cases that the sublime can be processed and integrated in the framework of experience. It can provide such an excess of sensibility and imagination that the finite human being cannot cope with it, and the origin of this Lyotard finds in the traumatic beginning of human life, which in turn entails a radical, that is, *unmasterable finiteness* of human beings (and thus, logically, their unmasterable incommensurability with the universe).[2]

1. I. Kant, *Critique of Judgment*, § 25, trans. J. H. Bernard (New York: Hafner, 1951), 88.
2. See J.-F. Lyotard, *Postmodern Fables*, trans. G. Van Den Abbeele (Minneapolis: University of Minnesota Press, 1997), 243.

Practicing cosmologists, however, who dare to speculate about the universe as a whole postulate the contrary, that is, that there is an intrinsic *commensurability* between their embodied intellectual abilities and the grand total of being that this intellect attempts to encompass. Here they follow Pascal, who in his *Pensées* anticipated a simple truth that in spite of the fact that the universe in its deep foundations and origin is hidden from the human grasp, to know the nature of things humanity must assume its proportion to nature: "It is a strange thing that [men] want to understand the beginnings of things and from that to progress towards the knowledge of everything: for that is a presumption as infinite as the object of their exploration. We surely cannot make plans like that without an infinite presumption—or an infinite capacity—like nature itself."[3] For most of working cosmologists this commensurability is taken as a premise, for otherwise their work could not even begin. In the history of philosophy of mathematics one can find an interesting trend of relating the finitude of human beings as creators of mathematics with its very origin.[4] Mathematics is seen as the mastery of the infinite so that only finite beings can make sense of the problem of the infinite; only they want to master it and thereby find themselves "confronted by the abyss of the unsurveyable, innumerable, and undecidable."[5]

The commensurability between humanity and the universe, assessed by the philosophers of existence, amounts to the fact that ". . . the physicist continues to think of himself as an Absolute Mind before the pure object and to count also as truths in themselves the very statements that express the interdependence of the whole

3. B. Pascal, *Pensées.* Selections, trans. and ed. M. Jarret-Kerr (London: SCM, 1959), 73.
4. See O. Becker, "Uber der sogenanten. 'Anthropologismus' in die Philosophie der Mathematik," *Phil. Anz. Jg.* III (1928): 369–87.
5. E. Ströker, *Investigations in Philosophy of Space* (Athens: Ohio University Press, 1965), 306.

of the observable with a situated and incarnate physicist."[6] Thus cosmologists exhibit that mode of thinking, which can be characterized as an anonymous and collective rational core, stripped of particular historical and personal incarnate situatedness in the world. The anonymous and collective nature of scientific thinking follows from the requirement that scientific knowledge and method must be "public," so that the results of science can be presented to a wide audience; this ultimately establishes science as linked with *episteme* (as being different from personal opinions). However, the requirement of "publicness" of science allows one to pose a question of its subject or subjects. It is clear that in order to achieve a universal communication and public structure of knowledge one must strip off all historically contingent and empirically individual characteristics of subjects, thus making them indifferent. In this case the truth established by one will be the same for another.[7] In other words, the commensurability between historical embodied consciousness and the totality of the universe is achieved by means of a radical reduction of all contingent properties of thinking (as related to the situation in history and science pertaining to it) and equating the scope of consciousness's receptacle to the totality of all articulated facts about the universe. Here the classical Cartesian constitution of the subject of knowledge is implied, namely, depriving this subject of any contingent characteristics makes this subject atemporal and ahistorical, attributing to its thinking capacity the qualities of absolute knowledge. The subject itself becomes universal and eternal so that it is here where the optimism of cosmologists lies: the knowledge of the universe is possible because of the universal and absolute nature of the subject of science. However, as it was pointed out by

6. M. Merleau-Ponty, *The Visible and the Invisible* (Evanston, IL: Northwestern University Press, 1968), 15.
7. Ibid., 15–16.

Kant in the context of the notion of the world, this optimism can be justified only as related to "thinking of" or "imagining of" the universe, not knowing it. A careful consideration of the Cartesian constitution of the subject of knowledge, historically corrigible in view of Kant's stance on it, tells us that this subject can only be achieved as a sort of eschatological reality, as an ideal that plays a regulative role with respect to particular scientific research. Then this implies that the representation of the universe as a whole as commensurable to scientific consciousness also represents an ideal, an asymptotic guideline that drives research, which will never exhaust this ideal in any particular stage of it.[8] Correspondingly, cosmological consciousness forgets the simple truth that its own facticity is related to embodiment in the physical, which, as science itself demonstrates, is not immutable and hence contingent. Thus there arises a general problem of reconciling historical contingency of forms of embodied consciousness with the alleged apodictic nature of its judgments about the universe.

Since cosmology pretends to deal with knowledge of the totality of the universe, it is legitimate to pose a question about the human capacity for such knowledge. This in turn implies an enquiry into the human condition in general. One of the basic assumptions of the European intellectual tradition (in its deviation from the characteristically premodern Christian stance on anthropology) is that human nature, or the human phenomenon, is part of the encompassing reality (that is, *nature*). The seeming efficacy of the methods of physics, for example, is transferred to anthropology and creates another conviction that one can exhaustively comprehend the meaning of human existence by methods of the natural sciences.

8. See an interesting discussion on the eschatological sense of the ideal of rationality in science in G. Goutner, "An Eschatological Aspect of Ethics and Epistemology," in *Scientific and Theological Thinking of Ultimate Questions: Cosmology, Creation, Eschatology*, ed. A. Grib (Moscow: St. Andrew's Biblical Theological Institute, 2008), 210–20 (in Russian).

What this means is that nature at large as well as human nature, as part of it, was implicitly conceived of as objects present-to-consciousness. In this case the reality of nature, as the collection of things and objects given to this anonymously present consciousness in their sheer facticity, had to be apprehended by science and knowledge through revealing the essential, universal, and common characteristics of these objects. Since human nature was considered a part of nature at large, it acquired the features of things (in the phenomenality of objects) within nature that reflected its shared physical and biological characteristics.

The major difficulty with this naturalistic stance is that it predicates human nature as being part of and determined by nature at large while this very nature can be attained only from within its particular fragment, that is, *human nature*. It becomes clear that a philosophical ideal for radical enquiry about the knowledge of nature (or being) as devoid of any presuppositions about what has been sought can hardly be achieved, because philosophy is created by human beings who cannot be removed from the central and initial point of any enquiry about nature (or being). Thus it is understood that the question of nature, and human beings in nature in particular, is most intimately connected with the question of *how this being can be attained*. Since it seems evident that the comprehension of being can only be reached from within the *experience* of what human beings themselves are, the attainment of nature (being) is always linked to that particular being which humans are. Thus if we refer to the universe as being a "part" or "mode" of being in general, one must admit that the being of the universe is always disclosed through the being of human beings.

Speaking differently, a *concrete* human existence (as incarnate existence in situation) becomes the root and source of access to the universe. But what is so particular about human existence? What does this existence mean? Existential phenomenology assigns to the term "existence" a special meaning by affirming, for example, that this "term" serves not to express that something actually belongs to the realm of existing realities, but to indicate *that mode of being which is proper to man and precisely constitutes him as human be-ing.* Existence in this sense is only intrinsic to human beings, and it is this existence that makes them a fundamentally special mode of being. Thus the *existence* of the universe in this view can only be understood as the transferral of the mode of human existence to what we call the universe. In a different language, the universe is manifested or subsisting through articulation by human beings.[9]

9. One can employ a theological terminology by saying that the universe is enhypostasized by human being. The Greek terms *enhypostatic* or *enhypostasis*, which were introduced in theology by Leontius of Byzantium in the context of Christological discussions of the sixth to seventh centuries, have meaning according to *A Patristic Greek Lexicon* as "being, existing in an hypostasis or Person," "subsistent in, inherent." Florovsky refers to the terms used by Leontius by saying that *enhypostasis* points toward something that is not self-contingent, but has its being in the other and is not contemplated as it is in itself. *Enhypostasis* is the reality in the other hypostasis. See, for example, G. Florovsky, *The Byzantine Fathers of the Sixth to Eighth Century* (Vaduz, Liechtenstein: Büchvertriebsanstalt, 1987), 191-203. An example, which illustrates what the existence in a hypostasis or person means, can be borrowed from a sphere of theological anthropology that asserts that "man is hypostasis [personality] of the cosmos, its conscious and personal self-expression; it is he who gives meaning to things and who has to transfigure them. For the universe, man is its hope to receive grace and to be united with God." P. M. Gregorios, *The Human Presence: Ecological Spirituality and the Age of the Spirit* (New York: Amity House, 1987), 83. The universe as *the expressed and articulated existence* is possible only in human hypostasis, that is, it acquires some qualities of existence if it is reflected in the personality of humanity. Using the words of Maximus the Confessor, every intellection about the universe inheres as a quality in an apprehending being (cf. his *Two Hundred Texts on Theology and the Incarnate Dispensation of the Son of God*, 2.3). The universe thus acquires qualitative existence in the being who apprehends it. Prestige, in order to illustrate how the apprehending knowledge becomes hypostatic existence, refers to Clement of Alexandria (*Stromata*, 4:22, 136:4) in order to articulate the point that speaking of knowledge, "apprehension extends by means of study into permanent apprehension; and permanent apprehension, by becoming, through continuous fusion, the substance of the knower and perpetual contemplation, remains a living hypostasis. This appears to mean that knowledge becomes so bound up with the being of the knowing subject, as to constitute a permanent entity." G. L. Prestige, *God in Patristic Thought* (London: SPCK, 1955), 176. The link between the universe as articulated existence and the apprehending being is not ontological, but rather hypostatic or personal. A patristic theologian would say that existence of the universe as the articulated existence is *hypostatic* existence, that is, the universe is *enhypostatic.*

Existential phenomenology considers human existence as a primordial phenomenon, as an initial fact of any further philosophizing about the world, which cannot be fully reduced to something else or demonstrated by reference to "the outside" of this existence. Theology in its "explanation" of the mystery of human existence asserts the creaturehood of humanity, as the radical otherness to God serving at the same time the basis for the unity and hence objectivity of all hypostatic (personal) humanity. The existence of a particular human person is not something that is inherent or latently present in the world, but represents an event that is initiated *in* creation but is not *of* creation. Christian anthropology argues that human subjectivity is inconceivable without embodiment, so that human existence is constituted by the unity of the body and soul. However, this unity is not an elementary fact of being-in-the-world, which can be easily taken as an initial point. In a Christian perspective the unity of body and soul is hypostatic, so that genuine existence is rooted in the mystery of the human hypostasis, which transcends the world. This existence has its ultimate source in the transcendent Divine, in whom human soul and body inhere in their hypostatic unity. Christian anthropology asserts that human beings are constituted not only through their "natural" predisposition to transcend their own subjectivity in the world, but they also have some ability and will to transcend the very relationship between their subjectivity and the world, that is, to transcend being-in-the-world;[10] this makes it possible to articulate *nature* in general (and human nature in particular) not only through its relationship to human intentional consciousness, but to position nature with respect to God. In this logic it is through this otherworldly dimension that humanity can have ability to articulate the world as a whole by

10. This is terminology from M. Heidegger, *Being and Time* (Oxford: Blackwell, 1998), 78.

effectively positioning itself outside the world; human consciousness is commensurable with the entirety of the universe due to the similarity of their foundations in the other of the world. In this sense *a* universe of a human being would be always existentially co-present and unique to this being by the fact of creation, although its context as related to living experience would become completely contingent and probably incommunicable to the other. This would create a problem of articulation of *the* universe as totality within history, as communion in community of human beings. Christian theology responds to this by asserting cosmic history as part of the history of salvation,[11] thus exercising a move similar to that of existential phenomenology.

It is clear that the views of existential phenomenology, which in its methods mimics that which has been asserted in theology, are stated as radically different in comparison with the scientific natural attitude that treats humanity as thing among other things. According to existential phenomenology all particular manifestations and all meanings of things in the world can only be understood and receive their foundation as correlates of the human hypostatic subjectivity whose mystery itself is never exhausted by means the of physical.[12] This implies that any scientific activity, including cosmology, cannot discard the tacit presence of human existence in its assertions of reality. Any attempt to speculate about the universe as if it is done in the name of anonymous and impersonal absolute consciousness becomes no more than a fallacious ambition of the human reason to produce syllogistic insights on the nature of the universe by disregarding the transcendental conditions of its presence in the

11. See O. Clément, *Le Christ, Terre des vivants. Essais théologiques.* SpiritualitéOrientale, n. 17 (Bégrolles-en-Mauges: Abbaye de Bellefontaine, 1976).
12. In this sense phenomenology reproduces a Christian stance on anthropology as being apophatic because it tells one exactly what the human person is not. See O. Clément, *On Human Being: A Spiritual Anthropology* (London: New City, 2000), 30.

world. This entails that cosmology, with respect to its claims about the universe, cannot be consistently understood without taking into account the nature of human beings who create cosmology. As it was expressed elsewhere: "Man and the universe are like two parts of the same book which can be understood only by means of one another."[13]

Subjectivity and Incarnate Existence

Existential phenomenology and existential tradition in general object to that stance of modern science which positions humanity as part of the outward reality (nature, world, universe) as if this reality existed independently of the constitution of this reality through human insight. Rather than beginning with nature and then seeing human life as its part, it argues for the reversal of procedure, that is, seeing the human reality (as existential events) as the primary fact of any enquiry so that nature is to be seen in light of it. This implies that not only human reality cannot be a subordinate part of the natural universe, but the methods that are used to study the universe are not applicable to the phenomenon of humanity and human reality, because the latter is never an "object" present to itself; that is, subjective reality can never be made an "object-thing" in spite of the fact that this subjectivity is always tacitly assumed in all modes of vaguely understood objectivity. Thus existential phenomenology insists that the ultimate meaning of the universe can be unfolded only by starting with human reality as separate and different in kind from the realm of the objectivized nature.[14]

13. A. Dondeyne, *Contemporary European Thought and Christian Faith* (Pittsburgh: Duquesne University Press, 1958), 10. Cf. H. Köhler, "The Relation between Man and the World," *Analecta Husserliana* CVII (2011): 37.
14. Cf. Kant's assertions on irreducibility of reason *Critique of Pure Reason*, A553–57.

The existence of humanity is intimately connected with the interiority of its intentional consciousness. But the affirmation of this interiority as a definition of the physically real human being leads naturally to the transcendence of the sphere of pure subjectivity through embodiment in order to become something special and concrete, that is, to be placed in a particular space and time. Human beings find themselves already in the world, in a particular place and at a particular time, and it is through subjectivity they try to find understanding of their meaning, which cannot be reached without transcending toward the world. This involves this subjectivity in a relationship with the world thus constituting the foundation of all meanings, associated with the articulation of the universe as a mode of this relationship that "produces the natural and antepredicative unity of the world and our life, being apparent in our desires, our evaluations and in the landscape we see, more clearly than in objective knowledge, and furnishing the text which our knowledge tries to translate into precise language."[15]

Existential phenomenology argues that the natural attitude with its opposition between subject and object must be overcome in order to discover the primordial field of the human being-in-the-world, in which the actual contact with the thing "awakens within me a primordial knowledge of all things, and to my finite and determinate perceptions' being partial manifestations of a power of knowing which is coextensive with the world and unfolds it in full extent and depth."[16] By being completely encompassed and permeated by the world, there cannot be anything in human beings that belongs to the world but does not belong to human beings.[17]

15. M. Merleau-Ponty, *Phenomenology of Perception* (London: Routledge, 1962), xx.
16. Ibid., 430.
17. This thought can be interpreted as *consubstantiality* between humanity and the universe. However, it does not preclude humanity from being incommensurable to the universe in terms of extended space and time. In a theological stance, however, this does not mean that human

The involvement of human beings in the world, the access to the world, and to its meanings, has its deepest ground in consciousness, which forms de facto the *universal and sole medium of access*. However, when one asserts the primacy of consciousness one means the perceptual consciousness that makes the contact with the reality that is not consciousness itself.[18] This is the sphere of immediate sense-perceptions and meanings, which forms the foundation one needs in order to construct the world through a scientific thematization and conceptualization. Consciousness being in an intricate link with the world does not represent itself entirely as an agency responsible for the constitution of the meaning; for, being responsible for the constitution of the meaning, it at the same time is perceptive of the meaning. Phenomenology doubts the legitimacy of any claim about existence of the ever-made preexistent world without human beings; but it also denies the possibility that the meaning of the world, as given together with the meaning of man, can have its foundation outside the bundle of the world and man.

beings cannot have a mode of existence different from the physical world. In addition to this, one must add that the term "consubstantiality" must be used with caution. The universe, as cosmology claims, consists of 96 percent of matter that has no direct physical interactions with those atomic structures that constitute the visible universe and human bodies, in particular. However, one can speak in this case of consubstantiality with the universe in a transcendental sense: the invisible parts of the universe are constituted through cosmological research as necessary cosmological elements for human existence. Thus the term "consubstantiality" reflects not only that link with the universe, which is empirically available, but has transcendental origin.

18. In G. Marcel's words: "My actual state of consciousness, which is bound up with the position of the organic body that it expresses, is the landmark in relation to which the infinite multiplicity of what can be thought by myself as existing is ordained. All existence can be traced back to this landmark, and outside of all relation to it, it is only by an abstraction that we can think existence. To think a thing as existing is to think oneself as the perceiver, it is to extend one's experience in such a way that it comprehends even that which it appeared to leave outside itself. This does not imply the kind of subjective idealism which attributes a privileged value to the immediate data of perception, but only the affirmation that existence supposes a relation to an immediate thought in general, that is to say to my thought." G. Marcel, *Metaphysical Journal* (London: Rockliff, 1952), 14.

The sense of this "bundle" can be explicated through the terms "dialogue" and "participation." This view implies that science does not represent an objectivizing description of the world on the side of an epistemological subject, which does not belong to the world and extracted from it. Cosmology, taken in its ontological meaning, is an *event* of the universe itself. In this sense cosmology is limited by an ontological status of the participant of the world. Here is an interesting explication of what is meant by the dialogue between humanity and the universe and participation in the universe's constitution. According to J. Kockelmans, the meaning of the world arises in the encounter between man and the world and "exists only in an interplay of question and answer. . . . Meaning arises in a dialectic relationship between man and the world, but it is not possible to say which of the two first begins the 'interplay' and which of the two first gives meaning to the other."[19] In similarity with this, J. A. Wheeler asserted a dialogism in relation between man and nature: physical reality reveals itself as an evolving complex of meanings in the course of the interplay between questions and answers that the human subject addresses *to* and receives *from* that "out there" which is articulated by human observers as the physical reality and nature. He writes: "Physics gives light and pressure-tools to query and to communicate. Physics also gives chemistry and biology and, through them, observer-participators. They, by way of the devices they employ, the questions they ask, and the registrations they communicate . . . develop all they know or ever can know about the world."[20] The world is not a clock-like machine which has been preconstructed and then discovered by human observers; it is a self-

19. J. Kockelmans, *Phenomenology and Physical Science* (Pittsburgh: Duquesne University Press, 1966), 53.

20. J. A. Wheeler, "World as System Self-Synthesized by Quantum Networking," *IBM Journal of Research and Development* 32 (1988): 5.

synthesized system, coming into existence through the articulation of impersonal reality "out there" via questions and answers processed by a collective of persons-observers who are capable of establishing the meaning and interpretation of their observation-participancy, which ultimately leads to the constitution of the integral view of nature.

By defining human existence as being-in-the-world, existential phenomenology asserts an inherent relationality between humanity and the world, the relationality that constitutes their ontology as relationship, not just simple epistemological coordination. Then "knowledge" can be treated as a special mode of this relationship. The world-pole of this relationship then appears only as its projection to human consciousness, that is, things are presented to subjects as they "look" at them, but not what they are.[21] Knowledge as a mode of relationship between humanity and the world can be described as a particular intentionality of the embodied consciousness toward the world, which treats this world as existing outside and independently of the sphere of subjectivity. This corresponds to the natural attitude that dissects the immediacy of being-in-the-world, extracting from it only the mode of its explicit or visible "presence" to consciousness. However, the exercise of the natural attitude presupposes a kind of "prescientific" knowledge, as awareness of the surrounding "medium" in which human incarnate subjectivity functions, but which "shows" itself in its empirical absence.[22] This "prescientific"

21. As was expressed by P. Florensky: "All that which is knowable by us is that which is grasped by us, and by us transformed into ourselves. . . . Through the act of knowledge all becomes our likeness." P. Florensky, "Macrocosm and Microcosm," in *Apology of the Cosmos* (St. Petersburg: Russian Christian Humanitarian Institute, 1994), 184 (in Russian). Things appear to humanity only within the context imposed by the very fact of its existence. In this sense the existence of things encodes the presence of humanity. By observing some things one can infer to the fact of existence of human beings (Heidegger, *Being and Time*, 88).

22. F. Nietzsche called this awareness "great reason of flesh." Cf. F. Nietzsche, "Von den Verächtern des Leibes," in *Also Sprach Zarathustra: Ein Buch für Alle und Keinen*, I, 4, in *Nietzsches Werke: Kritische Gesamtausgabe*, ed. G. Colli and M. Montinari (Berlin: Walter de Gruyter, 1967-91), trans. Thomas Common under the title *Thus Spoke Zarathustra*, vol. 11 of *The Complete Works of Friedrich Nietzsche*, ed. Oscar Levy (New York: Macmillan, 1919). Flesh does not mean the

knowledge can be described in terms of incarnation: "To be incarnate means to appear to itself as body, that is, this particular body . . ."[23] Or, in different words: "Incarnation is the situation of *a being who appears to himself to be*, as it were, *bound to a body*."[24] In Marcel the ontological event when "existence comes into being" is linked to incarnation, which is a primary and incomprehensible mystery: all attempts to state what incarnation means are approximate and ultimately inadequate. Marcel calls the situation of incarnate existence a "concrete reality" that is neither exclusively physical nor psychical but marks the limits of actions of an incarnate subject.

The body, as individualized flesh in empirical space and time, plays a central role by coordinating the incarnate conscious self, with the rest of what this self treats as the objects of its intentional grasp. This coordination can be expressed in the language of consubstantiality of the human flesh with the material content of the universe: "To say that something exists is not only to say that it belongs to the same system as my body . . . , it is also to say that it is in some way united to me as my body is."[25] The sensible universe then represents

body, which, extended in the space of the world, is found there perceived or rather sensed, but it means my unique body, which alone senses the bodies of the world. It is *my* flesh that senses bodies that themselves do not sense: it senses everything else only by sensing itself sensing. E. Levinas speaks of a similar mode of awareness as "non-intentional consciousness," which "accompanies all intentional processes of consciousness and of *myself* who, within this consciousness, 'acts' and 'wishes,' and has 'intentions.' This is awareness of consciousness, indirect and implicit, without an initiative to get back on myself, without aim; a passive consciousness as time which passes and makes me old without my involvement." E. Levinas, *Éthique comme philosophie première* (Paris: Payot & Rivages, 1998), 80.

23. G. Marcel, *Du Refus à l'Invocation* (Paris: Gallimard, 1940), 31.

24. G. Marcel, *Being and Having* (London: Collins, 1965), 16 (emphasis added), here defines existence in a way similar to that which was later used by E. Levinas, who spoke of *hypostasis* as an ontological event wherein "the existent contracts its existing." Levinas, *Time and the Other* (Pittsburgh: Duquesne University Press, 1987), 43.

25. Marcel, *Being and Having*, 15. Cf. P. Florensky: "Biologically, all that which surrounds us is our body, the extension of our body, the aggregate of our complementary organs" (Florensky, "Macrocosm and Microcosm," 184). W. Heisenberg expressed a similar thought by assigning to technology the meaning of extended biological functions: "Technology no longer appears as the result of a *conscious human* effort to extend man's material powers, but rather as a large-scale biological process in which man's organic functions are increasingly transferred to

the extension of the body of humanity in a very nontrivial sense: it manifests the ongoing incarnation of humanity.[26] Since incarnation, or "being in situation," cannot be objectified in terms of external constraints, the same is true with respect to space and time which, as part of one's being in situation, come together with this situation. Space expresses here some characteristic of dynamics of life, being thus a relational "entity" with respect to that human agent who makes room for itself as place and space. Making space constitutes a part of that creative development which accompanies any incarnation or being in situation, so that space acquires some specific forms of hypostatic expression of one's being, providing thus forms of communication of different persons as different "beings in situation."

The coordination with the world is understood as an inherent consubstantiality, and it refers to what human beings share with each other and with the substance of the universe in a transcendental (not only empirical) sense. This means that "matter" and "nature" have transcendental meaning as different expressions of that consubstantiality of the *flesh* with the whole universe. It is in this sense that one can argue that consubstantiality must not be understood only as sharing of nature and physical substance, but rather as a fundamental feature of humanity that relates itself to

his environment." W. Heisenberg, *The Physicist's Conception of Nature* (London: Hutchinson Scientific & Technical, 1958), 19-20. P. Heelan proposes the manifest image of nature as the totality of empirical horizons reached by human subjects through embodied intentions. In this case the body as subject is used by these intentions to extend itself into the environment and then to be adapted to any bodily extension. P. Heelan, "Nature and Its Transformations," *Theological Studies* 1 (1972): 497-501.

26. Marcel, *Metaphysical Journal*, 259. Certainly this thought can be extended by saying that the past is not only related to my body as such. It is related to my conscious body, which attempts to realize its past also in a strictly historical sense. To understand our past as history in general means to understand deeply and incorporate it in ourselves in order to become ourselves. One must attempt a breakthrough toward our past as origin from which all our present and future emerges. Cf. K. Jaspers, *Weltgeschichte der Philosophie, Einleitung* (München: R. Piper, 1982), 84, 117.

the universe. In this sense consubstantiality of flesh is related to the similarity in orders of creation.

Incarnation in flesh reveals itself *in* space through a particular body but is itself not *of* space: the incarnate consciousness manifests itself as nonlocal (that is, transcending the boundaries of a body) and stretching across the universe, thus carrying the whole world together with the body;[27] while being *in* a body it is not *of* the body, and this is the reason why it represents a typical situation when, in spite of its obvious personal *presence* as a body, the foundation of a sheer contingent facticity of this presence is unavailable to consciousness's grasp. Thus incarnation cannot be phenomenalized even to the reflective consciousness of the incarnate person (no access to the mystery of one's birth). This implies that every particular experience of the universe, as reflected and articulated through incarnation, cannot be studied in the same way as the universe is studied in cosmology in the phenomenality of objects. In cosmology the collective and anonymous subjectivity creates a picture of the universe where a concrete human consciousness acquires the features of a contingent epiphenomenon of the physical by disregarding the centrality of the multihypostatic humanity as being the center of disclosure of the universe from within the human, Divine-oriented history.

Humanity's Position in the Universe and the Paradox of Human Subjectivity

In spite of the simple philosophical truth that everything affirmed through observations and measurements receives its meaning and interpretation from within human subjectivity, modern science is

27. This is a premise of commensurability between consciousness and the universe based in consubstantiality.

still wrestling with the idea that reality, let us say physical reality, cannot be alienated completely from the acts of an apprehending intellect. Science pursues in many of its areas the methodology that is historically identified with the name of Descartes and in which it attempts to establish an ideal of objectivity of its theories by making a split between subjective conditions of any human knowledge and the meaning of the knowledge it tries to objectify. According to this view the object of knowledge is an objectively existing world, which is fully explainable and can be expressed in precisely formulated laws. The fundamental premise of the natural attitude of science is that the world exists in itself in its entirety and possesses a rationality that can be fully understood. Scientism, as a radical implication of the natural attitude, follows a definite criterion of objectivity, based on the principles of quantity: whatever can be quantified and mathematized according to certain rules, is objective by definition. All those aspects of "reality" that cannot be quantified are not objective and therefore meaningless, whereas the subject of knowledge is treated as pure consciousness, which is fully transparent to itself and faces the rational world objectively, that is, it grasps the universal aspects of reality (the world is reduced to our consciousness of it).[28] This ideal of knowledge was a certain reaction against embodiment and an attempt to free the knowing subject from the body's situatedness or localizing entanglements. In view of some trends in contemporary philosophy of science, this ideal of knowledge (which can be qualified as a sort of foundationalism) is

28. One should remind the reader that the natural attitude of exact sciences to what they called the "real" and "objective" stands in opposition to the *philosophical attitude*, which does not take for granted any presupposition of the given experience and anything pre-given as obviously existing. The latter considers it an error to assume that scientific knowledge is a pure mirroring of the world-in-itself, as well as the very assumption that there can be established an access to the world-in-itself in its absolute detachment from human senses and intelligibility. The overcoming of scientism and the natural attitude assumes thus that one cannot substitute the "objective" world of physical sciences for the fullness of experience of the living world by human conscious beings.

arguable. It is enough to point toward research on transcendental appropriation of modern physics in order to realize that the very criteria for objectivity changed by taking into account the historical and instrumental context of science, which brings with it the human factor.[29] This latter is ultimately linked to existential phenomenology's stance on embodiment, which avoids any lapsing in the foundationalism because it positions its view on the cognizing self as neither being grounded by the world nor being itself the ground of the world.

In order to illustrate the point that knowledge in modern cosmology is imbued with human presence, one may consider a simple example from popular scientific books that give an account about the place of humanity in the overall structure of the universe. If one tries (in the natural attitude) to demonstrate the whole grandeur of the world in terms of typical sizes of objects, putting atoms, molecules, DNAs, and so forth together with mega-objects like planets, stars, galaxies, clusters of galaxies, and even the whole universe, then humanity finds itself in a somewhat strange situation because the planet Earth inhabited by human life occupies a tiny portion of the volume of the visible universe; also the spatial scale of the human body (10^2 cm) is negligible in comparison with the radius of the visible universe (10^{28} cm). In a similar vein if the universe had a beginning 13.7 billion years ago, and then developed to its present state, it is not difficult to realize that the phenomenon of humanity came into existence at a very late stage in the history of the universe, so that the universe was devoid of human life (and hence devoid of self-expression) during the most part of its "history." It is then not difficult to realize that if the human presence in the universe is judged from the point of view of its spatial and temporal dimension,

29. See more details in M. Bitbol, P. Kerszberg, and J. Petitot, eds., *Constituting Objectivity: Transcendental Perspectives on Modern Physics* (Dordrecht: Springer, 2009).

human beings, considered as physico-biological bodies, turn out to be a contingent and *insignificant* part of the universe.

The paradox that is present here arises when one realizes that the very representation of the universe as a whole, including all different levels of its physical structure, and the very positioning of all objects in the universe against a spatial grid, is the product of human intellectual activity. The paradox is obvious: the finite, even insignificant embodied human agencies in the vast universe articulate the entire universe from a point-like position in space and time. Humanity actualizes in knowledge the totality of the universe as its intentional correlate and this manifests a fundamentally nonlocal essence of human presence, being a quality and a mode of being that transcends the finitude of its corporeal place, as well as all particular objects and laws associated with this corporeity.

One can see that if cosmology positions humanity in the vast universe, assuming that the universe (as entity) is preexistent (with respect to human intelligence), then humanity represents a particular type of "objects," passively dependent on the universe. The so-called "anthropic inference" in cosmology refines assertions about humanity's position in the universe by recapitulating consubstantiality of the universe and humanity in quantitative terms pertaining to specific embodiment.[30] In a way, this is a trivial observation that affirms self-consistency of human knowledge of the universe with the physical conditions of embodiment that make knowledge possible. However, the anthropic inference deals with the necessary conditions for physical and biological existence of humanity and does not cover the hypothetical realm of its sufficient

30. Literature on the "anthropic inference" in cosmology and associated fine-tuning is vast. See, for a classic example, J. Barrow and F. Tipler, *The Cosmological Anthropic Principle* (Oxford: Oxford University Press, 1986), as well as J. Barrow, S. Morris, S. Freeland, and C. Harper, eds., *Fitness of the Cosmos for Life: Biochemistry and Fine-Tuning* (Cambridge: Cambridge University Press, 2008).

conditions, related to humanity's intellectual capacity.[31] In this sense the famous and simplistic characteristic of humanity as "microcosm" turns out to be fundamentally insufficient.[32] The mystery of the sufficient conditions remains obscure in the same sense as the inability to account for the contingent facticity of all, including consciousness itself. In the natural attitude, where one attempts to explain the origin of consciousness as the epiphenomenon of the physical and biological one fails to recognize that it attempts to explain itself from itself. This fallacious logical circle originates in the fact that physics and biology operate in the framework of the already-given consciousness, but this very consciousness never becomes their subject matter.[33] A. Gurwitsch comments on this, accentuating the personal dimension of embodied consciousness: "What is of decisive and crucial importance

31. The sufficient conditions become actual in the present state of technology when humanity can control the factors of life's existence on the planet Earth from the side of so to speak "negative conditions": indeed it is in capacity to exterminate life on Earth so that the future continuation of life depends not only on the natural conditions and possible disasters which can terminate this life, but also on a conscious desire to have this life. This desire, however belongs to the sphere of the human morality and humanity's vision of its own destiny and that is why is not entirely controlled by the physical factors. In this sense the sufficient conditions of existence of humanity in the universe depend on humanity's own vision of its place in the universe, its importance or nonimportance for the fate of the universe itself. Correspondingly, cosmological research turns out to be important as contributing to the realization of cosmic goals of humankind. See discussion in A. Nesteruk, *Light from the East* (Minneapolis: Fortress Press, 2003), 195-208.
32. Being popular in classical Greek philosophy, the idea of microcosm was criticized in Christian literature because it did not take into account the intellectual abilities to disclose the sense of the universe. Consubstantiality is triviality and, according to Gregory of Nyssa, "There is nothing remarkable in Man's being the image and likeness of the universe, . . . in thinking we exalt human nature by this grandiose name (microcosm, synthesis of the universe) we forget that we are thus favouring it with the qualities of gnats and mice" (quoted in O. Clément, *On Human Being*, 34). Gregory's comparison of humanity with mice is remarkable, because it is very popular nowadays to advocate, on the grounds of biological evolution, that humanity did not progress too far from the animal-like state because humans and mice share 96 percent of their DNA.
33. B. Carr comments on this situation in physics: "That physics has little to say about the place of man in the universe is perhaps not surprising when one considers the fact that most physicists probably regard man, and more generally consciousness, as being entirely irrelevant to the functioning of the universe." B. Carr, "On the Origin, Evolution and Purpose of the Physical Universe," in *Modern Cosmology and Philosophy*, ed. J. Leslie (New York: Prometheus, 1998), 152.

is not whether the existence of consciousness is conceded or denied but rather that, even if this existence is conceded, consciousness and whatever pertains to it are considered as 'private' and thus not on principle subject to scientific investigation."[34] It is because science cannot accommodate the dimension of personhood that it has to abandon the reference to embodiment at all and to treat consciousness as a medium of access that is hypostatically uniform and thus nonobservable. Through such an oblivion human presence becomes irrelevant to the universe, whereas sciences themselves become obscure.[35] As it was eloquently put by Merleau-Ponty: "Scientific points of view, according to which my existence is a moment of the world's are always both naïve and at the same time dishonest, because they take for granted, without explicitly mentioning, it, the other point of view, namely that of consciousness, through which from the outset of a world forms itself round me and begins to exist for me."[36]

The ambivalence in assessing of humanity's position in the universe can be expressed in terms of a famous philosophical *paradox*

34. A. Gurwitsch, *Phenomenology and the Theory of Science* (Evanston, IL: Northwestern University Press, 1974), 133.
35. A. Gurwitsch described this state of affairs as the inability of science to give an account of its own possibility and efficacy: "All questions concerning human reason . . . are eliminated from the sciences; . . . if the human mind and human rationality are either overlooked or explained away in a naturalistic fashion, the sciences themselves become unintelligible; . . . the sciences appear as most ingenious technical devices which one may learn to use . . . but whose interior mechanism and functioning remain utterly obscure." A. Gurwitsch, *Studies in Phenomenology and Psychology* (Evanston, IL: Northwestern University Press, 1966), 399–400.
36. Merleau-Ponty, *Phenomenology of Perception*, ix. Apart from inadequacy in the comprehension of foundations of science, the whole stream of thought can be supplemented by a spiritual sentiment, namely that separating the world and the universe from the functioning conditions of human subjectivity, science based on the natural attitude, by using the words of Russian philosopher S. Bulgakov, acquires lifeless intentionality and orientates us in the kingdom of dead things. Bulgakov, *Philosophy of Economy* (Moscow: Nauka, 1993), 207 (in Russian). French phenomenologist E. Minkowski expressed in a similar way that everything science touches "becomes something immobile, dead nature." E. Minkowski, "Prose and Poetry (Astronomy and Cosmology)," in *Phenomenology and the Natural Sciences*, ed. T. J. Kiesel and J. Kockelmans (Evanston, IL: Northwestern University Press, 1970), 246.

asserting that while being *in* the universe, humanity is not *of* the universe, that is, in a certain sense, it transcends the universe by "holding" it through humanity's grasp. The dualism in human position in the world, which is present in this paradox, constitutes the inherent feature of any cosmological discourse that has to reconcile the locality and contingency of the cosmic position of humanity with its abilities to transcend this locality and encompass in theory the universe as a whole. Any naturalistic attempt to suppress or subvert the essential ambiguity of consciousness of being in the world[37] and at the same time, of the whole world, distorts a truly scientific interpretation of the universe.

The above-mentioned paradox was coined by E. Husserl as "the paradox of human subjectivity being a subject for the world and at the same time being an object in the world."[38] However, in its essence it was known since ancient times, and Kant, for example, expressed it in his *Critique of Practical Reason* as the difference in appreciation of the things that fill the human mind with "the starry heavens above and the moral law within": "The former begins from the place I occupy in the external world of sense, and enlarges my connection therein to an unbounded extent with worlds upon worlds and systems of systems, and moreover into limitless times of their periodic motion, its beginning and continuance. The second begins from my invisible self, my personality, and exhibits me in a world

37. In the words of A. Gurwitsch, "[Consciousness's] acts, on the one hand, depend functionally upon extra-consciousness facts and events, in this sense being effects of the latter, and, on the other hand, have presentational and cognitive function with regard to all mundane events and acts, including those upon which they depend causally." *The Field of Consciousness*, in *The Collected Works of Aron Gurwitsch*, vol. 3 (Dordrecht: Springer, 2010), 160.

38. Husserl formulated this paradox as follows: "Universal intersubjectivity, into which all objectivity, everything that exists at all, is resolved, can obviously be nothing other than mankind; and the latter is undeniably a component part of the world. How can a component part of the world, its human subjectivity, constitute the whole world, namely, constitute it as its intentional formation . . . ?" E. Husserl, *The Crisis of European Sciences and Transcendental Phenomenology* (Evanston, IL: Northwestern University Press, 1970), 179.

which has *true infinity*, but which is traceable only by the understanding, and with which I discern that I am not in a merely contingent but in a universal and necessary connection, as I am also thereby with all those visible worlds."[39] The paradox received numerous formulations and interpretations,[40] so we must provide the reader with a few clarifying references. M. Merleau-Ponty rephrased the same paradox in the context of the tension between two descriptions of the human condition: "On the one hand man is a part of the world; on the other, he is the constituting consciousness of the world"; this tension is to be overcome on the ways of existentialism.[41] E. Fromm, departing from a psychological dimension, gave to this paradox a status of "existential dichotomy" arising from the fact that, according to him, man emerged in being as an "anomaly" and "the freak" of the universe, whose being is in a state of constant and unavoidable disequilibrium, anxiety, dissatisfaction, and restlessness, which follows from being part of nature and transcending it.[42]

39. I. Kant, *Critique of Practical Reason*, Conclusion. ET: *Kant's Critique of Practical Reason and Other Works on the Theory of Ethics*, trans. T. K. Abbot (London: Longmans, 1959), 260 (emphasis related to "infinity" has been added). Kant's usage of the adjective "infinite" has a twofold meaning here: on the one hand he speaks of the "true infinity" as that which is related to what things really are, not simply to their appearances. On the other hand, speaking of the "infinite elevation," he de facto makes a statement of the incommensurability of human beings with all other natural things. The incommensurability, which in different words can be described as the unbridgeable gulf between human persons who articulate the things in the universe, and these very things, which have no power of self-reflection and articulation at all. More than that, the very notion of infinity as arising from freedom and inexhaustibility of the moral law is typically a human attribute in spite of a fundamental finitude of the human embodiment and the fact that it is because of this that anything received by us is conditioned, that is, finite. See more details in A. W. Moore, "Aspects of the Infinite in Kant," *Mind* XCII (1988): 205-23.
40. See D. Carr, *Paradox of Subjectivity* (Oxford: Oxford University Press, 1999).
41. "There are two classical views: one treats man as the result of the physical, physiological, and sociological influences which shape him from the outside and make him one thing among many; the other consists of recognizing an a-cosmic freedom in him, insofar as he is spirit and represents to himself the very causes which supposedly act upon him." M. Merleau-Ponty, "The Battle over Existentialism," in *Sense and Non-Sense* (Evanston, IL: Northwestern University Press, 1982),71-72.
42. In Fromm's words, "He [man] is set apart while being a part; he is homeless, yet chained to the home he shares with all creatures. Cast into the world at an accidental place and time, he is forced out of it, again accidentally. Being aware of himself, he realizes his powerlessness and

Similarly to Fromm, R. Ingarden describes the existential dichotomy as a very special and doubly complexioned result of man's feeling of being, on the one hand, quite alien to everything that happens in nature independently of him, so that he sees himself deprived by it of any kindly help and almost loses trust in fate; on the other hand, "in his pure and autonomous essence he feels himself to be something that stands out above nature, something that is so much more dignified than purely physical processes or what transpires in animals, that he cannot feel in solidarity with nature and live fully happily by being united with it in its domain."[43] According to Fromm's and Ingarden's insights, humanity, when it narrows its perception of the place in the universe to the status of a thing among other things, dooms itself to depression and anxiety of its own insignificance in the vast cosmos because life is enslaved and controlled by it. Contrary to this, the cosmos acquires some inward meaning if humanity sees itself as the center of its disclosure and manifestation. Then the universe receives intrinsic human qualities, thus being united to humanity: the question then is not of being positioned in the universe, but that of living here and now in communion with the universe. But this communion means much more than sheer consubstantiality. It means that a human being can "transcend" the universe while retaining its immanence with the universe. As was emphatically asserted by M. Scheler: "Only man, because he is person, can rise *above himself* as a living being and make all to be its subject of knowledge, including himself, as if he would be a single centre on *the other side* of the space-time world.

the limitations of his existence. He visualizes his own end: death. Never is he free from the dichotomy of his existence: he cannot rid himself of his mind, even if he should want to; he cannot rid himself of his body as long as he is alive . . ." E. Fromm, *Man for Himself: An Enquiry into the Psychology of Ethics* (London: Routledge & Kegan Paul, 1967), 40.

43. R. Ingarden, *Man and Value* (Washington, DC: Catholic University of America Press, 1983), 17–18.

But this centre of human acts appropriating the world, its own body and its psyche cannot be itself a 'part' of this world, that is, it cannot have any definite 'where' and 'when'; it can only be in the highest foundation of being. Thus man is a being which is above himself and the world."[44]

Hermann Weyl invoked this paradox as a riddle of the twofold nature of ego, which, according to him, is beyond the limits of science: "On the one hand I am a real individual man, born by a mother and destined to die, carrying out real physical and psychical acts, one among many. . . . On the other hand, I am 'vision' open to reason, a self-penetrating light immanent sense-giving consciousness or, however you may call it, and as such unique. Therefore I can say to myself both: 'I think, I am real and conditioned' as well as 'I think and in my thinking I am free'."[45] The paradox of human subjectivity reflects the fundamental existential dichotomy of the incarnate human condition as that primary reality from which any realistic cosmology and its philosophy must originate. And it is the inability of science to account for this paradox that leads inevitably to transcending tendencies to look for its foundation through the otherworldly connotations.[46]

Such connotations were implicitly present in theology and religious philosophy. As an example, we provide a few quotations from the earlier Christian literature as well as from twentieth-century religious philosophy to explicate the sense of the paradox. Here is a passage from St. Gregory the Theologian (Nazianzus, c. 329-389) with a characteristic formulation of the paradox: "Having decided to

44. M. Scheler, *Die Stellung des Menschen im Kosmos*, in *The Collected Works* (Moscow: Gnosis, 1994), 160 (in German).

45. H. Weyl, "Unity of Knowledge," in *Mind and Nature: Selected Writings on Philosophy, Mathematics and Physics* (Princeton: Princeton University Press, 2009), 197.

46. The detailed discussion of the paradox of human subjectivity in a theological context can be found in A. Nesteruk, *The Universe as Communion: Towards a Neo-Patristic Synthesis of Theology and Science* (London: T. & T. Clark, 2008), 178-84.

demonstrate this, the Artificer of the universe, the Logos, created man as a single living creature from both elements, that is to say from the nature of both the visible and the invisible worlds. On the one hand He took the body from already pre-existing matter, on the other He endowed it with breath from Himself, which Scripture terms the intelligent soul and the image of God (Gen. 1:27; 2:7). He sat man upon the earth as a second world, a great world in a little one, as a new kind of angel, adoring God with both aspects of his twofold being, fully initiated into the visible creation but only partially into the invisible, king of all that exists on earth but subject to the King above, both earthly and heavenly, both transient [by nature AN] and immortal [by the image AN], both visible [sensible by body AN] and invisible [intelligible by reason AN], situated between greatness [lord of the universe AN] and lowliness [slave of the universe AN], at the same time both spirit and flesh . . ."[47] In St. Maximus the Confessor (c. 580-662) the paradox was interpreted in the context of faith in God who created man in his own image and likeness, so that initially man was "like" God, that is, he was "all in all" (cf. Col. 3:11). For example, Maximus the Confessor described this presence of man in all things in terms of a potential unity of all creation, which was to be realized by man as originally created: " . . . man was introduced last among existent things, as the natural bond mediating between the extremes of the whole through his own parts, and bringing into unity in his own person those things which are by nature far distant from each other . . ."[48] Man was created in order to mediate between all divisions in creation, for example between the sensible (visible) and intelligible (invisible); he writes: "As a compound of soul and body he [man] is limited essentially by intelligible and sensible

47. *Oration 45, On Easter*, 7. ET: P. Nellas, *Deification in Christ: Orthodox Perspectives on the Nature of the Human Person* (Crestwood, NY: St. Vladimir's Seminary Press, 1997), 203.
48. Maximus the Confessor, *Ambigua* 41, PG 91, 1304-12B. ET: Nellas, 212.

realities, while at the same time he himself defines [articulates] these realities through his capacity to apprehend intellectually and perceive with his senses."[49] For Maximus, however, the dichotomy present in this affirmation was not a problem, for according to his theological position the fundamental nonlocality that is present in human insight about the universe originates from the human ability (as God-given) to comprehend the intelligible realm, which contains ideas about the universe as a whole. Another quotation is from a Russian religious philosopher N. Berdyaev: "There are in personality natural foundation principles which are linked with the cosmic cycle. But the personal in man is of different extraction and of different quality and it always denotes a break with natural necessity. . . . Man as personality is not part of nature, he has within him the image of God. There is nature in man, but he is not nature. Man is a microcosm and therefore he is not part of the cosmos";[50] hence "the place of man in the natural world is tragic. Man is not only an object in this world, first of all he is subject which cannot be deduced from an object. Taken with this the relation of man to cosmos is defined by its being microcosm; he enfolds cosmos and history. Man cannot be a part of something, he is the whole. Through the spiritual in him, man is not subordinated to nature and independent of it although natural forces can kill him. If man would be just a natural and finite being, his death would not be so tragic: what is tragic is death of an immortal being who aspires to infinity. Only from an object-perspective man is part of nature; from a spiritual perspective, nature is in him. He is a slave of nature and he is its lord."[51] This can be paralleled with P. Florensky, according to whom "nature and man

49. Maximus the Confessor, *Ambigua* 10:26, PG 91, 1153B. ET: "Various Texts on Theology, the Divine Economy, and Virtue and Vice" 5:71, in *The Philokalia*, v. 2, 277.

50. N. Berdyaev, *Slavery and Freedom* (London: Centenary, 1944), 94–95.

51. N. Berdyaev, "The Kingdom of Spirit and the Kingdom of Caesar," in his *Spirit and Reality* (Moscow: AST, 2003), 565–671 (in Russian).

are both infinite. And it is because of being infinite, that they are commensurable and can be parts of each other. . . . Man is in the world, but man is complex to the same extent as the world. The world is in man, but the world is also complex as man";[52] "Man is the recapitulation of the world, its summary; the world is the disclosure of man, its projection.[53] S. Bulgakov contributed to the same stream of thought: "On the one hand, man is potential all, the potential centre of the anthropo-cosmos, which, although, not yet realised but is being realised, on the other hand man is the product of this world, of the empirical."[54]

The implicit presence of the paradox in all scientific affirmations of the universe reflects the intrinsic split between the two different modes of intentionality. The self-awareness of its own transcendental nature happens when the intentionality has to deploy its means to cope with the constraints and pressures of the outer world (embodiment, for example). The more the universe attempts to crush human existence under the weight of astronomical facts, the more the egocentric intentionality prevails as a measure of resistance to it. The more the pressure of the outer world relaxes, the more the same intentionality relaxes and the transcendental "I" loses itself in the outer things. Thus the constraints of the constitution of the outer world, which escape clear-cut definitions and visibility, constitute the very intentionality to the extent that it cannot fully cope with these constraints.[55] Then one can see that the paradox of human subjectivity is not simply an epistemological conundrum, it reflects a genuine ambiguity or bipolarity of human beings, which must be balanced and *existentially* overcome.[56]

52. P. Florensky, "Macrocosm and Microcosm," 186.
53. Ibid., 187.
54. Bulgakov, *Philosophy of Economy*, 160.
55. Cf. J. Vialatoux, *L'intention philosophique* (Paris: Presses Universitaires de France, 1965), 34.
56. Cf. L. Leprince-Ringuet, *Science et bonheur des hommes* (Paris: Flammarion, 1973), 164.

The paradox of human subjectivity in the world can be understood in a different way. On the one hand, being inseparable from reality in virtue of their embodied intentional consciousness, human persons can exist only in the context of their immediate nondistance from reality (consubstantiality). On the other hand, being a hypostatic formation (humanity is endowed with an ability to inhere the universe in its own subjectivity through the fusion of knowledge, to form the meaning and act in the universe as its self-consciousness and self-realization, or, theologically speaking, as its hypostasis[57]), that is, being fundamentally different from other material things, human persons are "infinitely" ontologically distant from those other things (which makes it possible to neutralize an object within the primary relationship with subject and transform it into one-sided submission[58]). The ability to distance themselves from outer things (even, in abstraction, from one's own body) makes human persons equally positioned with respect to all objects in the universe, so that they can be articulated by human subjectivity as different and uniformly distant from it. Paradoxically, on the one hand, because of the infinite ontological distance from all things in the universe humanity is hypostatically commensurable (and thus equally close) with respect to all objects in the universe, including the universe as a whole, whereas on the other hand, being corporeally at nondistance from the universe, humanity is physically incommensurable with the universe.

The paradox of human subjectivity can be formulated in terms of space, that is, in terms of humanity's topological position in the universe (this will make the paradox even closer to the paradox of space in the Incarnation[59]). The formulation of the paradox in terms

57. See the explication of this in Nesteruk, *Light from the East*, ch. 7.

58. See more in C. Yannaras, *Postmodern Metaphysics* (Brookline, MA: Holy Cross Orthodox Press, 2004), 114.

of space is achieved through a metaphor of the container and the contained: on the one hand by its physical and biological parameters humanity is contained in the universe; on the other hand the universe itself is "contained" by human subjectivity as its intentional correlate (that is, enhypostatically). In this formulation the ontological centrality of humanity is contraposed to its cosmographic mediocrity (cosmological principle). The distinction between two worlds is accentuated here: the one is affirmed by cosmology as existing whole and scientifically thematized in terms of elements and essences, and another one, associated with the immediate life of consciousness, the so-called life-world, the medium of indwelling into which every human being is brought into existence. For every particular being their life-world, being "here and now," is linked to the planet Earth and thus is geocentric. Earth is ontologically central in a spiritual sense,[60] in the sense of "wherefrom" manifestations and disclosure of the universe do originate. In spite of the fact that astronomy and cosmology deal with Earth as an object and ascribe to it a movement in space, both of them were produced by human beings on Earth, and it was here, on this planet, that scientific thought developed the definitions of motion, rest, and space understood in a general objective sense. Cosmologists' statements concerning the indifferent position of Earth in cosmic space (cosmological principle) receive their meaning from experiences acquired here, on the planet Earth. The *here* which is the place of this initial experience *is not* therefore a place in space, since it is itself a place of origin of a notion of space.[61] In this sense the cosmological principle, as a philosophical hypothesis

59. See on the topological paradox of incarnation T. F. Torrance, *Space, Time and Incarnation* (Edinburgh: T. & T. Clark, 1997); Nesteruk, *Light from the East*, 226-36; also Nesteruk, "The Cosmos of the World and the Cosmos of the Church: St. Maximus the Confessor, Modern Cosmology and the Sense of the Universe," in *Knowing the Purpose of Creation Through the Resurrection*, ed. Bishop Maxim (Vasiljevic) (Alhambra, CA: Sebastian, 2013), 323-28.

60. Cf. V. Lossky, *Orthodox Theology: An Introduction* (Crestwood, NY: St. Vladimir's Seminary Press, 1997), 64.

articulating the uniformity of space at large, enters into contradiction with the singular and unique "here" which is radically incomparable with any "there," thus predetermining the non-homogeneous topology in any ideation about space at large. The nontrivial nature of this last comment follows from a phenomenological stance on space as not a preexistent objective out there originating from the subject's passive contemplation of it, but in terms of the subject's comportment "in" it. This so-called attuned space becomes an initial instant and a medium of disclosure of that "objective" space through relation to which this subject is constituted as corporeal existence in the universe. However, this relationship is manifest of a paradox similar to that of the container and of the contained put in an interrogative form: How can one grasp the relationship of a particular being (subject) as if it is "in" space when this being is essentially constituted by being "over against," and hence beyond space?[62]

What is obvious, however is that the constitution of space, first of all of the attuned space, is intertwined with and not detachable from the fundamental aspect of human embodiment or corporeity, where the latter manifests itself as a living being in relation to other beings and to the world, in whom this relation is announced and articulated in a way of its sense-reaction and its comportment, or its action in situation. In this sense the constitution of space in all its varieties (from attuned space to the mathematical space of the universe) represents the modes of explication of embodiment or corporeity through which human beings interact with the world.

61. This is theologically similar to what Christ as the Logos experiences with respect to the whole universe: the universe, being a spatial extension with respect to his human nature, is beyond space, that is, not in space, with respect to his divine nature.

62. E. Ströker, *Investigations in Philosophy of Space* (Athens: Ohio University Press, 1965), 15. This reminds one of a Kantian stance on human being as being simultaneously phenomenon and noumenon: on the one hand space is an a priori form of sensibility that allows a subject to order its experience; on the other hand this form of sensibility is unfolded not from within that space which is depicted by it, that is, it comes from beyond any possible spatial presentation of experience.

Thus the lived body entails a kind of lived space that bears the character of self-givenness "in the flesh." In other words, the initial point of any discourse on corporeity and associated spatiality implies a kind of knowledge as presence "in person" or "in the flesh" as a mode of givenness of an object in its standing in front of the functioning corporeity.

In cosmology, by articulating the entirety of the universe human beings remain corporeal, so that their corporeality as relationship to all things contains in its facticity the very premise of being incommensurable physically and commensurable hypostatically to the totality that humanity attempts to reveal in cosmology. The attitude to this totality is twofold: on the one hand humanity attunes to it through belonging to it; on the other hand, through a non-egocentric intentionality humanity positions itself as if it were beyond the universe, "looking" at this universe as an object and depicting the latter as something being present over against "the flesh" and in person. However, since humanity cannot abandon its position of corporeal existence in situation on the planet Earth, all cosmological models contain the elements of this given embodiment, even in those cases when they predicate the universe in transhuman or even nonhuman (the early universe or multiverse) terms. In other words, the commensurability with the universe, as the transferral of some human qualities to it, is not *of* space, but originates *in* space.[63]

And finally, a brief note on the temporal dimension of the paradox.[64] Despite science's claims that intelligent humanity is an

63. For human beings to achieve the sense of commensurability with the universe one must be in space as a delimiter of their embodiment. Interestingly, this conclusion is similar to a Christian theological stance on space in the context of knowledge of God. It is because the incarnation of the Logos of God took place in rubrics of space and time, that no knowledge of God is possible outside the ways of Christ in space and time. (See, for example, Torrance, *Space, Time and Incarnation.*)

64. Cf. N. Berdyaev, "Attempt at Eschatological Metaphysics: Creativity and Objectification," in his *Spirit and Reality* (Moscow: AST, 2003), 523 (in Russian).

insignificant and accidental evolutionary artifact in the material universe,[65] philosophers object to this by pointing out that human consciousness cannot be explained in terms of cosmic factors as well as in terms of evolutionary theories (both cosmological and biological) that themselves are mental creations. This brings a temporal dimension to the paradox by pointing out that any speculation about the world as it existed prior to emergence of conscious humanity is a dubious enterprise, for what is affirmed as existing in prehistorical time still has features of that consciousness which is limited by the conditions of embodiment at present.[66] This implies that the very history of the universe, given through the display of its frozen past, has sense only through the historical consciousness of human beings.

If human history is treated only as an *epiphenomenon* or as a continuation of the natural history of the universe, then the emergence of the phenomenon of humanity in the late history of the universe is merely a contingent aspect of cosmic and biological evolution and thus has no philosophical meaning. Any question about mankind's significance or insignificance has meaning only if the whole of natural history is seen through "teleological eyes," which are themselves not an integral part of the scientific attitude. If, on the contrary, human history is not only distinguished from the natural history of the universe, but actually understood as incorporating natural history as the unfolding constitution of the world and humanity within human history, then human history ceases to be a part of cosmic determinism and acquires some features of a "trial," an

65. Some scientists, however, claim that the universe is intrinsically imbued with life, so that its outcome is not an accident. See, for example, in this respect C. De Duve, *Vital Dust: Life as a Cosmic Imperative* (New York: Basic Books, 1995); see also Barrow et al., *Fitness of the Cosmos for Life*.

66. The discussion of problems connected with the description of those eras that were not lived through by human consciousness can be found, for example, in R. Aron, *Introduction à la philosophie de l'histoire* (Paris: Gallimard, 1938), 39-46.

"event" (the "humankind-event"[67]) in which man's intrinsic freedom and a dualistic standing in the universe are encapsulated.[68] The centrality of humanity's position in the universe (as a subject of history of this universe) thereby acquires teleological connotations: the universe needs humanity in order to be explored and thus transfigured (through knowledge) through the acquisition of the *sense* contrary to its seeming nonsensical contingent facticity. This point gives another dimension to the stance of phenomenological philosophy that the world is radically human in a nontrivial sense: "[I]f man is attached to the world, the world likewise is attached to man, in such a way that it is no longer possible to speak about a world-without-man."[69] The world, being a container for humanity, is contained by human persons not through the power of physical forces (consubstantiality), but through the power of intellect, which makes humanity as conscious persons equally, that is, qualitatively *distant* from the world in its apprehending capacity to articulate the world.

The Paradox of Human Subjectivity and Personhood

The paradox of human subjectivity in the universe can be further explicated as pointing toward the different positions human subjectivity can adopt with respect to ontology of being. On the one hand there is an explicit treatment of the world in terms of thinghood, that is, in terms of things pre-given in order to be

67. This is terminology from my *Light from the East*, ch. 7. Cf. E. Morin, *Science avec conscience* (Paris: Fayard, 1982), 120.
68. Cf. with Orthodox cosmology's claim that cosmic history is part of the history of salvation related to humanity, and not vice versa. See O. Clément, *Le Christ, Terre des vivants. Essais théologiques. Spiritualité Orientale*, n. 17 (Bégrolles-en-Mauges: Abbaye de Bellefontaine, 1976), 80.
69. W. A. Luijpen, *Existential Phenomenology* (Pittsburgh: Duquesne University Press, 1960), 25.

recognized by thinking consciousness. In this sense the universe preexists as substance and the ultimate ontology of being is thought to be the ontology of this substance. Then the facticity of human beings in the universe is treated as the variation of this substance. In this case, humanity being a part of the universe experiences fear that the laws of the universe with their contingent, but fine-tuned outcomes at some stage can remove the phenomenon of humankind from it. On the other hand there is a different intuition that things which are out there, objects and entities in the universe, appear not as external and hostile environment but as the manifestation of the living presence of humanity in the universe, which actually makes all these things *beings*. The making of the universe must not be understood as manufacturing things from some pre-given material, but rather as *creating* things in a rather different sense. To create in the sense of personhood means not to physically dominate the pre-given, but to create such an ontological situation where all so-called things acquire the "presence" relevant to the totality of existence understood not in terms of substance but in terms of *hypostasis* (personhood). Humanity itself becomes present and manifest through transferring its hypostasis to being. This hypostasis is not something that "preexists" in substance or in nature; it is not an impersonal combination of the worldly elements or Platonic forms, but the center and the ultimate beginning of all articulated existence.[70]

The paradox of human subjectivity can thus be explicated as the tension between ontologies of being based either on substance or on hypostasis. For example, in modern physics and cosmology the grandeur of the world is understood through particles, fields, space-time, planets, galaxies, the whole universe, but, as we have pointed out above, there is no place for human subjectivity and personhood.

70. See more details on the explication of the meaning of "hypostasis" in my *Light from the East*, ch. 7.

The personal characteristics of those who create the physical picture of the world are remarkably missing from the very result of their activity. One can say that while being tacitly present behind the works of its own creation, personhood as the source of this creativity, is absent from its own creation. The picture of the universe is the manifestation of personal presence in the universe, but persons who created this picture are not explicitly found in it.[71] Cosmology in its outward expression creates such conditions for unconcealment of the universe, which takes place at the expense of concealment of persons.

Personhood is missing from the natural sciences because they approach human beings in the same way as they approach other things, that is, in the phenomenality of objects: one needs to "mortify" human beings and reduce them either to "walking dust" or to impersonal physico-biological robots in order to affirm their presence by means of observation and rational induction. But personhood as existential events escapes scientific grasp by transcending either materialistic definitions or idealistic beliefs. Personhood manifests itself as that givenness which cannot be subjected to any constraints of matter or categories of the understanding. This is the reason why it is impossible to define personhood in the way one defines things. Things can be defined

71. In terms of a historical reference one can point to Erwin Schrödinger, who in his work "Mind and Matter" empathically exposed a state of affairs in "the world of science" (contemporary to him) as "becoming so horribly objective as to leave no room for the mind and its immediate sensations." E. Schrödinger, *What Is Life? The Physical Aspect of the Living Cell* with *Mind and Matter & Autobiographical Sketches* (Cambridge: Cambridge University Press, 1992), 120. In the same paper he wrote: "Mind has erected the objective outside world of the natural philosopher out of its own stuff. Mind could not cope with this gigantic task otherwise than by the simplifying device of excluding itself—withdrawing from its conceptual creation. Hence the latter does not contain its creator" (ibid., 121; see also 122). To make his arguments stronger, Schrödinger quotes similar passages from C. G. Jung ("The flood of external objects of cognisance has made the subject of all cognisance withdraw to the background, often to apparent non-existence," 120) and C. Sherrington ("mind, the anything perception can compass, goes therefore in our spatial world more ghostly than a ghost. . . . It remains without sensual confirmation and remains without it forever," 121).

because they can be possessed and operated, but it is impossible to possess persons.[72]

However, in spite of all the insufficiency of science to deal with the problem of personhood, persons do not disappear but reveal themselves in a rather dramatic way. Since humanity as personhood is not content with the presence of beings in the world as they are given to it empirically and studied scientifically. As we pointed out in the Introduction, humanity wants to recognize beings not so much according to their nature, but as the results of *free will*. Thus by subjugating that truth which is gained on the grounds of the scientific to the desire for truth of the whole created existence, humanity exhibits its hypostatic essence, that is, its personhood. Humanity as personhood prefers to express its own presence by appealing to the belief in the transworldly source of this existence in the conditions of its incapacity to overcome the absence of personhood in scientific articulations of the world. Humanity makes this effort as a resistance to being contained by the universe and being comprehended in the phenomenality of objects. It does not want to be manipulated through circumscribability and individualization, which are inherent in spatio-temporal representations of the universe. It is in this sense that humanity as personhood longs for truth of existence, which is *in* this world but not *of* this world. This longing forms the spiritual motives of humanity and points toward the *telos* of explanation of all, in which the paradox of the human hypostatic existence as presence in absence will have to be finally resolved.

The reinstatement of personhood implies that one must turn to the foundations of the sciences, their origin in constitutive acts of

72. Clément, *On Human Being*, 30. Knowledge of persons is possible only through love, so that the Cartesian "Cogito ergo sum" must be replaced by "Amo ergo sum." G. Marcel, in a similar vein, develops an idea that love of a person precludes possession of this person in any possible sense (Marcel, *Being and Having*).

subjectivity. The absence of personhood in the resulting scientific picture of the world must be subjected to the phenomenological scrutiny in order to recover those intentionalities of human subjectivity that led to the development of the worldview in terms of efficient physical causality. We have here a kind of a phenomenological reversal in attitude to cosmology: to look at it not from the point of view of the content of its theories and their alleged reference to the physical world, not to enquire into the meaning of concepts, such as, for example, the universe as a whole, its origin, and such, but in fact, to use cosmology as a hermeneutical tool for understanding humanity itself, to use the human image of the universe as a kind of mirror through which human subjectivity and persons constitute themselves. It is through this shift in attitude that the sense of cosmology can be reversed: it can be seen as that activity of the human self, which through its outward look establishes itself and brings out (according to its will) the absence of personhood in the mathematized science to its explicit philosophical presence. The phenomenological reversal to the noetic pole of cosmological research is thus dictated by the work of intentionality. Seen this way, cosmological research exhibits an interesting interplay between the elements of the natural and human sciences that we discuss below.

The phenomenological reversal of such a construct as the universe as a whole (which served for the naturally oriented mind as the ultimate objective background of all facticity of life), reveals this construct as a certain structure of the incarnate transcendental subjectivity. If in the natural attitude science affirms the explicit presence of the universe at the expense of the absence of personhood, in the philosophical attitude the universe as an intentional correlate of human subjectivity does not possess qualities of "out there" measured in terms of distance. It is not the "other" as object here or there, above or below, right of left, near or far. The universe in all its entirety is posed as existent in the human hypostasis, but since this entirety is not available to any empirical acquisition, this *enhypostasization* manifests the universe's presence in absence. This result is not surprising, for as human personhood escapes complete definitions by manifesting itself through "presence in absence," the universe, being a mirror of the human reason through which humanity constitutes itself also escapes complete definitions, thus acquiring a mode of "presence in absence," that is, a mode of personal "opposite" of dynamic ecstatic reference. When one articulates the universe in terms of measurements of distance one loses personhood; when one brings the universe to being a personal "opposite" of ecstatic reference one loses the sense of the universe as extended space and time. The universe can then be understood as a kind of otherness of personhood that is present in the event of a person's self-affirmation.

The positioning of human beings in the universe in terms of extended space and time represents a mental abstraction from the living experience (personal ecstatic reference) as that context where human beings conceive themselves as unique forms of existence. There is an immediate living context common to all human beings that makes their life meaningful and valuable regardless of what scientific books write about human insignificance on a cosmic scale.[73] But this living context, the life-world in the terminology of Husserl, is not taken into account by science (as an ingredient of its own constitution). This indicates that the scientific picture of the universe, and hence its assertions about the insignificant place of human beings

73. The articulation of insignificance of humanity in the universe plays a strange role in contributing toward the disappearance of the existential function of modern science. The statement of this insignificance represents a scientific truth that is valuable for science itself. It does not affect at all the "man on the street" who "feels" in some manner that science literally "has nothing to say to him." Consequently, "the man of science is incapable of telling him how to live, what is the significance of his job, what are those historical events which have grave consequence for his life." S. Strasser, *Phénoménologie et sciences de l'homme. Vers un nouvel esprit scientifique* (Louvain: Publications Universitaires, 1967), 211. It is in this sense that science as activity does not affect the life-world as the sphere of existential events in their sheer facticity.

(as physical bodies) in it, represent abstractions from the living experience. The abstraction of the universe as the measurable and extended distance manifests the ideal of classical rationality to objectify the meaning of things (events) by means of paralleling the immediate experience of these things through their intelligible image in an abstract "space" that contains universal structures accessible to every consciousness.

This implies mathematics as a language of description that allows everyone to position a physical event (as an existential event) in the abstract mathematical space and time. Then one can realize that the "physical" reality that is articulated by conscious beings in the intelligible world contains in itself nonempirical elements, rooted in the capacity of human subjectivity to exceed sense-impressions. This implies that "nature" (or the universe), understood classically as something that is independent from acts of consciousness, cannot function anymore as a logically consistent notion, for it involves human consciousness in its definition as constitution. This is, in fact, a new saying of that which W. Heisenberg advocated long ago, namely that the natural sciences are created by men, and that the function of the sciences is to be a part of the interaction, or relationship between nature and human beings.[74] This entails the conclusion that scientific *reality* is not only the realm of the outer world, but also the realm of consciousness's work in scientific discourse. Reality becomes a relation, an *encounter*, a coexistence of oneself and the world, an ordination of the world to one and an opening of oneself to the world, at once oneself and that which is other than oneself.[75] Correspondingly, physical theories of the

74. W. Heisenberg, *Physics and Philosophy* (London: Penguin, 1989) 69; *The Physicist's Conception of Nature*, 24 (emphasis added).

75. De Waelhens, "Science, Phenomenology, Ontology," *Cross Currents* 17 (1957): 168. Cf. J. A. Wheeler's ideas on the dialogical character of study of the universe, when the resulting picture of reality appears as a cumulative result in the chain of questions to and responses from that

universe can be interesting not only for the sake of physics and cosmology *per se.* While studying physical nature we also study the relationship between us and nature, and as a result, we study human nature. Cosmology, being initially a natural scientific aspiration toward the whole universe, thus becomes a "cosmology" of the interior cosmos of the human mind. Cosmology becomes a science of human affairs whereas the universe, as science portrays it, becomes a "mirror" of the human embodied soul, the vision of humanity inside out. By paraphrasing a passage from M. Merleau-Ponty, one can suggest that scientific thinking of the universe, a thinking that looks on from above, and thinks of the universe as an object, must return to the "there is" that precedes it, that is, to the site, the soil of the humanly produced and modified world, as it is in our lives and for our actual bodies that we call ours, this sentinel standing quietly at the command of our words and acts.[76] The same can be reformulated differently.

Since from a phenomenological point of view the structures of the disclosure of embodiment are part of the same cognitive life that extends to things like the categories of thinking, the exact sciences, logic and mathematics, the very paradox of human subjectivity in the universe contributes toward the constitution of the person understood as a transcendental subjectivity that faces its own corporeality through the empirical structures of the world. In other words, the very constitution of corporeality as an intentional correlate of consciousness takes place through the positioning of conscious bodies in the background of all cosmic things. But this implies that cosmology, as a special physico-mathematical thematization of this corporeal background, contributes indirectly

which is sout there. See, for example, J. A. Wheeler, *At Home in the Universe* (New York: American Institute of Physics, 1994) and bibliography therein.

76. Cf. M. Merleau-Ponty, "Eye and Mind," in *The Merleau-Ponty Aesthetics Reader: Philosophy and Painting*, ed. G. A. Johnson (Evanston, IL: Northwestern University Press, 1993), 122-23.

toward the problem of the hypostatic corporeality, that is, "to the site, the soil of the humanly produced and modified world" as that personal "opposite" of dynamic ecstatic reference.

Here we come to a threshold in our discussion of the meaning of the human dimension in cosmology by formulating a thesis that cosmology is such a discipline where the demarcation between the elements of the natural and human sciences becomes very loose: cosmology functions at the crossroads of the natural and human sciences. Correspondingly, our next move is to elucidate this thesis.

The Dilemma of the Object-Noematic and Act-Noetic: The Paradox of Subjectivity and Cosmology's Untestability

The problem of the interplay between the dimensions of the natural and human sciences in cosmological discourse has its origin in the paradox of human subjectivity formulated phenomenologically as positing that the universe of physical cosmology is produced by means of special intellectual operations rooted in the life-world and thus is specifically *different* from the latter; at the same time, the universe apparently proves to be part and parcel of the life-world.[77] In other words, on the one hand the physical universe, as an "object" of the natural and scientific constitution, is different from and alien to the human world, thus transcending and encompassing human reality itself; on the other hand, being a mental accomplishment, the universe exhibits itself as cultural reality, thus being contained within the life-world.

Since the thesis of the paradox creates the sense of a possibility to find the foundation of human existence in the physical universe, physics sees its task to develop the theory of the universe in order

77. Cf. Gurwitsch, *Phenomenology and the Theory of Science*, 147.

for humanity to have no escape from its law-like-necessities. If this ideal could be achieved, the paradox would be removed because the foundation of a scientific explanation of the universe would be part of the universe itself. The question is then whether cosmology can sustain this ideal and thus prove its right to be qualified as a natural science in the strict sense. In order to answer this question let us analyze the line of object-noematic interpretation of cosmology appealing to its particular feature, that is, its lack of testability.

In the conclusion to his survey of philosophical issues of modern cosmology, George Ellis stated the thesis that uncertainty constitutes a key aspect of cosmology: "Scientific exploration can tell us much about the *universe* but not about its *ultimate nature*, or even much about some of its major geometrical and physical characteristics. Some of this uncertainty may be resolved, but much will remain. Cosmological theory should acknowledge this uncertainty."[78] It seems that what is implied here is a distinction between "the universe" as it appears through study and construction by a cosmologist and that which can be termed as the ultimate, underlying sense of its contingent facticity.[79] The uncertainty of cosmology lies in the fact that it is functioning in the framework of what is already given but, nevertheless, is incapable of accounting for this givenness and hence its own possibility as a fact of life. One can say, using terminology of J.-L. Marion, that scientific cosmology, in what concerns the universe as a whole, is functioning in the conditions of the "positive incertitude," that is, asserting things provisionally, which will always be amended and complemented by new experimental and theoretical results, or even replaced by a more powerful scientific paradigm.

78. G. Ellis, "Issues in the Philosophy of Cosmology," in *Handbook of the Philosophy of Science, Philosophy of Physics*, Part B, ed. J. Butterfield and J. Earman (Amsterdam: Elsevier, 2007), 1274 (emphasis added).

79. It is just a different way of naming of another perennial distinction between "appearance" and "essence."

Characteristically, this positive incertitude forms the condition of cosmology's progress at all. In contrast, if one approaches cosmology philosophically, it is clear that philosophy was "certain" about limits of reason in predicating the universe as a whole for centuries. (It is enough to point to Kant's analysis of the rational cosmological idea in order to realize that its actuality has not been extinguished after nearly a century of scientific cosmology.) The perennial problem of the universe's facticity, that is, its being, has not been removed from the philosophical scene despite desperate attempts by apologists of the multiverse and the so-called "theory of everything" to remove this problem on scientific grounds. In this sense philosophy works in conditions of the "negative certitude," recognizing that there are only certain aspects of experience of the universe that are subject to discussion and explanation. The universe as donation and event in any life, as the beginning and end, as being and meaning, cannot be explicated in terms of positive definitions. In this sense Ellis's recourse to uncertainty and untestability in cosmology represents a natural intellectual move toward the "negative certitude" that the universe as a whole will never be exhaustively understood.[80]

Unlike other sciences (with respect to which one can assert a similar thing, namely that they do not account for their own foundations), cosmology has a particular status because it pretends to deal with the unique and all-encompassing object—the universe as a whole, which by definition includes not only what is observed here and now (that is, in a particular spatial location—home-place—and in a particular historical period), but that totality to which human thinking assigns the sense of reality independent of space and time (this totality as a generic mathematical view can have a sense of

80. See on the concept of "negative certitude" J.-L. Marion, "Je suis un sceptique instruit," *Philosophie Magazine* 39 (2010): 58-63, as well as his book *Certitudes négatives* (Paris: Bernard Grasset, 2010).

a Platonic idea and thus being devoid of space-time propensities). Correspondingly, its intended object, that is, the universe as a whole, cannot be constituted as accomplished here and now but, on the contrary, represents an asymptotic ideal (the universe as a whole) reached through a self-correcting advance of knowing. In approaching this ideal cosmology invokes ideas and intuitions (related to constructs and theories) introduced on the level of cosmologists' *intentionality* (supported by beliefs, such as, for example, the inexplicability of the universe) and not under the pressure of evidence based on empirically accessed physical causality. Thus, naturally, these ideas cannot be tested on the level of physical causality, and are, probably, untestable in principle in a finite span of time.[81]

The accentuation of *uncertainty* and *untestability* by Ellis can give the impression that cosmology, in those parts that do not refer to direct astronomical observations, is not what is usually classified as a "natural science." For the natural sciences it is typical to bring unknown aspects of nature to their explicit presence, which is confirmed by observations and tests. Theories typical for the natural sciences aim to refer to empirical reality by means of rules of correspondence, that is, through tests and procedures established by the scientific community and accessible to public comprehension on the grounds of common sense. Mathematization plays an important role in a theoretical advance of the natural sciences, but mathematics, associated by its historical origin with the natural sciences, remains a tool, a method, ultimately created from within the conditions of

81. Ellis's wording of "untestabillity," "unprovability," or "improbability" is used in both cases, whether Ellis speaks about our universe (e.g., "the deduction of spatial homogeneity follows not directly from astronomical data, but because we add to the observations a philosophical principle that is plausible but *untestable*" (Ellis, "Issues in the Philosophy of Cosmology," 1226)), or when he refers to the so-called multiverse (other worlds: e.g., "multiverse proposals are unprovable by observation or experiment . . . ," [ibid., 1263]).

the life-world in order to appropriate this world further through articulation and by means of theoretical thematization.[82] The aim of the natural sciences is to explain facts by arranging them in the framework of physical causality as a sort of human control. In this case, even if for a philosopher (a philosopher of science, for example) a cosmological text (a book, an article, a conference talk) represents an interest of an independent order pertaining to the human sciences, there still remains a certain correspondence between this text as a fact of life, and that reality which this text describes. The validity and relevance of this text to what it describes as reality is exactly established through references to experiments and tests that are a part of this text.

If now cosmology is perceived as an enterprise that bases itself in nontestable assumptions, there is a question that arises on the nature of this enterprise and the sense and validity of its epistemic claims. Do cosmological theories indeed act as intentional correlates to physical reality, posed as existing in itself, so that cosmological constructs provide a set of convergent approximations to it? Or does, alternatively, cosmology, being a mode of cultural activity, create an intellectual environment with a corresponding narrative on humanity's place in the universe without any ontological commitment? It seems reasonable to conjecture in this case that any narrative about the universe is based on empirical extrapolations and intellectual conjectures that in some cases are untestable *now*

82. One is reminded of Husserl's famous criticism of the radical mathematization of nature on the grounds that mathematics, as a method providing a certain mathematical representation of nature, was taken for the reality of this nature. Husserl writes: "Mathematics and mathematical science, as a garb of ideas, or the garb of symbols of the symbolic mathematical theories, encompasses everything which, for scientists and the educated generally, *represents* the life-world, *dresses* it up as 'objectively actual and true' nature. It is through the garb of ideas that we take for *true being* what is actually a *method*—a method which is designed for the purpose of progressively improving, *in infinitum*, through 'scientific' predictions, those rough predictions which are the only ones originally possible within the sphere of what is actually experienced and experienceable in the life-world" (Husserl, *The Crisis of European Sciences*, 51-52).

and some are untestable *in principle*. Nevertheless, these untestable conjectures contribute to the wholeness of the narrative, thus becoming indispensable constituents of the idea of the universe. The natural origin of these conjectures is related to their historical contingency, which entails cosmology's *positive incertitude* and thus the "natural" presence in it of some untestable or eventually falsifiable and theoretically redundant statements. One can generalize this thought by asserting the historical contingency of the conditions of disclosure in cosmology, that is, of the historical relativizing a priori as relative to a certain situation of cosmology changing from one step to another of its history. It is this positive incertitude of cosmological research that guarantees its progress and at the same time makes it radically different in comparison with a "negative certitude" of philosophical insights in the topic of the universe or the world. The historical relativizing of the conditions of cosmological knowledge could extend the Kantian understanding of constraints on knowledge, as transferred from ahistorical a-priories to ever-corrigible conditions of constitution of reality related to the evolving human condition.[83]

The problem is that by definition *the universe as a whole* cannot be thought as contingent "formation" (unless in a theological sense as *created out of nothing*) because it assumes a spatio-temporal totality as its apodictic identity. This invokes a conflict with the historical contingency of cosmology as an activity of knowing. The question is: Can the historically contingent discipline of cosmology (as related to contingent events of human subjectivity) pretend to encompass the universe as a whole, which by definition exceeds all spatially distinct realms and eras and thus all aspects of created contingency?

83. Cf. Bitbol et al., *Constituting Objectivity*, 13.

One can comment on the sense of the historical contingency of cosmological facts. By studying astronomical objects through their image in the two-dimensional celestial sphere we study de facto *free phenomena*, that is, those that contain in themselves or establish from themselves the reasons for their being given to us. Their contingent facticity consists in that they are self-caused related to historically concrete and individual acts of consciousness. Their appearance in the living history of astronomical observation is free from any underlying causes in the following sense. Astronomical observations can be interpreted as caused by the contingent factors of human history; however, one cannot construct a trans-historical "trajectory of knowledge" as if it would be driven by a sort of law in the "space" of all possible knowledge and which as a potential possibility led to the given outcome of observations. Or, in different words, one cannot assume the existence of a universal "observer" that could anticipate the path of knowledge independently and apart from the acts of knowing as their first-time happening. Knowledge and experience are accumulated in consequent observations in time from different perspectives where the conditions of knowledge, the very fashion in which this knowledge is collected, are not controlled by us; we do not know all factors that influence the course of that existential manifestation which is called "knowledge." Up to some extent this is true even in the case when the experiments are anticipated through theory and thus planned in advance within a given paradigmatic constraint. The resistance of the universe to its disclosure makes these experiments amendable and corrigible because their outcomes are still contingent. This rephrases the sense of the contingency of knowledge as that process which is not exhausted or only driven by its subject matter: the choice of what to observe (or whether to observe or not) and of the notions for the description of the universe is based not in understood physical laws and interconnections, but remains hanging within a *free intentionality* of researchers. Correspondingly, the untestable conjectures naturally contribute to this intentionality. Untestability of some conjectures about the universe thus reflects the contingency of the conditions of constitution of cosmology's "object," a process that is not one-to-one related to causal developments in the object of knowledge and sources of experience. One implies here a change in the conditions of constitution that are not related to the instrumental requirements and means of accessibility. As an example one can refer to changes of theoretical paradigms that have philosophical motivations; for example, they follow from a philosophical desire to explain away the problem of the contingency of the initial conditions, which is attended in inflationary cosmological scenarios or ideas of the multiverse. The uncertainty of cosmology witnesses to the fact that the *intentional* acts of cosmological research which allegedly attempts to reveal causality in physical processes do not lead to the fulfillment of intentions, and this, as we argue, constitutes an irreplaceable and unavoidable feature of any knowing about the universe as a whole.

Being intrinsically historical, the narrative about the universe that contains untestable propositions is subject to ongoing change and renewal. In terms of research practice, this means that theories change rapidly and the survival of theories against the flow of data (that is, testable propositions), let us say within a couple of years, is treated as its success. In cosmology, especially related to ultimate questions, this renewal does not only follow the logic of revealing new causal connections among facts of nature, but is driven by an intentionality that reflects the aspirations and advances of the human spirit, its "infinite tasks."[84] In this sense some conjectures about the untestable

84. Husserl defines humanity as mankind with infinite tasks as a carrier of philosophy, conceptions of ideas through which "man gradually becomes a new man," man who "lives toward poles of infinity." Infinite tasks are associated with the theoretical attitude to the world. In this sense

aspects of the universe as a whole (being acts of human will) represent *existential events* (manifestations and assertions of that which does not present itself through physical causation) as elements of intentions, motivations, and goals of a historical cosmologist-actor, through which a personal experience and intuition of the universe attempts to express itself outwardly in the scientific narrative. Cosmological narrative thus spontaneously appeals to the idea of a wider cosmic context of human existence, beyond this planet, by shifting home-places not only in visible space, but beyond it, toward the intelligible universe. This happens at the expense of leaving out the sphere of the empirical and sensible experience and making a transition to stable and allegedly nontransient intelligible entities that are beyond the reach of any straightforward testability. In some extreme versions of such a narrative, its apologists appeal for the removal of all "human baggage" behind the underlying theories. As an example, one can point to a hypothesis of the mathematical universe,[85] which advocates an extreme view of mathematics as an underlying structure of physical reality, which is stripped of all aspects of the human presence. The claim that the universe is mathematical is tantamount to an assertion of the universe's epistemic exhaustibility (that is, knowability). This contrasts with the general philosophical understanding that mathematical concepts have no intuitive content and are very poor in donation, thus leaving behind all aspects of experience of the universe that are so powerful in donation that they block the discursive reason from being able to constitute the phenomenon of the universe at all.[86] Contemporary studies on

science itself signifies the idea of the infinity of tasks. See Husserl, *The Crisis of European Sciences*, 277–79.

85. See M. Tegmark, *Our Mathematical Universe: My Quest for the Ultimate Nature of Reality* (London: Allen Lane, 2014), in particular chapters 10 and 12.

86. See in this respect J.-L. Marion, *Being Given: Toward a Phenomenology of Givenness* (Stanford: Stanford University Press, 2002), 179–247.

transcendentalism in mathematical physics also point out that the grasp of reality can only be achieved through computational synthesis as a process in time, so that any attempt to claim that subjectivity can get hold of the ultimate structure (as computable) would contradict a simple fact of the physical incommensurability between the finite human incarnation and a potentially infinite universe. The fact that cosmology involves a "computational synthesis" of the observable astronomical phenomena implies that physical objectivity of these phenomena as well as of the universe as a whole cannot be tantamount to an ontology of some independent reality. The possibility of a complete mathematical reconstruction of such an ontological reality would ascribe to the human mind excessive intellectual capacities that transcend its finitude following from the limits of embodiment. Correspondingly, according to H. Weyl's qualification, cosmology here concedes to idealism (or transcendental idealism) in the sense that "its objective reality is not given but to be constructed . . . , and that it cannot be constructed absolutely, but only in relation to an arbitrarily assumed coordinate system and in mere symbols."[87] The totality of the universe and its actual infinity, being the source of the computational incommensurability, is, however, not beyond its reach in a sense of manifestation. It is always hidden within the range of its formulations by cosmology. To the extent, therefore, that a particular cosmological theory is functioning adequately even in the conditions of untestability, one can regard the structure it articulates as constituting the intelligible pattern of the universe as a whole. The untestability thus is the constituting element of the apophatic intelligibility of the infinite universe.

87. H. Weyl, *Philosophy of Mathematics and Natural Science* (Princeton: Princeton University Press, 2009), 117.

Cosmology cannot avoid dealing with hardly testable conjectures simply because it is not an experimental science: according to Ellis's theses "The universe itself cannot be subjected to physical experimentation" and "The universe cannot be observationally compared with other universes."[88] In this sense the presence of speculative mathematical elements that cannot be directly related to empirical reality must be accepted with a sort of humility, and in no way as the end of realistic commitment in research, assuming that reality stands here not for a preexistent antecedent entity, but as being unceasingly constituted. The escaping absence of the physical correlate to that which is intuited through mathematics rather represents an invitation to continue the scientific quest when the reality of what is called the universe is withdrawn from a simplistic empirical grasp. The presence of unattained goals as imaginatively projected standards encourages and activates in a scientist a different intentionality that is based on existential inspirations and not fully articulated insights.[89] The initiation of this intentionality allows a philosopher and theologian to appropriate cosmology not as that type of knowing which delivers ultimate truth but rather as a particular account of the human encounter with the universe. In this case, the demand for cosmology follows from existential orientation, that is, from asserting the sense of human existence and its *telos*.[90]

88. Here is a full version of his theses A1 and A2: (A1) "The universe itself cannot be subjected to physical experimentation. We cannot re-run the universe with the same or altered conditions to see what would happen if they were different, so we cannot carry out scientific experiments on the universe itself"; (A2) "The universe cannot be observationally compared with other universes. We cannot compare the universe with any similar object, nor can we test our hypotheses about it by observations determining statistical properties of a known class of physically existing universes" (Ellis, *Issues in the Philosophy of Cosmology*, 1216).

89. These existential inspirations initiate a generic cosmological mythology that has already been described in terms of cosmogenesis, that is, how the world was made, how the present universe that stands before our eyes developed from what went before, from the non-universe, the formless. Cf. J. Ladrière, *Language and Belief* (Dublin: Gill & Macmillan, 1972), 153.

90. This thought has been expressed in numerous ways by philosophers and scientists. See in this respect, for example, Minkowski, "Prose and Poetry," 244; M. K. Munitz, "Kantian Dialectic

Seen in such a way, cosmology in its status and function in human knowledge does not fall exactly under the definition of natural science. In the latter, scientists make claims and debate about them: they can reject claims on the grounds of their irrelevance to the reality of things; by this, scientists express their attitude to epistemic claims contained in theories. In experimental sciences these claims are corrigible to a sort of limit, which must ground them in reality. The situation in cosmology, in what concerns the universe as a whole, is different because the constant corrigibility of its epistemic claims implies that one cannot hope for their convergence, which would imply a certain ground in physical reality as the ultimate correlate of these epistemic claims (although one cannot prevent anyone from exercising a belief in such a convergence).

There are some elements in cosmological research that make it similar to a work of a historian.[91] The cosmological narrative can receive a historical interpretation that vindicates in a different way the presence of untestable elements in cosmology. For example, by studying the history of astronomy (as a certain introduction into cosmology proper) one must take a neutral position with respect to the epistemic claims of Ptolemy or Giordano Bruno. Certainly from a modern view their cosmologies were incomplete and contained untestable elements. But a *historical perspective* stops us from such an evaluation, accepting, rejecting, or amending their epistemic claims concerning the knowledge of the universe unless we evaluate their theories through the eyes of the present era, and our understanding of history is fundamentally conditioned by the present scientific views. The same can be applied to modern cosmology. Indeed, if cosmology

and Modern Scientific Cosmology," *The Journal of Philosophy* 48, no. 10 (1951): 338; P. Brockelman, *Cosmology and Creation: The Spiritual Significance of Contemporary Cosmology* (Oxford: Oxford University Press, 1999), 42; J. Primack and N. Abrams, *The View from the Centre of the Universe* (London: Fourth Estate, 2006), 280-90.

91. Cf. Ellis, "Issues in the Philosophy of Cosmology," 1221.

is seen as an ongoing narration about the universe that is intrinsically historical and hence contingent, then the presence in its structures of untestable elements as well as the overall uncertainty cannot be used to qualify the epistemic unfoundedness of cosmology. On the contrary, these untestable elements should be seen as *cultural artifacts* that must be accepted as existential events, that is, as historically motivated actions. Then one must abstain from judgments on the truth of cosmology's epistemic claims and keep neutrality with respect to them. In this case the presence of untestable hypotheses and assumptions in cosmology can be treated as that inevitability which contemporary methodology of cosmology has to face. From the point of view of physics this can be a misfortune, but seen phenomenologically this turns out to be a fact of the human condition in which knowledge of the universe takes place.

In fact, the presence of untestable elements in cosmology indicates an intrinsic apophaticism in knowledge of the universe (that is, inability to exhaust knowledge of the universe through its signifiers) as a feature of the limitations following from the human condition. It is the desire to overcome these limits originating in the human free will that provokes cosmologists to appeal to intelligible realities (which are not necessarily physically testable) or philosophical principles (rooted in existential beliefs) in order to "restore" their commensurability to the universe. One can see that this appeal is driven by *intentions* of human subjectivity, and not by the necessities of physical observations themselves. In this sense the introduction of untestable conjectures points rather to mechanisms of functioning of human subjectivity when it faces phenomena that exceed their capacity of constitution. This in turn contributes to a generic thesis that the study of the universe contributes toward the study of man.[92]

92. Cf. P. Heelan, "Nature and Its Transformations," 486. According to Primack and Abrams: "We can enhance the sense of ourselves, as we've been successfully doing with our other senses, by

However, one must not treat the last thesis in a simple naturalistic fashion, as if epistemic questions are reduced to questions of cognition. What we want to say is that the epistemology underlying cosmology as allegedly natural science is in many aspects free from determinations induced by the physical causality and expressed in natural language, as well as from the over-formalization that takes place in mathematical cosmology. Thus cosmology can be seen as explicating the general functioning of human subjectivity as embodied in the universe.[93]

By making explicit the workings of human corporeal subjectivity, cosmology places itself within the cultural world, thus exhibiting some features of the cultural and human sciences along the lines of the arguments used, for example, by Husserl: namely, that the cultural or human sciences reveal themselves as all-encompassing, since they also comprise the natural sciences and a mathematized nature; they are themselves mental accomplishments, that is, cultural phenomena.[94] Cosmology acquires the meaning of a cultural science in the sense

means of a scientific but nevertheless metaphorical telescope- a new cosmological lens through which we can see how the expanding universe really works and how astoundingly special our place is in it" (*The View from the Centre of the Universe*, 282). Placed in a cosmological context the following quotation from F. O'Murchadha points to the same thought: "The otherness of nature is beyond us, beyond our humanity, beyond history. Yet it is in a place carved out of nature that we dwell. In dwelling—as ethical being—we become what we are. Hence, *how our dwelling is conceived, how the relation of the history of this dwelling to the natural environment, through which it is carved out, occurs, relates directly to who we are.*" F. O'Murchadha, "Nature as Other: Hermeneutical Approach to Science," in *Continental and Postmodern Perspectives in the Philosophy of Science*, ed. B. E. Babich et al. (Aldershot, UK: Avebury, 1995), 189.

93. One can make a parallel between our phenomenological interest and a similar approach to philosophical issues of physics which is developed in the movement called "formalized epistemology." See, for example, F. Bailly, "Remarks about the Program for a Formalized Epistemology," in *Quantum Mechanics, Mathematics, Cognition and Action: Proposals for a Formalized Epistemology*, ed. M. M. Mugur-Schächter and A. van der Merwe (Dordrecht: Kluwer, 2002), 3-8. However, unlike the "formalized epistemology" that aims to explicate the cognitive structures underlying science, this research attempts to relate these structures to fundamental existential conditions predetermining their facticity.

94. E. Husserl, *The Crisis of European Sciences*, 237. However, the converse is not true, that is, the cultural sciences cannot be given a place among the natural sciences. See in this respect A. Gurwitsch, *Phenomenology and the Theory of Science*, 148–49.

that it deals not only with the disclosure of "objective" reality of the universe (as humanity's natural environment) but encodes human understanding of its own history as well as the sense of existence, for example, by measuring the universe through the standard of human life.[95] It is in this sense that cosmology exhibits itself not as a monological questioning of the universe as if it was "out there," but rather as a dialogue with the universe as a noematic pole through which humanity enquires in itself.[96] The fact that this noematic pole is not fixed and escapes its ultimate grasp constitutes an inalienable feature of the noetico-noematic link in the cosmological enquiry leading to the formation of the *ego* of a cosmologist, exactly to the extent that it is not capable of grasping the donating intuition of the universe. The *ego*'s subjectivity is evolving through the invitation by the universe to disclose it in certain limits. This confirms a simple truth that in any form of speech about the universe one does not imply the already-given and independent object of disclosure.[97]

95. There are different connotations that can be pointed out in this respect. First of all, the views on the universe were always important for traditional communities to draw some remote expectations having ethical character: ethics depends on the idea of environment and its developmental perspective. See in this respect F. Mathews, *The Ecological Self* (London: Routledge, 1991), 3-6. Secondly, from a Christian point of view cosmological history is part of the history of salvation, that is, of the history of relations between humanity and God. See Clément, *Le Christ, Terre des vivants*, 80. And thirdly, if salvation is a *telos* of human existence, the study and knowledge of the universe also have a teleological sense: the aspiration of cosmology to disclose the origin of the universe turns out to be a *telos* of cosmological explanation.
96. The constitution of one's ego through knowledge so that the "object" of this knowledge is formed together with its subject represents a crucial feature of the human sciences. According to Russian philosopher M. Bakhtin, the object that is studied in the human sciences belongs to the same realm as the subject who studies, and thus it is no less active than the knowing subject. M. Bakhtin, *Aesthetics of Verbal Creativity* (Moscow: Iskusstvo, 1979), 349 (in Russian). In this context it is interesting to make reference to J. A. Wheeler, who affirmed a similar thing with no recourse to the human sciences. For example: "In giving meaning to the universe, the observer gives meaning to himself, as part of that universe." C. M. Patton and J. A. Wheeler, "Is physics legislated by cosmogony?," in *Encyclopaedia of Ignorance*, ed. R. Duncan and M. Weston-Smith (Oxford: Pergamon, 1977), 31. That is, by disclosing the universe the observer forms its own structures of subjectivity that apprehend the universe.
97. The opposite would imply sheer idealism, resulting in the universe being an intelligible entity graspable through the already-formed intellect. Its grasp would imply mysticism as communion

Correspondingly, one intuits the universe as related to the continuity of human conscious experience of the ongoing embodiment in the universe. Seen in this way, cosmology could acquire a teleological sense if one relates it to humanity's "infinite questions" of the beginning and the end of the universe as having connotations with enquiries into the sense of humanity's own beginning and its own consummation. It is because of the infinitude of its tasks that human subjectivity, while attempting to comprehend the sense of the beginning and the end of the universe, is being formed and comprehended by itself. It is through the unlimited donation of the universe (as the field of possibilities) that humanity awakens to its own finitude in spite of a potentially infinite consciousness (based in the infinite human will), that is, to the issue of the "beginning" and "end" of consciousness itself, that is, its contingent facticity. This sense of finitude entails another mode of incommensurability, that is, the infinite can only be represented through symbols (apophaticism), which by their origin function not on the level of physical causality.[98] In this sense cosmological theorizing as an ongoing symbolizing of the universe contains in itself logic and necessities that are not directly related to its subject matter, but rather to the sense of goals of

with the realm that is beyond the empirical (a mysticism similar to the one envisaged in the Platonic philosophy of mathematics). Even if one would assume the possibility of a complete computational synthesis of this intelligible universe, this would, as we mentioned above, exceed human finite capacities limited to the conditions of embodiment. The desire to know the universe in its totality ("all in all") brings to mind some eschatological connotations, as if thinking of totality of the physical universe is equivalent to the anticipation of the eschaton in which the overall transfiguration of creation will allow one to see the universe from the perspective of the Divine: the postlapsarian loss of the sense of the world as "all was in all" (1 Cor. 15:28) will be overcome.

98. H. Weyl linked the human longing for the sense of the infinite with that moral and human standard by which all human deeds are judged: "[M]ind is freedom within the limitations of existence; it is open towards the infinite. . . . The completed infinite we can only represent in symbols. From this relationship every creative act of man receives its deep consecration and dignity." H. Weyl, "The Open World: Three Lectures on the Metaphysical Implications of Science," in his *Mind and Nature*, 82.

humanity itself. Here again cosmology manifests itself as the science of human affairs.

Cosmology and Human Will

In the background of what we have said so far there still remains a question (related to Ellis's comments on the presence of untestable and unprovable assumptions in cosmology) as to why in spite of the a priori understanding that cosmology will never achieve the fullness of explanatory power and adequacy with truth, the urge for cosmological search and narration continues. Indeed, even for a reader unexperienced in all the subtleties of the methodology of science, the claim that cosmology is based in a fundamental uncertainty must make the staggering impression that cosmology forms an exotic set of trans-scientific ideas and intuitions that, by virtue of popular science and mass culture, have acquired the status of a stable social *belief*. However, cosmology is not a new mythology[99] resembling cosmic ideologies of the ancient past where humanity aspired to be dissolved in cosmic immensities, that is, ideologies that were strongly dismissed by the early Church Fathers as pagan and gnostic.[100] Cosmology is not a sheer imagining, but has its own logic and drive, which reflects the sense and value (as well as *telos*) of

99. Here one must agree with Ellis that the fact that cosmologists write about contentious issues in cosmology "is proof that they consider it meaningful to argue about such issues" because their quality emerges naturally from knowledge of the physical universe (Ellis, "Issues in the Philosophy of Cosmology," 1272).

100. Cosmic ideology, or the so-called *cosmism*, is a philosophical and spiritual longing for fusion with cosmic entities. Cosmism means not only the affirmation of our commonalities with nature and our contingence upon its laws and accidents but also a much more sophisticated kind of spirituality, which longs for fusion with "cosmic life" and its mystery and which is ecstatic in its essence. N. Berdyaev called this spiritual tendency "the lure of the cosmos" and described it as man's slavery to cosmos (and nature in general), as opposed to the freedom of hypostatic existence in the divine image. See N. Berdyaev, *Slavery and Freedom* (London: Centenary, 1944), 93–102.

communion with the universe conditioned by necessities of nature and, at the same time, pertaining to human freedom. This communion is not a dissolution in the cosmos but rather is a state of *apatheia* as transcending the natural causes that are beyond human control and, at the same time, transcending the immanence to the universe by a sheer fact of human presence, through which the universe is transformed by human will (that is, brought to self-articulation), its *energeia* and operations of cognition. Thus the persistence of cosmological research comes not only from the logic of cosmological gnosis, but from other factors originating in the human condition.[101] Indeed, finite human beings, because of their paradoxical standing in the universe are not content with the presence of things in the universe as they are given in their empirically contingent facticity. Cosmologists, by invoking the idea of the universe as a whole, intend to understand the meaning of finite things (astronomically observed objects or earthly phenomena) not only through their *nature*, that is, through that which is subjected to physical causation, but through the purposes and ends of these things as they stand with respect to the universe as their ultimate foundation. But this intentionality is sustained by humanity's aspirations not only to be commensurable to the universe but, in fact to be above it, to transcend it and thus to encompass it through the power of intellect rooted in human will. It is because of its paradoxical position in being, which causes existential discomfort,[102] that humanity appeals

101. J. Moltmann, by formulating the quest for a sense of cosmic utopia that creates in us the interest to know the universe, suggests that the whole of science-fiction can be summed up as the infinite survival of humankind and the unlimited development of human consciousness. J. Moltmann, *Science and Wisdom* (Minneapolis: Fortress Press, 2003), 72.

102. E. Fromm, for example, speaks about existential and historical dichotomies of human beings when they question the sense of their existence: "Man is the only animal which can be bored, that can be discontented, that can feel evicted from paradise. Man is the only animal for whom his own existence is a problem which he has to solve and from which he cannot escape." These are dichotomies about life and death, loneliness and relatedness, individuality and sociality. Because of their free will human beings can attempt to annul historical contradictions through

to the idea of the universe as a whole as an alternative to being contained by finite natures, that is, being comprehended only as an "object" among other objects in the universe. Existentially it does not want to be manipulated through circumscribability and individualization, which are inherent in spatio-temporal forms of the finite cosmos, and correspondingly, humanity longs for the truth of its existence *in* the space-time rubrics of this universe as if it is not *of* the universe as it appears to us. Here humanity wants to recognize things not according to their compelling givenness, but as results of humanity's free will realized in its intentionality. This naturally leads to the transcendence of the empirical and the invocation of intelligible entities (sometimes untestable and unprovable), which serves as a pointer and invitation to further research rather than its end and ultimate certainty.[103] The presence of free-will agents in the universe imposes certain constraints on the nature of the universe: it must contain the necessary conditions for them to exist[104] (or, as it is argued elsewhere, the universe must be moral).[105]

their actions, but it is futile to overcome existential contradictions; humans remain dissatisfied, anxious, and restless. Then "there is only one solution to his problem: to face truth, to acknowledge his fundamental aloneness and solitude in the universe indifferent to his fate, to recognise that there is no power transcending him which can solve his problem for him." Fromm, *Man for Himself*, 41-44.

103. As eloquently expressed by H. Küng, there is no intellectual compulsion in questions beyond empirical reality, but *freedom* dominates in them. H. Küng, *The Beginning of All Things* (Grand Rapids: Eerdmans, 2007), 78. Here an analogy comes to mind: that the invisible foundation of the universe falls under the rubric of aesthetical idea (rather than rational idea) which, being qualified as an unexpoundable presentation of imagination, functions according to "*free play*." Kant, *Critique of the Power of Judgment*, § 57, Comment 1, trans. P. Guyer and E. Matthews (Cambridge: Cambridge University Press, 2000), 217.

104. This argument corresponds to what is generally called the anthropic inference in cosmology, accentuating a delicate interplay between the physical and biological parameters of human existence and large-scale properties of the universe as well as fundamental physical constants. This inference does not account for the facticity of humanity's existence, because it does not cover the realm of sufficient conditions, which rather belong to the sphere of human morality and conscious will.

105. This statement of the morality of the universe advocated by N. Murphy and G. Ellis, *On the Moral Nature of the Universe* (Minneapolis: Fortress Press, 1996) (see in particular p. 207), does not have any straightforward scientific reference, for "free will" as well as the very facticity

The perception of cosmology as that block of insights which involves deeply human anxieties and correspondingly persists as an existential quest invokes a different stance on the ontological commitment exercised by those who are engaged in the popularization and adoration of cosmology and who usually claim that whatever is theoretically and mathematically possible is physically real and true, although nonobservable and untestable.[106] The countersense put forward by the human sense of cosmology doubts not the legitimacy of the cosmological narrative (comprising theories of nonobservable entities) *per se* but the validity of epistemic justification adopted for its realistic interpretation.

Indeed, if many cosmological hypotheses and inferences are not testable, that is, the correspondence principle between theory and empirical reality as epistemic justification does not work, there is a way of interpreting cosmological propositions about the nonobservable and invisible by assigning to the universe the sense of a mental accomplishment but achieved through the idea of *coherence*, where "coherence" stands not only for the clarity of theoretical explication and cohesion of mathematical calculations, but for the "collaborative agreement"[107] among cosmologists who effectively *hypostasize the notion of truth* related to the universe and postulate the ways of epistemic justification which lead to it.[108] In this case

of consciousness in the universe cannot be accounted for through any reduction to physics or biology. One can add to this that the very existence of cosmology as a free and creative exploration of the universe is thus inherent in the fact of free human choice to explore the world at large. The initiation of cosmology lies in the freely made decision to act and transcend the limitedness of the given. And this free decision as such is not subject to a scientific account.

106. Regardless of any possible stance on whether the mathematical exhausts the whole of reality, one must recognize that since the mathematical models are constructed in *this* universe, they contribute in a way to the articulated content of this universe in a generic epistemological sense.

107. N. Rescher, *The Coherence Theory of Truth* (New York: University Press of America, 1989), 318-19, 333.

108. See J. Bowker, *The Sacred Neuron* (London: I. B. Tauris, 2005), 118-48, where its author argues on the importance of coherence considerations in the science-religion dialogue as a different form of justification in comparison with the correspondence principle. On the limited

the implied truth of cosmology cannot be an ontological truth (that is, physical truth as allegedly existing in itself) but is a human-dependent hypostatic constitution of truth possessing the qualities related to the corporeity of human beings.[109] In this interpretation many cosmological constructions appear as historically contingent but coherent mental accomplishments (based in beliefs) whose truth contributes toward the spiritual goals (*telos*) of community but obviously does not exhaust them. However, the question of whether the locally established truths are subject to convergence with ultimate truth remains beyond scientific scope and represents in turn a belief motivated by trans-scientific convictions. In this option the validity of cosmology's claims is dictated not by a direct reference to reality but through the adoption of a consistent and creative set of beliefs that themselves constitute the sense of reality, although contingent as related to the goals of the community of cosmologists. (One can see a certain analogy between the forming of sense in cosmology and theology: indeed theology forms its sense of truth not through empirical references to the Divine [which are characteristically absent], but through experience of God as elaborated and established by ecclesial agreement.[110])

Knowledge of the universe interpreted as a collaborative construction cannot be treated as independent of human insight, so that cosmology's alleged status of following the standards of a

application of the idea of coherence of epistemic justification in cosmology see Nesteruk, *The Universe as Communion*, 244–46; A. Nesteruk, "Transcendence-in-Immanence in Theology and Cosmology: A New Phenomenological Turn in the Debate," *Studies in Science and Theology* 12 (2010): 179–98; A. Nesteruk, "The Universe as a Construct: Epistemic Beliefs and Coherence of Justification in Modern Cosmology," *Journal of Siberian Federal University. Humanities & Social Sciences* 5, no. 9 (2012): 1304–35; A. Nesteruk, "Coherence of Epistemic Justification versus the Principle of Correspondence in Modern Cosmology," *Izvestiya: Herzen University Journal of Humanities and Sciences* (St. Petersburg), no. 156 (2013): 59–67. See also more details below in chapter 4.

109. Cf. Bitbol et al., *Constituting Objectivity*, 4.
110. See Nesteruk, *The Universe as Communion*, 244–46.

natural science (as that in which the "object" of study can be entirely separated or detached from the subject) is not achievable. Cosmology, in contradistinction to astronomy and astrophysics,[111] is rather a "universology"[112] that deals with a single, unique totality of all, which not only cannot be treated as an object and hence subjected to experimentation, but also cannot be made devoid of the delimiters of human insight. This, as we mentioned before, implies that human beings, as part of the universe, cannot position the universe as a whole in front of their consciousness, unless as a mental abstraction.[113] If such a mentally constructed universe nevertheless were to be identified with the physical totality, this would imply a sort of impossible transcendence of the actual physical universe as if one were able to "look at it" from the outside, hence transcending one's embodied existence. The inseparability of humanity and the universe as their consubstantiality entails that all speculations about other worlds remain intrinsically immanent, being noematic correlates of an embodied subjectivity that is an irreducible element of being of this universe. Thus the universe as an intentional correlate of cosmological consciousness represents a mental accomplishment and

111. G. Ellis underlines the essential characteristic of that subject matter which pertains to cosmology proper: "[I]f we convince ourselves that some large-scale physical phenomenon essentially occurs only once in the entire universe, then it should be regarded as part of cosmology proper; whereas if we are convinced it occurs in many places or times, even if we cannot observationally access them . . . then study of that class of object or events can be distinguished from cosmology proper precisely because there is a class of them to study" ("Issues in the Philosophy of Cosmology," 1219). This careful distinction is related to the universe as a whole and makes a clear-cut demarcation line between cosmology *proper* and other celestial sciences like astronomy and astrophysics.

112. S. Jaki, *Is There a Universe?* (Liverpool: Liverpool University Press, 1993), 1-2.

113. This would correspond to a Platonic treatment of the construct of the universe. In this case cosmology would face, in analogy with the general Platonic stance in the philosophy of mathematics, a serious problem of justifying the interaction between the universe as an intelligible entity and its empirical appearance to embodied consciousness, an interaction that would imply a sort of mystical communion. See in this context a discussion on the epistemic status of mathematical objects in R. Tieszen, *Phenomenology, Logic, and the Philosophy of Mathematics* (Cambridge: Cambridge University Press, 2005), 46-68.

cultural achievement[114] exhibiting features of the sciences of human affairs.[115]

114. Cf. Gurwitsch, *Phenomenology and the Theory of Science*, 44–45. See also Husserl, *The Crisis of European Sciences*, 227, on "nature" as correlate of a universal abstraction.

115. See A. Nesteruk, "Cosmology at the Crossroads of Natural and Human Sciences: Is Demarcation Possible?" *Journal of Siberian Federal University. Humanities & Social Sciences*, Part. 1, vol. 4, no. 4 (2011): 560–76, Part 2, vol. 4, no. 5 (2011): 644–66. A similar observation, with no recourse to phenomenology and the concept of intentionality, has been made by Russian philosopher V. Rosin. See, for example, V. M. Rosin, *Types and Discourses of Scientific Thought* (Moscow: Editorial URSS, 2000), 81 (in Russian). In another paper Rosin writes: "The object of cosmology (in analogy with the objects of biology, cultural sciences and sociology) cannot be described within a single scientific discipline. . . . From the point of view of the philosophy of science the universe represents an ideal object of theories pertaining to the human sciences, based in its construction in facts (astronomical observations and their interpretation) and related to the process of realization of cosmologists' values and approaches, as well as to the discourse of the human sciences (for example the treatment of astronomical observations as characteristic texts and activity of the Cosmos) . . ." V. Rosin, "Towards a Problem of Demarcation of the Natural and Human Sciences, and Where to One Must Relate Cosmology," *Epistemology and Philosophy of Science* XI, no. 1 (2007): 111–28 (in Russian). Rosin makes his claims on the human sciences' nature of cosmological knowledge by referring to works of another Russian philosopher, V. Kazyutinsky, in particular to his paper "Worlds of culture and world of science: An epistemological status of cosmology," in *A Socio-Cultural Context of Science* (Moscow: Institute of Philosophy of the Russian Academy of Science), 1998, 101–18 (in Russian). See his paper, "No, Cosmology Is a Physical Science and Not a Human One," *Epistemology and Philosophy of Science* XI, no. 2 (2007): 125–29 (in Russian), which admits that indeed cosmology "exhibits dimensions pertaining to humanities." V. Kazyutinsky, "Physical Cosmology and Culture. Introduction," in *Cosmology, Physics, Culture*, ed. V. Kazyutinsky (Moscow: Institute of Philosophy of the Russian Academy of Science, 2011), 13 (in Russian). He quotes an eminent Russian astrophysicist, D. Varshalovich, that cosmology is a science not only of nature, "it is at the junction of the human and natural sciences" (ibid., 4). An interesting criterion for cosmology to assert itself as a natural science was suggested by A. N. Pavlenko: he claims that the extent to which cosmology expunges from its theories any notions of infinity (e.g., an infinite universe) can be considered as a criterion that cosmology indeed has the right to be called a natural science (for otherwise the infinite variables of its theories cannot be observed and lead to intrinsic contradictions, which undermine cosmology's status as natural science). However, Pavlenko does not qualify cosmology with its tendencies to deal with infinites as a human science; he rather assesses it as being in the middle, between a natural science and a mathematized metaphysics. See A. N. Pavlenko, "The Principle of 'Observability': Why the Theory of the Infinite Universe Cannot Be Sustained," in *Philosophical Problems of Cosmology: The Universe from "Nothing" or the Universe from "Non-Being"?* (Moscow: URSS/Librocom, 2012), 141–42.

The Explication of the Interplay between the Dimensions of the Natural and Human Sciences in Cosmology

Let us articulate further the sense of interplay between the elements of the human and natural sciences in cosmology. Cosmology is a scientific activity of human beings: it is because of this that, in its constitution, it is a human science in a trivial sense.[116] This claim comes from a noetic pole and implies that the epistemic and socially significant achievements of cosmology just are cognitive and manipulative achievements of human beings.[117] To say that "cosmology is human science" is to say that the doing of cosmology is an existential characteristic of human beings, their mode of being-in-the-world. However, if one looks at the interplay from the point of view of their noematic poles, one must admit that the difference between them is always understood in terms of the radically distinct object domains outlined by the faculties of cognition. The natural sciences are characterized by the conviction that their subject matter is always "an object," and, in particular, a nonhuman object (whose principle of existence is not related to subjectivity and personhood), so that its reading does not require any mutual agreement or reciprocity apart from common substance-based connotations.[118] It is in this sense that if cosmology pretends to be consistently a natural science, it must fulfill the major requirement: the "object" of cosmology (allegedly the universe as a whole) must be "at a distance"

116. Cf. C. W. Harvey, "Natural Science Is Human Science. Human Science Is Natural Science: Never the Twain Shall Meet," in Babich, *Continental and Postmodern Perspectives*, 122.

117. Cosmological research is driven by cultural and social factors, even by fashion. See, for example, R. Penrose, *The Road to Reality* (London: Vintage, 2005), 1017-20.

118. E. Husserl accentuated a feature of "corporeity" that physics wants to see in that world from which this same physics originates, that is, the life-world: "The natural science of the modern period, establishing itself as physics, has its roots in the consistent abstraction through which it *wants* to see, in the life-world only corporeity. Each 'thing' '*has*' corporeity even though, if it is (say) a human being or a work of art, it is not merely bodily but is only 'embodied,' like everything real." Husserl, *The Crisis of European Sciences*, 227.

from subjects of knowledge and thus, in a way, must be inhuman; its contingent existence manifests *itself from itself*, and is not conditioned by the constituting human subjectivity, that is, it is devoid of noetic carriers. However, this requirement creates a tension with the fact that the very constitution of any object is performed by a particular operation of reason, which in spite of its imposing detachment from an object still remains behind it. In other words, this requirement immediately invokes a countersense to the noetico-noematic inseparability that is asserted by phenomenology, thus leading to a dissonance with our previous observations that cosmology, being a cultural activity, cannot avoid speaking of humanity itself. Correspondingly, the noetic pole in predications of the universe can be removed only in some applications of cosmology (its astronomical part) dealing with distinct physical objects, which as such do not have immediate impact on the being of humanity (remote planets, stars, or galaxies). Such objects are characterized by persistent identity through a span of historical time and their appearance in human experience is not a construction, but an empirical fact. It is this identity that gives them the status of objectively existent entities. Contrary to these, such cosmological "objects" as clusters of galaxies are constructions because they are observed and hence are constituted wholes only from this particular location and cannot be treated as objects independently of this fact: clusters consist of "galaxies," which are at different distances from us and thus at different times, so that the question of the status of the cluster of galaxies as a distant and distinct *"object" with fixed spatio-temporal characteristics* (this is usually implied in physics and natural sciences) does not make sense—this "object" is a mental construction. While introducing a construct of a "cluster" of galaxies on the basis of the manifested phenomena, a different intentionality is invoked that unifies different galaxies in one "physical object," assigning to it such an existence *as if* it were based on physical causation between its parts. The fact that it is intentionality that plays a pivotal role here can be realized through an observation that a cluster of galaxies as a correlate of this intentionality remains unfulfilled on the level of physical causality. A cluster cannot be conceived as a physical system or object whose components are in a physical interaction constitutive of this object. There is no "body" of the cluster of galaxies in the same way as there is a body of a passing train whose appearance through its front entails the assurance in its physical objectivity as a "solid body" localized in space and time. One can conjecture that the question of existence of such objects as clusters of galaxies is established through insistence on their epistemic identity and not space-time attributes.[119] Here one can see that the language of intentionality (pertaining to the human sciences) cascades toward the language of physical causes (pertaining to the natural sciences). This shows that the ideal of the natural sciences is not only problematic on the scale of the whole universe, but, in fact, on the scale of its "elementary constituents" such as clusters of galaxies.[120]

What happens then is that a similar shift from intentionality to causality takes place in creating the idea of the universe as a whole, when the appearance of filaments of clusters of galaxies through observations and their scientific articulation is referred to the universe as a whole understood as a singular entity unified on the basis of physical causality. Here the language of intentions (in this case a *belief* in existence of the overall physical totality) is transformed into the language of physical causality. The intention is to unify (on the basis of a successive synthesis) the causally disconnected regions in the sky in one single whole. But this unification cannot be achieved

119. Cf. Primack and Abrams, *The View from the Centre of the Universe*, 171.

120. This resonates with Ellis's qualification that "cosmology is both a geographic and a historical science combined into one: we see distant sources at an earlier epoch, when their properties may have been different." Ellis, "Issues in the Philosophy of Cosmology," 1221.

as an accomplished phenomenalization: one cannot unify the various empirical display of the universe because one cannot grasp its facticity. The universe as a construct is poor in intuitive donation in the same sense as all mathematical constructions are. Consequently the construct of the universe, while contributing to the constitution of the universe, does not explain the facticity of this particular constitution as contributing toward the facticity of the universe in general. Correspondingly, the intentional objective is to produce a model where the unity of "all in all" in the universe would be explained in terms of *physical causality* but related to the common origin of the universe (through which all causally disjoint regions are unified). Here the work of intentionality presupposes transcendence of visible appearances toward a nonvisible, that is, toward that which is not reducible to the visible and yet that which is the condition of it. Intentionality thus implies a speculative transcendence similar to that of creative mythology inherent in the very human condition and which may be having an evolutionary importance. The appeal to the "past" of the universe, which "unites" the phenomenal in some unformed and undifferentiated matter, means at the same time the invocation of the original time which in its actuality is infinitely far away but still active and present as an open-ended fulfillment. Being nonhuman, this "past" is in radical discontinuity with this world, but it serves as a productive act in relation to the visible world, that is, as the world at distance, that distance which can be crossed over but not immanently overcome.

The "natural" appeal to the past of the universe in order to interpret its contingent present manifests an *epistemic causation* from "intentions" to "physical causes." The work of intentionality reveals itself as a desire to remove this contingency and replace it by an apodictic substantiality with a characteristic identity. Such a unification of all different aspects of the universe's appearances in one substance[121] implies, however, not so much causal connection based on physical processes (contemporary cosmology is clear about the fact that the universe consists of space-time disconnected sections[122]), but connectedness as belonging to the same underlying foundation (be it the overall encompassing space-time structure or substance, the "all in all" of the universe). It is this type of consubstantiality that is implied when (in some textbooks on cosmology) the universe containing, according to relativity, the infinitely many causally disconnecting regions, is depicted in a single diagram meant to symbolize the totality of all.[123] However, this "consubstantiality" does not entail physical causality; it has rather a transcendental character referring to the conditions of knowledge of the universe as a whole. It is not difficult to grasp that this type of transition from the language of intentions to the language of physical causality cannot be made on strictly scientific grounds; rather it implies *faith in the existence* of the universe, or the world as such. It is this faith that delivers us the sense of the given when we use the term "universe" in the conditions when the givenness pertaining to the natural object is unattainable. Using, by analogy, Husserl's words, "It is this *universal ground of belief in a world* which all praxis presupposes, not only the praxis of life but

121. One must not understand substance straightforwardly in the style of an old-fashioned metaphysics. When, for example, we talk about the unity of the universe in the Big Bang, we assume a sort of unified field that contains potentially all differentiated forms of observable material objects. This assumption is similar, for example, to an old Greek idea of "water" as that underlying agency which gives rise to all varied forms of matter.

122. This fact is related to the potentially infinite geometry of space and the finitude of the speed of light. Since we observe the universe along the past light-cone, which imposes constraints on the maximal distance causally connected with the earthbound point of observation, we are not only detached from all regions beyond this light-cone, but even within this light-cone we effectively receive signals from regions that, according to the standard model of cosmology, have been disconnected in the past. This constitutes a famous horizon problem, whose alleged solution was assigned to inflationary cosmology. See, e.g., S. Weinberg, *Cosmology* (Oxford: Oxford University Press, 2008), 205-6.

123. The examples of such diagrams can be found in many standard books on cosmology; see, for example, E. R. Harrison, *Cosmology: The Science of the Universe* (Cambridge: Cambridge University Press, 2000), 345-55. See also chapter 4.

also the theoretical praxis of cognition. . . . *Consciousness of the world is consciousness in the mode of certainty of belief*."[124] Husserl's usage of the term "world" does not correspond exactly to what is meant by the universe in cosmology. He does not reduce the meaning of the world to the all-encompassing extended spatiality and temporality, but rather means the world as an irreducible context of all experience, as the "horizon of all horizons" in all intentional acts.[125] Those beliefs that Husserl is talking about, if one translates them into the language of intentions, correspond to that which in the natural attitude can be described as empty and never-fulfilling intentions. Then the very tendency to transform the language of intentions into the language of physical causes in the context of the universe as a whole represents an attempt to make the universe a target of ongoing but unfulfilled intentions.[126] The universe acquires here the features not of that which is manifest (*the manifest*), but of *the manifestation* related to humanity. Correspondingly, the universe as the world cannot be considered as an astronomical concept (which by definition assumes the fulfillment of intentions at least to a certain extent); it is rather a transcendental notion related to our immediate experience of life in the universe and with the universe.

On the one hand the notion of the universe comes from astronomical observations and theories based in the ideal of physical causality; on the other hand we have some stories of the Big Bang and the universe's facticity expressed in philosophical enquiries and scientific-mythological narratives guided by the language of intentions and having origination in the human condition. This

124. E. Husserl, *Experience and Judgment* (London: Routledge & Kegan Paul, 1973), 30.
125. See in this respect a classical paper by L. Landgrebe, "The World as a Phenomenological Problem," *Philosophy and Phenomenological Research* 1, no. 1 (1940): 38–58; also M. Farber, "First Philosophy and the Problem of the World," *Philosophy and Phenomenological Research* 23, no. 3 (1963): 315-34.
126. Cf. R. Sokolowski, *Introduction to Phenomenology* (Cambridge: Cambridge University Press, 2000), 43.

all suggests that either region in cosmological discourse (both its observational base and eidetic extrapolations) encompasses the other; in other words, each type of cosmological understanding—be it the physical part with its idea of objectivity and causality, or a purely speculative part on the global aspects of the world—accounts for the hidden unity of both intentionalities dwelling in one and the same human person who discloses the universe. Put differently, cosmological theories need inputs from existential faith and hence from philosophy (regulative ideals in a Kantian sense), whereas philosophical imagination in the creation narrative borrows and exploits, for its "visualization," physico-mathematical images, thus offering a metaphysical extension of physics.[127]

One now anticipates that any attempt of totalizing the worldview, that is, making a unique and consistent whole in our perception of the universe, is doomed to fail. All these attempts start from within the life-world associated with a geocentric world, and it is the life-world that remains patchy and incoherent through different articulations including not only philosophical and scientific, but also religious ones. In other words, the life-world does not allow its totalization either through the language of matter and body or that of spirit and soul, through physical causes or through human intentions, because it is the world of historically contingent *events* whose instantiation is not subject to the physical or purely spiritual. As was argued by J.-L. Marion by referring to Kant, unique occurrences

127. In fact, all diagrams that recapitulate the wholeness of the universe and its link to the articulating humanity represent such a metaphysical extension. As an example one can point to a famous "closed circuit" in J. A. Wheeler's writings symbolizing the world as a self-synthesizing system of existence (see, for example, "World as a System Self-Synthesized by Quantum Networking," 5), or a picture of the so-called "Cosmic Uroboros" (see Primack and Abrams, *The View from the Centre of the Universe*, 160, 284). These diagrams mean to stand for such a unity of the world in which the historically formulated physics is raised to the level of apodictic structure of being. One then understands the apophatic sense of these representations providing signifiers of the universe's manifestation with no pretense for the exhaustion of the sense of that which is signified.

typical for historical phenomena (events) do not fall under the rubric of *analogies of experience*, which can deal only with a reduced and incomplete phenomenality typical for the objects constituted by the sciences, a phenomenality that is poor in intuition.[128] The reality of the life-world is too existentially complex to allow a simpleminded reduction to either one of these. This implies that cosmology is doomed to deal with this intrinsic dualism between its orientation toward the natural sciences and, at the same time, its dependence upon the dimensions of human life. One must then expect that the language of physical causation as a mode of thematization of the life-world will go side by side with the language of *communion* dealing with the excess of intuition over reason, so that neither of them will be reducible to the other. Every attempt to semantically transgress the normalized sense-borders of everyday intentional life in the conditions of inescapable communion with the universe, by the use of causally reductive language that sees the universe in stages of evolution and hierarchies of objects, issues in a countersense. The same holds for attempts to transgress the causal domain with intentional language: the intentional language is useful to analyze and refer the very causal language to existential motivations of research, but not to the truth of a *fact*, which the causal language attempts to affirm.

One can refer to the analogy between the phenomenon of birth (as coming into being of a new hypostatic existence) and the idea of the Big Bang as an example of how the language of intentionality takes control over the language of causality.[129] In both cases the "event" of birth and the origin of the universe are phenomenologically concealed because one cannot transcend one's own life or the

128. See J.-L. Marion, "The Saturated Phenomenon," in D. Janicaud et al., *Phenomenology and "The Theological Turn": The French Debate* (New York: Fordham University Press, 2001), 204.
129. See a detailed discussion below in chapter 6.

universe in order to "look" at their "origin" from outside. By referring to the mystery of origin of every particular personal existence (its contingency), a physical problem of comprehending the temporal origin of the universe is transformed into a philosophical problem of the contingency of this origin. If one thinks of the ontological condition for the possibility of processes in time in the universe, then no appeal can plausibly be made either to a set of fixed entities or to an ontic first cause since, within the material universe, neither the former nor the latter can "originally," that is, ontologically, precede what is caused (nonoriginary origin). The notion of "cause" is acting here as a pragmatic fiction, which disguises the fact that "that which causes" is only something that is changing into something else. In different words, "causing means giving," implying that the cause is going out of itself as an effect, while the effect is wholly "from" the cause in which it eminently abides. The sought "explanation" of the facticity of the universe thus implies the inevitability of "givenness," that is, that donation of the universe which we receive. Correspondingly, all conventional mythologies acting on the premise of imitating the otherworldly "origin" of the universe as a precondition for the unfolding temporal flux fail to address this contingency in an "explanatory" manner: they just imitate the mode of donation of the universe through the appeal to creative power of imagination, which allegedly refers to realities beyond the facticity of the donation. In other words, the donation itself is explicated in terms of layers and strata of this same donation, which is not perceived as manifested.[130]

The analogy with the phenomenological concealment of the origin of personal consciousness clarifies the sense of communion with the universe. It is exactly because there is no access to one's

130. Ladrière, *Language and Belief*, 153.

origin and one has to accept life as it is given, every personal consciousness (as an event of existence) treats itself as indefinite and commensurable with the universe, through an intuition of coexistence with the universe, which is not fulfilled through acts of reason in terms of spatial and temporal distinctions. G. Marcel wrote at the beginning of the twentieth century: "The universe as such, not being thought of or able to be thought of as an object, has strictly speaking no past: it entirely transcends what I called a 'cinematographic' representation. And the same is true of myself: on a certain level I cannot fail to appear to myself as contemporary with the universe (*coaevus universo*), that is, as eternal."[131] Marcel asserts here that the sense of communion with the universe (as a fact of life) makes existentially irrelevant any notion of evolution and stage-by-stage description of the universe (a "cinematographic" representation).[132] Communion is here and now and it is absolute in its personal ecstatic phenomenality as an event of life; in other words, it is the facticity of life that retains transcendence of that with which and in which life is as communion. If the universe in its phenomenality appears to be a stable and enduring background of existence, then a human being as a communicant with the universe realizes itself as commensurable with it and hence also coeternal with it (not in a theological sense). One can say that human beings as long as they are alive experience apophatically the immanence of the actually infinite through their finitude. The life of a human being is then an act of communion, an event whose fullness does not need any acknowledgment of history of the universe. Within the fact of

131. Marcel, *Being and Having*, 24.
132. The so-called "cinematographic representation of the universe" (terms that ascend back to H. Bergson) is irrelevant in the approach of communion even in spite of a trivial observation that all material constituents of communicants contain elements of the historical past of the universe (human bodies are made of stardust and ultimately originate in the Big Bang). This past is encapsulated in human bodies (being present in us in every cell and every breath) and thus is recapitulated through communion with no reference to spatial and temporal extension.

life it is the universe that becomes a part of existential history of every human being and not vice versa.[133] Thus the alleged "identity" of the universe, which receives its fulfillment in acts of communion with hypostatic creatures, transcends any phenomenalization of the universe (as its representation) through cosmology.[134] However, theologically, the sense of immanence with God's creation, as such, does not remove those moral tensions in the created world which are implied in the way of salvation.[135] Correspondingly, the experience of immanence with the world as God's creation does not entail the actualized presence of the Kingdom of God.[136] In other words, the cosmological gnosis of creation does not necessarily entail knowledge and communion with the Divine.

What happens in cosmology is that physicists' intentionality pertaining to the natural attitude breaks the noetico-noematic (subject-object) inseparability and explicates the event of communion with the universe through creating theoretical models of the universe. By being in communion the knowing subject, a cosmologist,

133. This observation corresponds not only to a phenomenological stance on the sense of history as constituted by transcendental ego. See E. Husserl, *On the Phenomenology of the Consciousness of Internal Time*, trans. J. B. Brough (Dordrecht: Kluwer, 1991); P. Ricoeur, *Husserl: An Analysis of His Phenomenology* (Evanston, IL: Northwestern University Press, 1967), 145-50, but also to a Christian conviction that cosmic history is included in the history of salvation.

134. The epistemological dichotomy between communion and discursive representation of the universe implied here can be expressed as the opposition between *subjective absolute* and *objective relative* mentioned by H. Weyl in reference to M. Born: "The immediate experience [as communion, AN] is *subjective and absolute*. . . . The objective world, on the other hand with which we reckon continually in our daily lives and which the natural sciences attempt to crystallise . . . is of necessity *relative* [this relativity is first of all related to its historical contingency, AN]; it can be represented by definite things (numbers or other symbols) only after a system of coordinates has been arbitrarily [that is, historically contingently, AN] carried into the world. . . . Whoever desires the absolute must take the subjectivity and egocentricity into the bargain; whoever feels drawn toward the objective faces the problem of relativity" (Weyl, *Philosophy of Mathematics and Natural Science*, 116).

135. This point was clearly articulated by Maximus the Confessor in his teaching on mediation between divisions in creation: the overcoming of moral divisions in creation and between the created world and God is to take place in the conditions of immanent hypostatic presence of God in creation without trespassing the boundaries between creation and the uncreated.

136. Contrary to what is suggested by Comte-Sponville in his *L'esprit de l'athéisme*.

articulates the universe as a sort of "out there" that follows the objective laws of nature and thus is independent of a cosmologist's insight. Such a scientific notion of the universe naturally falls under the phenomenological critique, which reinstates a simple fact that any statements about the universe represent an articulated *truth* related to human beings and their bodies, which are consubstantial to the universe and commune with it. However, this intertwining between humanity and the universe does not deprive the universe of independence from the conditions of its expression by human subjectivity. The universe as communion, denying any complete synthesis in its phenomenalization, always retains that overwhelming presence which cannot be conditioned by the rubrics of subjectivity, inducing an excess of intuition over any attempt to see or constitute them. Thus the universe appeals to humanity as a *saturated phenomenon* always retaining its own transcendence with respect to all attempts to grasp the sense of its facticity.[137] This transcendence reflects the fact that humanity experiences the all-unity of the universe in the conditions of the archetypical memory of "all in all" retained since the event of the Fall. There is the sense of immanence with the universe through created consubstantiality. However, because of the moral tensions between parts of creation, its wholeness as originating in the Divine is not available to humanity as *presence* thus manifesting itself as ever-escaping transcendence.

Thus there is an obvious and probably unavoidable tension between the representation of the universe as an object in cosmology and its presence in existential communion. This tension is of a

137. The notion of the "saturated phenomenon" was introduced by J.-L. Marion in his paper "The Saturated Phenomenon," and developed further in his books *Being Given* and *In Excess: Studies of Saturated Phenomena* (New York: Fordham University Press, 2002). The idea of the universe as a saturated phenomenon was coined by the present author in the paper Nesteruk, "Transcendence-in-Immanence in Theology and Cosmology." Its further development can be found below in chapter 7.

theological kind related to the human condition attempting to restore its lost Divine image. To avoid this tension remaining in the postlapsarian condition, one could step into a dispassionate *phenomenological* description of the universe as it discloses itself in the natural attitude in the language of causes on the one side and in the language of intentional immanence through communion on the other side. One might say that the *ontological commitment* should be left out and a certain phenomenological calm must be adopted with respect to various languages used for assessing the universe as a whole. This means that one can use discoveries achieved by natural sciences as well as by philosophical insights and communion without committing semantically or ontologically to one region's priority over the other. The same idea can be expressed differently: the cosmological narrative follows either the logic of physical causes and is shaped by mathematics aiming at pure objectivity or, alternatively, the "logic" of life and inseparable communion with the universe—which is not subject to intellectual persuasion and thus is open to free-willing employment of artistic expression, ever incomplete metaphorically, and fundamentally open-ended. It is because of this dichotomy that one must learn how to live with the incomplete wholes, partial and shattered totalities—totalities requiring different languages although, after all, belonging to one culture.[138] Correspondingly, the objective of a theological and

138. Cf. C. Harvey, "Natural Science Is Human Science," 133. See also L. Papin, "This Is Not a Universe: Metaphor, Language, and Representation," *Proceedings of the Modern Language Association* 107, no. 5 (1992): 1253-65. The implied diversity and plurality as a valid approach to knowledge of the universe brings one to a characteristic conclusion that in cosmology, all signifiers of experience of the universe do not exhaust that which is aimed to be signified. As was conjectured by D. Bohm and D. Peat, each scientific theory bears the inscription "this is not a universe" (based on the analogy with R. Magritte's famous paintings with characteristic titles "This is not a pipe," or "This is not an apple"), meaning that "every kind of thought, mathematics included, is an abstraction which does not and cannot cover the whole of reality," and this is why "perhaps every theory of the universe should have in its fundamental statement 'this is not a universe.'" D. Bohm and D. Peat, *Science, Order, and Creativity* (New York: Bantam, 1987), 8-9.

philosophical insight in cosmology is not to find a unified language for understanding the universe, but rather to realize that in the human approach to its totality, always initiated in the life-world, we progress by the various ways given to humanity. The reality of the universe then is much more than is met by the discursive mind; it forms a mysterious sense of identity, which is intuited, but never completely grasped by the mind: it bedazzles us, while constituting our own sense of identity to the extent that we cannot circumscribe the universe in the rubrics of thought. In its perennial leap toward understanding the sense of the universe, humanity stretches its capacity to grasp itself. Indeed, ontologically, the universe we disclose through embodiment involves us as disclosing it. Given this, our personhood achieves its status as a "place" that permits disclosure of the universe through what this personhood is. The "I" as person discloses, by being structured by disclosure itself, that is, in the words of M. Merleau-Ponty, by being a "concrete emblem of general manner of being."[139]

The question of the very possibility and facticity of this disclosure remains a theological mystery. In this sense all modern cosmological theories, while explicating and articulating this mystery, do not challenge a proper theological stance on creation of the universe. The ontological otherness of all creation remains that saturating givenness which is not affected by the advance of science. Its sense can only be appropriated experientially through communion with the good God through the universe as his good creation.

139. Merleau-Ponty, *The Visible and the Invisible*, 147.

3

Constituting the Universe

Transcendental Delimiters and Apophaticism in Cosmology

> *For cosmic thought to be possible, thought must find the means to separate itself from things, to gaze beyond their appearance, to transcend the visible towards a non-visible that is irreducible to the visible and yet the condition of it. A speculative transcendence is required.*
>
> *–Ladrière, Language and Belief, p. 150*

In this chapter we analyze the general epistemological conditions for knowability and explicability of the universe as a whole. Both knowability and explicability imply an anthropological dimension in cosmology, for both these requirements have human origin: to speculate about the universe one must experience its presence through the instantaneous synthesis associated with conscious life. It is argued that the cosmological principle acts as reduction of this synthesis to a mental spatial uniformity of the universe. However, the contingent facticity of the cosmological principle leads to the necessity of its explication in terms of generative steps required to grasp the sense of the universe in terms of its origin (similar to the origin of any hypostatic existence). This implies that the study of the universe presupposes the transcendence of the visible and forms the construct of the universe in the realm of intelligible forms. The latter characteristically rearticulates the dual constitution of humanity, articulating the universe, pointing this to the inseparable nature of cosmology and anthropology.

Introduction: The Universe's Knowability and Explicability

It is a fact of contemporary science's sociology that cosmology reveals better than any other physical study a constant advance of theory and observations that rapidly extend and revise our views about the universe. Cosmology makes all its statements using affirmations that attempt to approach the universe in its totality in terms of references, imported from the earthly physical world and astronomical cosmos. However, it seems reasonable to suggest that because of the physical incommensurability between human subjects and the universe as a whole, this set of defining statements is open-ended and cannot hope to terminate at any final stage.[1]

Small details about the universe, the technicalities of its theories, contribute toward our perception of "pieces" or "moments" of the universe as a whole whose *identity* is anticipated by cosmologists as the *pre-given*, inherent, and nonrelational (with respect to these

1. Cf. M. Munitz, "Approaches to Existence," *Philosophy East and West* 25, no. 3 (1975): 337-38. L. Brisson and F. W. Meyerstein, developing a parallel between ancient Greek cosmology and modern physical cosmology, assert a similar thought that modern science shares the same predicament as it was implied by Plato in his cosmological account in Timaeus, namely that "every theory is at best a provisional explanation of some phenomenon, which remains open-ended, since a theory can only resist falsification, but can never be verified" (Inventing the Universe. Plato's Timaeus, The Big Bang, And the Problem of Scientific Knowledge (Albany: State University of New York Press, 1995), 12.

"pieces" and "moments" as well as to anything hypothetically "outside" the universe) in itself, having a character of either an entity or a logical subject. The notion of the universe as a whole implies some particular features of what is predicated, which immediately puts this notion in a sort of difficulty. First of all the universe as a whole, by definition, constitutes a *principle of existence* (and in this sense a foundation), which does not allow one to use any definitions based on the ontology of distinct physical events taken as parts and pieces of the universe as a whole. Secondly, it is impossible to talk about the universe as a whole in terms of its origin (one implies here not temporal origin, as in physical cosmology, but an absolute ontological origin as facticity of being), because there is nothing beyond the universe (in terms of space and time) that could determine its origination (one can speak of the nonoriginary "origin" of the universe). By historical analogy one can appeal to the witness of Descartes who, while making a careful distinction in application of the terms of "infinity" and "limitlessness" to the universe, notices that "we must not be so presumptuous as it seems we would be if we supposed that the universe has any limits without being assured of it by divine revelation or, at least by very evident natural reasons; because it would [mean] that we want our thoughts to be able to imagine something beyond that to which God's power has extended itself in creating the world."[2] Following Descartes' thought, one could add that in speculating about the origin of the universe's facticity human beings transcend their embodied consciousness, and enquire into the ultimate sense of things, as if they could possess the "mind" (or *logos*) of that agency which brought the universe with its physical laws into being. That is why the universe as a whole (*intellected*, not through its manifestations in pieces and moments) is

2. Descartes, *Principia Philosophiae*, part 3, § 2, quoted in A. Koyré, *From the Closed World to the Infinite Universe* (New York: Harper & Brothers, 1958), 109.

nonspatial, atemporal, beyond space and time. Thirdly, the universe as a whole is not accessible to our grasp as part of a causation from one "thing" to another in time because even the visible universe represents a frozen compendium of temporal eras that are linked to each other but cannot be related to anything beyond them. In other words, *relational logic* cannot be applied in order to affirm anything about the universe as a whole if one attempts to do so through a series of causations starting in the astronomical cosmos. Ages, times, and places belong to the category of *relationship*, and consequently no object necessarily associated with these things can be other than relative. The universe as a whole transcends the category of *relationship*; for nothing else whatsoever is necessarily associated with it. Relational logic can be applied to all things that are in space and time; it can also be applied to our, human, relationship with the universe. But since the universe is beyond any relationship, our participation in it does not affect its being (whatever this could mean); in other words, our intellection about the universe as a whole does not provide us with any knowledge of the universe as it is in itself. We are in relationship with the universe, but this relationship does not allow us to penetrate its mystery independently of our participation in it.

As a logical subject, the universe (or its, so to speak, identity) appears, for example, as an *idea*, similar to the idea of the *world* in a Kantian sense, that is, as an accomplishing term of the series of causation in the realm of the conditioned "jumping" from the empirical "pieces" and "moments" to the inferred mental fullness. This jump exhibits a belief that there must *be* a united and unconditional all to which one belongs and in which one participates. However, this "must be" originates in our mind because we are *given* the conditioned (finite things), and we believe that we must, in the same way, have a capacity to access the unconditioned.[3] This is not the case, as the unconditioned itself can only be taken as an infinite series of conditions that is not given to us. The universe as a whole is not in space and time and cannot be encountered in experience. Correspondingly, the universe's ontological status becomes uncertain (as Kant would say: the universe does not exist in itself as if it would have a determinate magnitude).[4] Hence the universe as a whole "emerges" as a *rational* idea (as distinct and different, in a Kantian classification, from an *aesthetical* idea, that is, an idea of a beautiful arrangement—*cosmos*) with a certain regulative use. This idea has a minimal perceptible intuition and never becomes cognition (that is, a never-fulfilled intentionality). The famous Kantian antinomies of reason associated with the notion of the world, which never disappear if the universe is approached through the categories of the understanding, demonstrate that if cosmology dealt only with such an abstract (a priori) notion of the universe, all talk about the identity of the universe would have a crypto-theological meaning, implying a language that does not have references in the mundane world. In this case the universe would be defined in an abstract way something like this: the universe as totality is one, incomprehensible, possessing completely the total potentiality of being, excluding notions of when and how related to its wholeness, and not to be known through natural image.[5] In spite of the fact that Kant's insistence that cosmological questions can only be settled a priori, that is, with no recourse to experience, seems now to be inadequate (because some cosmological hypotheses on the age and size of the universe can be subjected to empirical verification or refutation), still the question

3. I. Kant, *Critique of Pure Reason*, 2nd ed., trans. N. K. Smith (London: Macmillan, 1933),A332/B388, A409/B436, A497/B525. Kant's critical assessment of this view can be found inA498/B527.
4. Ibid., A503-5/B532-34.
5. These definitions are not surprisingly reminiscent of the definitions of God. See, for example, Maximus the Confessor, *Two Hundred Texts on Theology and the Incarnate Dispensation of the Son of God* I:1-10, in *The Philokalia*, vol. 2, 114-16.

of the universe's identity as related to its whole remains effectively unchanged through the progress of science. The advancement of science teaches us exactly that which Kant was asserting on general metaphysical grounds, namely, that scientific progress will never accomplish the disclosure of the universe's identity (this is a *negative certitude* of philosophy); the idea of the universe as a whole with a certain identity has a perfectly legitimate regulative use that can lead us to search for the conditions of the universe's explicability and increase our knowledge and understanding of it.[6] In this case the antinomial skepticism could be overcome and Kant would approve such a usage of the notion of the universe and its identity.[7] The universe, treated as "a particular" ever-escaping precise definition and exhaustion through its signifiers, becomes a subject not of the *understanding* but of *judgment*. To understand the universe one must judge whether it exists or not. Since judgment with respect to the universe can only be *reflective* (and not *determinate*),[8] it is this kind of reflective judgment that vindicates cosmology as the research of the universe as a whole.

Correspondingly, in actual cosmological research the sense of identity present behind the term "universe" is rather an expression of that gradual substitution of the material accumulated from observations and theories for an inherent intuition of the universe (related to reflective judgment) as totality of all, thus following a path to objectivity through coordinating phenomena and theoretical constructs into strata of invariants across a variety of subjective and

6. Kant, *Critique of Pure Reason*, A307/B364, A508–10/B536–38.

7. Cf. ibid., A223/B380, A670–72/B698–700, A685/B713.

8. Here is Kant's description of reflective judgment: "The reflective judgment thus works with given appearances so as to bring them under empirical concept of determinate natural things not schematically, but *technically*, not just mechanically, like a tool controlled by the understanding and the senses, but *artistically*, according to the universal but at the same time undefined principle of a purposive, systematic ordering of nature." I. Kant, *First Introduction to the Critique of Judgment*, trans. J. Haden (Indianapolis: Bobbs-Merrill, 1965), 18.

instrumental circumstances. In this case, together with M. Munitz, one can formulate a thesis that the universe appears as a constructive achievement, rather than discovery.[9] This thesis confirms a general conviction formulated in the phenomenological stance on science, namely that there must be made a distinction between nature as it appears in primary perceptual experience and nature-for-physicists (that is, "nature"), which is a *mental accomplishment* ("hypostasis of mental creations") as an ideal limit of convergent sequences of "images of nature" that are constructed by physicists in the course of history.[10] In other words, in physical cosmology, the "identity of the universe" is not an a priori inherent and immutable in-itself, but involves constitution. It is because of the finitude of the human scientific understanding that the notion of the universe as a whole acquires features of an open-ended construct[11] in the conditions of a hidden belief in ultimate convergence of the correspondence rules relating constructs to the alleged reality.

There is one particular feature in the constitution of the universe that implies for the universe to have a sort of identity: namely, *its identity through time*. Indeed, for cosmological research, in order to assert that it deals with the universe as physical reality, there must be some *stability* in the "object" studied through time. Then the

9. See, for example, M. Munitz, *Question of Reality* (Princeton: Princeton University Press, 1990), 141.
10. Cf. A. Gurwitsch, *Phenomenology and the Theory of Science* (Evanston, IL: NorthwesternUniversity Press, 1974), 41-46. One must also mention the physicalistic attempt of J. A. Wheeler to promote a similar view that the universe is not a pre-given clock-like mechanism, but the world of existences that is constituted by human observers-participants. See, for example, J. A. Wheeler, *At Home in the Universe* (New York: American Institute of Physics, 1994), 112-31, 295-311, and numerous papers cited therein.
11. By constructs, we understand, in analogy with H. Margenau, *The Nature of Physical Reality: A Philosophy of Modern Physics* (Woodbridge, CT: Ox Bow, 1977), 69-72, the entities through which the empirical perceptions and presentations receive their theoretical explication. Constructs, sharing in their function something from concepts and something from ideas, submit themselves to logical procedures to a much fuller extent than to the immediate astronomical data.

question is: What is that referent in physical reality which sustains such a stability? The answer is nearly obvious: the material content of the universe, a physical composite of fields, particles, macroscopic structures, and other theoretically predicted but allegedly physical components. However, one cannot claim that this substance is that which preexists the very process of investigation of the universe because it itself demands constitution through the irrational, experientially identified, resistance to any attempt to disclose its identity. We refer here to the fact that modern cosmology predicts that 96 percent of its material content is formed by the so-called dark matter and dark energy, whose physical nature is not known yet: theoretical prediction experiences resistance from nature to be disclosed in terms of experimentally observed and identified fields and particles. This is the reason why the required stability of cosmological constituents has to rely on mathematical assertions. For example, "dark matter" and "dark energy" are mathematical predictions (following from the demand of consistency in cosmological theories and observations) but their ontology remains, up to now, hanging on some implicit belief-based commitments to the realistic nature of their mathematical constructions. In this case the ontological status of the identity of the universe receives elucidation from the conviction that its intelligible (mathematical) pattern is devoid of mutability. Since mathematical physics appears as a leading technique of thematization and objectification of the universe, it pretends to fulfill the function of bringing the alleged identity of the universe to unconcealment (despite the fact that there are probably other ways of communication with the universe).[12] Indeed, since the universe

12. Here one mentions that this question resonates with a general issue of mathematization of nature and whether it exhausts the sense of reality. In a historico-philosophical perspective one must refer to E. Husserl, who critically assessed such a mathematization in his *Crisis of the European Sciences* (Evanston, IL: Northwestern University Press, 1974). See also some secondary sources in this respect: J. Kockelmans, "The Mathematisation of Nature in Husserl's Last

as a whole is thought as being beyond any worldly causality and temporality, it is primarily posed by consciousness as an intelligible entity stripped (by means of phenomenological reduction) of any specific facts and theoretical predications. However, the presence of the universe in consciousness is not only a matter of constitution through mathematics; it is the fact of conscious life in general. The universe appears to humanity as the unconditional *coming to truth* of the *visible* (sensible) and *invisible* (intelligible) *being*, manifested in life itself. Paraphrasing P. Brockelman, the universe is not this chain of mountains or that galaxy or solar system, but the continuous eruption of myriad forms, the active *that-ing* or *is-ing* of everything which emerges into experience of life.[13] Before being constituted the universe is intuited as an aesthetical idea (not as a rational idea), and this means that different means of communion with the universe are invoked.

Being reminiscent of the ancient approach to the universe as beautiful cosmos, contemporary physical cosmology, before its

Publication, *Krisis*," in *Phenomenology and the Natural Sciences*, ed. T. J. Kiesel and J. Kockelmans (Evanston, IL: Northwestern University Press, 1970), 45-67; A. Gurwitsch, "Galilean Physics in the Light of Husserl's Phenomenology," in *Galileo: Man of Science*, ed. E. McMullin (New York: Basic Books, 1967), 395-401. A discussion of the mathematization of nature can be linked to the question of mathematical realism as it is understood in modern philosophy of mathematics. As an example one can point toward a phenomenological position with respect to mathematical objects that is similar to the idea of "weak objectivity" (R. Tieszen, *Phenomenology, Logic, and the Philosophy of Mathematics* [Cambridge: Cambridge University Press, 2005]), and to another trend called structural realism advocated in M. Resnik, *Mathematics as a Science of Patterns* (Oxford: Clarendon, 2005), as well as in S. Shapiro, *Thinking about Mathematics: The Philosophy of Mathematics* (Oxford: Oxford University Press, 2000). This latter structuralist trend asserts that the physical domain of cosmological theory represents an instant of a mathematical structure; see in this respect G. McCabe, "The Structure and Interpretation of Cosmology," *Studies in History and Philosophy of Modern Physics* 1, no. 35 (2004): 549-95; 2, no. 36 (2005): 67-102. In a way, the structuralist view positions itself as a radical form of mathematical realism, which asserts that everything is mathematics; see, for example, M. Tegmark, *Our Mathematical Universe. My Quest for the Ultimate Nature of Reality* (London: Allen Lane, 2014) chapter 10 and 12.

13. P. Brockelman, *Cosmology and Creation: The Spiritual Significance of Contemporary Cosmology* (Oxford: Oxford University Press, 1999), 79.

intention to study the universe as a structure extended in space and time, anticipates the universe as a singular unity that possesses some inherent *logos*. Thus any reasoning about the identity of the universe implies that the *reality* of this identity must be secured by bringing out the fact that it is different from its manifold presentations and showing that, despite its uncertain status, it is truly a component of what we experience: the universe is presented to us through its "pieces" and "moments" (places and eras depicted in numerous cosmological diagrams[14]), but the identity of the universe (not only as the *name* of its facticity[15]) is empirically *absent*: it escapes rubrics of extended space and time, and is thus trans-temporal, that is, not belonging to the temporal flux.

One may clarify the sense of the identity of the universe as it stands in the context of "presence in absence" by bringing out some particular phenomenological stance on the identity of objects. The identity of an ordinary object appears as the unity of its profiles and impressions available to the public mind.[16] One particular feature of constituting the identity of an ordinary object is that it can be formed through consciousness of its absence, that is, its potential nonexistence. For example, while experiencing the beauty of a flower one appreciates it without clear understanding as to why this particular kind of beauty came into being if anyone were not there to appreciate it. The identity of this flower (apart from an obvious synthesis of pieces and moments) can implicitly enter consciousness from the perspective of its possible nonexistence, so that the very

14. See, for example, a typical diagram called "The Cosmic Spheres of Time—Our Visible Universe," in J. Primack and N. E. Abrams, *The View from the Centre of the Universe: Discovering Our Extraordinary Place in the Cosmos* (London: Fourth Estate, 2006), 135.

15. Munitz, *Question of Reality*, 175.

16. On a general phenomenological analysis of identity in absence, see R. Sokolowski, *Husserlian Meditations* (Evanston, IL: Northwestern University Press, 1974), 8-56. See also a simplified version of the same exposition in R. Sokolowski, *Introduction to Phenomenology* (Cambridge: Cambridge University Press, 2000), 17-21.

factual beauty of it (as identity) is weighed against its potential (eidetic) nonexistence. Unlike in ordinary objects, in cosmology, any attempt of constituting the notion of the universe as potentially nonexistent is intrinsically contradictory, for it eliminates with itself that embodied consciousness which thinks of the universe. Here an analogy can be drawn with the theology understood as experience of God which claims that the elimination of God in consciousness leads to cessation of this consciousness: one refers here to the famous § 58 of Husserl's *Ideas I*,[17] where Husserl argued that the notion of a transcendent God who allegedly transcends both the world and the field of "absolute" consciousness and thus is posed in the natural attitude as existing objectively out there, must be subjected to the transcendental *epoché* (phenomenological reduction) so that the question about its reality is suspended and thus the very theology of a transcendent God is brought to a methodological halt. It is important to realize here that "transcendency pertaining to God"[18] takes place not through an ascending series of the world phenomena but through observing "marvellous *teleologies*"[19] in nature that characterize the activity of consciousness. Then arises the question about the ground of facticity of this same constitutive consciousness, which naturally does not have the sense of a physical-causal reason.[20] By proclaiming the transcendence of God to consciousness, his next step is to *reduce* the thusly affirmed God and to remove the whole subject of the Divine from the phenomenological project. The major problem in this procedure is that it does not differentiate between God as a mental construction that is subject to any possible operation of

17. E. Husserl, *Ideas Pertaining to a Pure Phenomenology and to a Phenomenological Philosophy*. First Book. *General Introduction to a Pure Phenomenology*, trans. F. Kersten (Dordrecht: Kluwer, 1998), 133-34.
18. Ibid., 134.
19. Ibid., 134.
20. Ibid., 134.

consciousness, such as reduction, for example, and the living God of faith whose presence in consciousness is exactly that ontological link which makes this consciousness possible at all and which can be cut off only in abstraction. It is clear that the "absolute" consciousness of Husserl, which through the very mode of its being expresses our existential participation or communion with God, cannot bracket or reduce the living God because by so doing this consciousness attempts the impossible: to deprive itself of its own foundation and hence, de facto, to destroy itself as consciousness *in* God. Here, by using a religious language, consciousness degenerates and falls into an illusion of its might. J.-L. Marion writes in this respect: "Husserl submits what he names 'God' to the reduction only in so far as he defines it by transcendence (and insofar as he compares this particular transcendence with that, in fact quite different, of the object in the natural attitude); and yet in Revelation *theo*-logy, God is likewise, indeed especially, characterised by *radical immanence to consciousness, and in this sense would be confirmed by a reduction*."[21] This is the main point of our analogy with cosmology: human beings are immanent to the universe through consubstantiality, so that the hypothetical removal of the universe must inevitably lead to the cessation of the conditions of consciousness's embodiment. In this respect one must point out that in a different treatise, his "First Philosophy,"[22] Husserl himself was preoccupied with a problem of nullification of the world, namely whether the pure ego, the subject, disappears if the world as such is reduced to nothing through a so-to-speak epistemological nullification. His answer was "not," that is, what is subject to nullification is the empirical ego which is indebted to the world. However, if one supposes the existence of a nonphysical

21. J.-L. Marion, *Being Given: Toward a Phenomenology of Givenness* (Stanford: Stanford University Press, 2002), 242–43 (see also footnote 4 at p. 343).
22. E. Husserl, *Erste Philosophie* 2 (The Hague: Nijhoff, 1959), 73.

center of human existence, the pure "I" of experience, which is beyond the worldly delimiters, then this "I" does not disappear and, in fact, it is the reduction of the world that allows one to proceed from here to the discourse of the "I."[23] Finally, one can repeat that the identity of the universe cannot be acquired by discursive thinking either through the rational idea of the world, or through its eidetic nullification.

Whereas Kant was critical of the notion of the world as a whole by pointing out that the difficulties of reasoning about it clarify the limits of reason, thus contributing toward the constitution of subjectivity, one can say that the ideation of nonexistence of the world creates a similar difficulty by not only revealing the absolute preconditions for any reasoning in rubrics of embodiment: there must be the universe in order to think of it, for it is through consubstantial embodiment that the existence of reason is linked to existence of the universe. In view of this, the nullification of the world is possible only if this world is considered in the phenomenality of objects. In other words, what is reduced in the very modality of objectification is the world as object. However, the world as existential givenness remains through the saturation of the intuitive content that cannot be reduced because it cannot be held by the discursive thinking. Correspondingly the above-mentioned disclosure of the "I" results in this "I" being constituted by the intuitive donation of the world exactly to the extent this "I" cannot grasp the sense of the world's facticity. In this sense, the net result of the eidetic nullification of the world is the same as its predication through intelligible series: it reveals the limits of subjectivity thus structuring it. Finally, an attempt to construct the identity of the universe through its eidetic nullification seems to be existentially contradictory: the world comes

23. See more discussion in M. Farber, "First Philosophy and the Problem of the World," *Philosophy and Phenomenological Research* 23, no. 3 (1963): 315–34.

into existence together with the divine gift of life, so that it is only God who is potentially capable of nullifying the world together with the consciousness of this world.

The contemplation of the identity of the universe in the conditions of its empirical absence constitutes the core of the human condition, pointing again to the twofold presence of humanity in the universe and its particular intentionality, which has its origin in the singularity of every human being-in-the-world (Heidegger), existence-in-situation (Marcel), or hypostasis (Levinas). The very mode of conscious life implies the sense of immanence *with* the universe. The universe appears through an *instantaneous intuitive synthesis* that is encapsulated in the very fact of life and forms pre-predicative experience. The anticipation of the identity of the universe at the pre-predicative level represents, in fact, a *transcendental* requirement not only for its knowability, but also for its accessibility in general. This requirement originates in the trivial fact that the identity of the universe is not a "regular particular" and escapes the exhaustion through signifiers pertaining to the understanding. When this conceptual item extends human thought beyond the limits of possible experience, theoretical knowledge becomes impossible, and one must look for the special conditions that adjust our thought to fit this idea. The understanding is not precisely suitable for this purpose, for it determines (through application of its categories) only those particulars that are given in sensible intuition. Judgment (which, according to Kant, is a faculty distinct from understanding and reason), however, requires that all particulars be subject to understanding, that they be rendered intelligible and explicable under some law or other. In respect to the universe, our limited cognitive capacity makes it impossible to fulfill the demands of judgment in any theoretical way: knowledge remains apophatic, that is, incapable of exhausting the reality of that which it signifies. In spite of this there is a pre-predicative belief (relying on the indemonstrable power of mathematics) that the universe is organized so that we will be able to understand it. This presumption, or better, a certain belief-based commitment, establishes in a characteristic way the intrinsic *teleology* of research aiming toward the concept of the universe as a whole, its alleged identity, as a *telos* of explanation in cosmology.

Nevertheless, reason must discipline subjectivity in apprehension of the universe as a whole, and the intuition of the identity of the universe as a teleological commitment exactly fulfills this function: it not only provides us with the assurance that the universe *is*, but it ultimately ordains the methodology of physics to be applied to its study. Physics and mathematics based in the analytical faculty of reason heroically overcome the lack of empirical evidence for the identity of the universe by employing creative imagination (variation of the empirical carriers of essence) and extrapolating sensible images of reality, as well as local physical laws through space and time, summarizing them in a kind of unity, which is intended by cosmologists as potentially graspable.[24] In this case the intentionality pertaining to physics and mathematics, being based in the pre-predicative instantaneous synthesis, serves as a delimiter in knowledge of the universe: it does cover only certain aspects of the universe. Characteristically, the limitation of knowledge follows from an attempt to express symbolically the unity of the actually infinite universe.[25] Indeed, the visible universe is limited by cosmological

24. The teleological commitment, which is implanted in the belief that the whole of the universe can be grasped, is manifested in particular through diagrams pretending to symbolize the universe as a whole. What is characteristic for all these diagrams is that they depict the universe as if this universe is somewhere outside of the reflecting consciousness, as if this universe were a geometrical object that could be analyzed outwardly. According to Kant, knowledge of the universe (judgment of it) is only possible if the presupposition that the universe's empirical laws are related in the form of a deductive system is fulfilled. See I. Kant, *First Introduction to the Critique of Judgment*, 215. Cosmological diagrams manifest in their imagery the fulfillment of such a deductive system.

horizons[26] whose very existence is inferred from the supposition of the global space-time structure (as a symbol of the universe's unity) exceeding this horizon. The universe as totality is seen as uniform space filled with a uniform "cosmological fluid" (made of clusters of galaxies) disguising all differences among particular physical objects. This description of the universe as uniform in space and in terms of matter can be treated as a symbol of identity of the universe, as a construct and intellectual achievement based in an instantaneous synthesis (judgment) and not in a successive induction (that is, analytic derivation). The uniformity of space and distribution of matter, as we argue below, is a major transcendental delimiter in cosmology, following from a *judgment* that the universe must be knowable.

In this sense, in spite of its "anti-teleological" pathos,[27] the cosmological principle represents a sheer manifestation of a teleological commitment in study of the universe that ordains the mechanistic-like explanation in cosmology. Teleology here refers first of all to the *telos* of explanation, which is the unity and integrity

25. The limits of knowledge that proceed from human finitude contain in themselves a potential for grasping the infinite, at least at the level of a simple thinking or imagining this infinite. Indeed, if one's own finitude sets limits on how much one can take in in the sense of how much this one can be affected by that which is out there, then the taste of the infinite, that is, of the self-explanatory and unconditioned (the universe as a whole or its identity), can only be achieved if one is *a priori* aware of it, so that this one has to become infinite as being absorbed by the infinite. This thought reproduces the paradox of subjectivity expressed here as a paradox of grasping the infinite in the conditions of finitude. Cf. A. W. Moore, *The Infinite* (London: Routledge, 2001), 231.

26. Whatever we see in the sky represents a frozen image of the past, since the signals we receive come from the different distant objects. In this sense what we observe as the universe is limited to the so-called past light-cone which, in its contingent facticity, constitutes one particular observable configuration among the infinite number of possible, subject to the condition that the universe is considered as extremely large in its actuality. It is interesting that most of possible observable configurations are causally disconnected apart from the very early universe, where the smallness of the universe and its short-lived history had not allowed it yet to decompose into physically disjoint regions.

27. See E. McMullin, "Indifference Principle and Anthropic Principle in Cosmology," *Studies in History and Philosophy of Science* 24, no. 3 (1993): 359–89.

of the universe. The diagrams that represent the wholeness of space and time thus can be seen as pictorial symbols of the sought identity of the universe having not an inferential, but intentional origin.[28]

From Cosmological Principle as a Transcendental Delimiter in Knowledge of the Universe to Apophatic Cosmology

One sees that the anticipation of the universe as a whole, or its identity, implies a particular transcendental assumption,[29] known as the "cosmological principle." Its transcendental dimension originates in the observation that if the universe changed radically in *space*, one could not apply physical principles discovered locally to other parts of the universe. To guarantee the universal validity of physical laws, at least related to the observable forms of matter, the universe must be uniform in space, or in different words the universe must look isotropic from every possible location in it. This principle postulates the uniformity of the universe in space, as if one could reposition oneself from one point in space to another and see the universe as statistically the same. The effect of this postulation is that the

28. See more on these diagrams in the next chapter.

29. This assumption differs from assumptions of the authors of the so-called Steady State cosmological model, which is commonly rejected nowadays on observational grounds, who argued that in order for physical laws be universally valid and thereby guarantee that the laws of physics discovered here and now could apply to the distant past of the universe one must postulate the "perfect cosmological principle," which demand the uniformity of the universe not only in space but in time. See H. Bondi and T. Gold, "The Steady-State Theory of the Expanding Universe," *Monthly Notices of the Royal Astronomical Society* 108 (1948): 252-70; F. Hoyle, "A New Model for the Expanding Universe," *Monthly Notices of the Royal Astronomical Society* 108 (1948): 372-82. See also the discussion of the transcendental nature of this reasoning in Y. Balashov, "A Cognizable Universe: Transcendental Arguments in Physical Cosmology in *Constituting Objectivity: Transcendental Perspectives on Modern Physics*," ed. M. Bitbol, P. Kerszberg, and J. Petitot (Dordrecht: Springer, 2009), 269-77. For a discussion of the methodological foundations of the Steady State cosmological model see Y. Balashov, "Uniformitarianism in Cosmology: Background and Philosophical Implications of the Steady-State Theory," *Studies in History and Philosophy of Science* 25B (1994): 933-58. For a history of the big bang-steady state controversy, see H. Kragh, *Cosmology and Controversy: The Historical Development of Two Theories of the Universe* (Princeton: Princeton University Press, 1996).

uniformity of the universe thus becomes an all-encompassing circumscription of the universe encapsulating in a discursive manner the instantaneous synthesis. As already mentioned, cosmology as a scientific discipline would not be possible without this principle: we could not construct a physics that would describe different, non-homogeneous parts of the universe.[30] Alternatively, based on the observed isotropy of the universe from the planet of our habitation, and in order to avoid any inclination to *teleologism*, related to the selected nature of our position in the cosmos, one postulates the same isotropy from all other possible positions in the universe, which is tantamount to its spatial uniformity. We argue below, however, that the cosmological as such still represents a certain trend of teleology related to the process of research. Since the aim of research is to explicate the whole universe, there must be an assurance that there is an object of research as well as its possibility related to the harmony between the varied manifold phenomena and their integration in human subjectivity. The cosmological principle plays exactly this role: it effectively postulates the possibility of knowledge of the universe and in this sense it also functions as a transcendental principle. Finally, it seems reasonable to claim that in all possible scenarios the cosmological principle emerges as a natural consequence of the transcendental requirement of the explication of the wholeness of the universe, its anticipated (through the instantaneous synthesis) identity, where the instantaneity is effectively reduced to the mental spatial uniformity of the universe. Thus the universe's identity reemerges through the cosmological

30. As was asserted by D. W. Hogg, "Is Cosmology Just a Plausibility Argument?," arXiv: 0910.3374v1 [astro-ph.co], (18.10.2009): 9, the idea that the universe is not homogeneous makes no quantitative predictions and explains nothing, so that it cannot be a scientific contender with the present observations: "An inhomogeneous universe is so intractable that there is almost no near-term future in which we are likely to be able to either observe or compute anything interesting in this area."

principle acting as a transcendental requirement for the explicability of the universe.[31] Here two comments are required: historical and theological.

The cosmological principle as a transcendental assumption naturally appears in the Kantian stance on space as an a priori form of the sensible experience. His apriorism demanded that space is Euclidian. The cosmological principle in the Kantian view of the universe follows simply from the fact that the Euclidian space is uniform. As to the distribution of matter, its uniformity, independent from that of space, follows from the Newtonian gravitational dynamics in infinite space. One must be careful, however, in understanding the aprioristic overtones that naturally accompany any reference to Kant. The cosmological principle is not a priori in a Kantian sense of a truth established before observations. Rather, as inferred from the observational isotropy in the distribution of matter as it is seen from the Earth, this principle is rather a posteriori. However, historically, being implemented in the scientific practice of cosmologists, this principle acquires some features of "methodological inevitability," thus quietly transforming in the "law of thought," that is, a delimiter of a transcendental kind. A similar thought related to attempts of physics to acquire an axiomatic form starting from the most general principles (similar to the cosmological principle) was expressed in the work of E. Whittaker[32] and later quoted by J. Jeans.[33] Whittaker called these general principles "the principles of impotence," that is, statements that assert the impossibility of achieving something. For example, in the case of the cosmological principle, one cannot achieve knowledge of an absolute position in

31. See, for example, S. L. Jaki, "Christology and the Birth of Modern Science," *The Asbury Theological Journal* 45, no. 2 (1990): 69.

32. E. Whittaker, "Some Disputed Questions in the Philosophy of the Physical Sciences," *Proceedings of the Royal Society of Edinburgh* 61 (1941): 160–75.

33. J. Jeans in *Physics and Philosophy* (Cambridge: Cambridge University Press, 1945), 80.

space: space is uniform, so any information about such a position is lost.[34]

The second, theological comment, originates in that, paradoxically, the cosmological principle has a geocentric origin. The centrality of the Earth for the overall comprehension of the universe is characteristically accentuated in patristic theology, in particular in the concept of the incarnation of the Son-Logos of God, on the planet Earth in the human flesh. The paradox of the incarnation, formulated by early Greek Fathers, explicated a nontrivial topological relation between the universe which was created by and through the Logos and the universe as it was accessible to Jesus Christ in his earthly place. Indeed being in a body on the Earth, the incarnate Logos was present everywhere in the universe because he did not give up his place at the right hand of the Father. Being at the planet Earth he controlled the whole universe by retaining his divine nature. But the latter implies that the universe, as seen through the "eyes" of the Logos, does not have any extension and is uniform because of its "absolute simultaneity" for the Logos. From a human point of view the presence of the Logos in all places of the universe indicates that the universe is at least *theogenic*. A theological affirmation of the unique position of the incarnate Son of God though being in a body at a given point in the vastness of cosmic space and, at the same time still being coinherent with every point in space, provides an implicit principle of order in the universe, which ensures that every place in the universe, as a place of the "presence" of the Word, is coinherent with the place where God was bodily incarnate, that is, with the Earth.[35] This in turn implied, in the view of the Christian scientists of the time, that there was a uniformity in the laws of nature

34. On the importance of this discussion in building the methodology of cosmology, see G. Gale and N. Shanks, "Methodology and the Birth of Modern Cosmological Inquiry," *Studies in History and Philosophy of Modern Physics* 27, no. 3 (1996): 290–93.
35. T. F. Torrance, *Space, Time and Incarnation* (Edinburgh: T. & T. Clark, 1997).

(which were known from their experience on earth) throughout the whole of the cosmos. This intrinsic rationality in the world, according to Athanasius of Alexandria, is maintained by the creative Logos of God, which is not an immanent principle of the world, but the transcendent artificer of order and harmony in created existence, which is thus contingent upon the transcendent rationality of God (see, for example, Athanasius, *Contra Gentes*, 40:1). Two implications of this theological development for physics were realized by a Christian thinker, John Philoponus of Alexandria (died c. 570). He recognized that any true order in the universe must be universally valid, and he inferred from the colors of the stars that the same laws govern the stars and bodies on earth.[36] It is because of this that one can conjecture that historically the cosmological principle could be supported by the conviction that the universe is *theogenically* uniform. The question of on what grounds cosmology can proceed from the *theogenic* uniformity to the cosmographic uniformity (cosmological principle) cannot be answered apart from pointing out that it is the Divine image in human subjectivity contemplating *theogenic* uniformity at the pre-predicative level (that is, in rubrics of faith) that guides reason in its appropriation of the external world in the phenomenality of objects through imposing the principle of spatial and material uniformity upon it.[37]

Coming back to the main line of our discussion, we mentioned above that the cosmological principle allows one to use an imagery of the universe, depicting it through diagrams representing the universe as a whole. These diagrams become another way of expressing

36. See, for example, Jaki, "Christology and the Birth of Modern Science," 69.
37. See more on the interplay between the paradox of incarnation and the cosmological principle in my paper A. Nesteruk, "The Cosmos of the World and the Cosmos of the Church: St. Maximus the Confessor, Modern Cosmology and the Sense of the Universe," in *Knowing the Purpose of Creation Through the Resurrection*, Proceedings of the Symposium on St. Maximus the Confessor, Belgrade, October 18-21, 2012, ed. Bishop Maxim (Vasiljevic) (Alhambra, CA: Sebastian, 2013), 297-333.

outwardly the instantaneous synthesis of the universe giving an image to the identity of the universe. For example, if the universe is thought to be closed and finite in space and time, it is depicted as a curvilinear cylinder with two apices symbolizing Big Bang and Big Crunch. Here the universe acquires a kind of identity as being "created" by consciousness: the identity of an image is supposed to originate from the identity of an "artist" who produced this image. One can speak of reduction of the identity of the universe from personal communion to its intersubjective and impersonal representation following from the transcendental requirements for its explicability. By its constitution within these transcendental limits the identity of the universe functions only as a *signifier* of that which is meant to be achieved through the instantaneous synthesis. However, since such a representation of the identity of the universe does not extend beyond its signifiers, it has the sense of an *apophatic* identity: it tells us what the identity of the universe *is not*, and this is the *negative* reason why this image is *valuable*. The *apophatic* stance in cosmology does not mean that thinking of the universe in the negative certitude values physical cosmology only for being, de facto, a *negative cosmology*. The characteristic feature of the apophatic approach consists in that, while employing the language of physics, it honestly states that physics has been used as a tool, exercised in its extreme, in order to express the human aspiration and hope for the things unseen. But, *apophaticism*, above all, is an attitude of the thinking intellect that refuses to form concepts about the universe as a whole as accomplished truths: it rejects the claims of exhaustibility of knowledge sometimes made in scientifically "enframed" cosmology that would adapt all mysteries of the universe to human ways of thought.[38] The "apophatic knowledge" brings the person who

38. See on the principle of apophaticism in modern metaphysical extension of science in C. Yannaras, *Postmodern Metaphysics* (Brookline, MA: Holy Cross Orthodox Press, 2004), 84. The

exercises this knowledge to a direct experience of the universe as communion, so that the intellectual knowledge of the universe is ultimately grounded in the experience of the universe in life.

One can go even further and trace the origin of the cosmological principle as a transcendental delimiter in knowledge of the universe to the initial fact of life as finite embodiment. The adjective "finite" plays here a pivotal role: any attempt to think of the universe places this thinking in the conditions of the impossible, namely to access the actual infinity of the universe from within its finite formation. On the objectivistic side, if this access is thought as possible, the universe must be thought as having in itself that originary foundation which makes possible the infinitely progressing self-representation from within its own finite formation. On the epistemological side, this access becomes a transcendental requirement following from the postulate of the universe's explicability: to know the universe through the locally and historically contingent fixed astronomical phenomena and established physical laws one must postulate a nomological uniformity of the universe. This uniformity bears an intelligible character: all parts of the universe are subject to similar laws and thus, as a uniform intelligible entity, the universe is accessible to the human intelligence as a stable and self-identical pattern. The postulate of the

general definition of this principle which historically originates in theology can be found in C. Yannaras, *On the Absence and Unknowability of God* (London and New York: T. & T. Clark, 2005), 59-60. According to Yannaras, apophaticism can be defined "as the abandonment of all claims to an 'objective' assessment of truth, or the denial that we can exhaust the truth in its formulation. Abandonment or denial does not mean here a rejection or overlooking of the possibilities for knowledge represented by a rational formulation of knowledge. Apophaticism is not to be identified with irrationalism, or indifference to the rules of logic in the formulation of knowledge—for these rules represent the possibility of communicating and sharing in knowledge. Nor may apophaticism be confused with self-centred mysticism, the flight to private emotional certainties. The apophatic way or position presupposes the prior acceptance of the methods of philosophical epistemology—the acceptance, for instance, of both the way of affirmations and the way of denials—as potentialities for attaining knowledge. It is precisely the emphasis on the *possibility* of knowledge that sets apophaticism apart from any *positivism* about knowledge, that is to say, from any form of absolutizing of the rules or presuppositions needed for ascertaining the validity of any formulation of knowledge."

uniformity of the universe demonstrates the transcendental ability to displace itself in the intelligible space in order to stretch consciousness across the whole universe. However, this displacement (interchange of home-places), being eidetic in its essence, does not imply any interchange of spatial hypostatic embodiments. Thus the cosmological principle makes the universe uniformly intelligible. In this sense the personal instantaneous synthesis of the universe as a whole through the living communion on the planet Earth is replaced by the anonymous and impersonal discursive representation of the universe as uniformly extended in space and time.

The intuition about the identity of the universe as its instantaneous synthesis can thus be treated as a deposit of personhood in perception and articulation of the universe. This does not imply that the content of this synthesis is somehow exhausted by a person on their own, as if personhood mimics some innate and nonoriginary (with respect to the universe) modes of apperception. It just says that this instantaneous synthesis, as an ecstatic movement toward being, receives its manifestation in a human person, and in no way is the sense of the identity of the universe reduced to the rubrics of the "I," nor does this identity acquire explicitly any hypostatic features different from those deposited by the carrier of its enhypostasization. The universe does not become hypostatic despite its being presented through the properties, which, by their constitution, are *enhypostasized*. However, one cannot exclude that cosmological research, as human activity, forms the idea of the identity of the universe in a "quasi-hypostatic" sense, when research becomes an interaction with yet unknown and fundamentally open-ended "being." Then, seen from a philosophical perspective, a theoretical exploration of the universe can be interpreted as an attempt to enter communion with another quasi-hypostatic being. The situation here becomes analogous to comprehension of a human person. Namely, that human person cannot be fully understood and knowable as a physical or biological object. Metaphorically the discursive method of apprehension of human beings in terms of labels and external parameters provides access only to their bodies, or, as some philosophers said, to their corps. Human persons cannot be known and the sense of their existence exhausted though knowledge as subjugation to the judgment of the other. Indeed it is possible to *communicate* with persons, but it is impossible to dominate them.[39] In this case the fundamental irreducibility of quasi-hypostatic being to its "pieces" and "moments" (one cannot know a person through analysis and dissection) confirms our previous intuition that the cosmological research (as related to the universe as a whole) is a fundamentally open-ended, and thus an apophatic enterprise. In the same way as any other human person (as a modus of unique and incommunicable being) cannot be known by using syllogistic faculties of the cogito, the universe, being perceived as a quasi-hypostatic being (in theology enhypostasized by the Logos of God[40])

39. This situation with respect to the person of the other was characteristically formulated long ago by Gabriel Marcel: "The other as other exists for me only in so far as I am open to him (insofar as he is a thou), but I am only open to him insofar as I cease to form a circle with myself within which I somehow place the other, or rather, the idea of the other; for in so doing, the other becomes the idea of the other, and the idea of the other is no longer the other as such, but the other qua related to me, as dismantled, as dismembered or in the process of being dismembered." G. Marcel, *Être et Avoir* (Paris: Éditions Universitaires, 1991), 75. Here is another quote from Olivier Clément elucidating this point: "To know something of the mystery of the person, we must go right beyond its natural context, beyond its cosmic, collective, and individual environment, beyond all the ways in which it can be grasped by the mind. Whatever the mind can grasp can only be the nature, never the person. The mind can grasp only objects, whatever is open to inspection. But the person is not an object open to inspection, any more than God is. Like God it is incompatible, inextinguishable, fathomless." O. Clément, *On Human Being* (New York: New City, 2000), 30–31. See a vast discussion of this issue in C. Yannaras, *Person and Eros* (Brookline, MA: Holy Cross Orthodox Press, 2007), as well as in N. A. Berdyaev, "'I' and the World of Objects," in *Spirit and Reality* (Moscow: AST, 2003), 126–34 (in Russian).

40. See details of this concept in A. Nesteruk, "The Universe as Hypostatic Inherence in the Logos of God" (Panentheism in the Eastern Orthodox Perspective), in *In Whom We Live and Move and Have Our Being: Reflections on Panentheism in a Scientific Age*, ed. P. Clayton and A. Peacocke (Grand Rapids: Eerdmans, 2004), 169–83.

cannot be exhaustively known by means of simple observation, analysis, and theorizing. In other words, what cannot be known is the intrinsic way of existence of the universe. Contrary to this, in cosmological syllogistic thinking the identity of the universe is posited as that aspect of the universe which allegedly can be described outwardly.

One can clarify this point by saying that the identity of the universe for a cosmologist forms a subject of his intended ideal: to find the ultimate formula or set of formulae (ultimate theory) that will express the physical law behind the origin, uniqueness, and concreteness of the universe we perceive through the symbolism of the night sky; in this sense cosmology is naturally prone to foundationalism, as a tendency to establish the ground-grounded relationship between the visible universe and some underlying reality. This desire reflects humanity's anxiety about the contingency of its own existence and desire to rely on some stability and assurance in the midst of the natural world. The anticipation of a certain stability of personal existence as following from the integrity of the world is always *present* in our epistemic horizon, but has a deficit of explication so that the ways of expressing this integrity in scientific cosmology represent mere metaphors and symbols of the unknowable; the search for the sense of hypostatic existence is transferred to the universe as that other under the pressure of which the incarnate ego is constituted.

The issue of identity of the universe as linked to its instantaneous synthesis in subjectivity may be approached from a different angle. The instantaneous synthesis makes any description of the universe in terms of its gradual stages of formation psychologically irrelevant: the sense of the entirety of the universe as co-present to a living human subject enters into an existential contradiction with an object-like vision of the universe's parts and phases of development. The instantaneous synthesis suspends the perception of the universe as being extended in space and time. This mode of communion can be paralleled with thinking of a person who is not empirically present, that is, "present in absence," so that any notion of space as separation is suspended, for there *is no place or space* can be allocated to such a "presence." This presence can be rather described as hypostatic, that is, as belonging to that who articulates the universe as a whole. Seen phenomenologically, the act of thinking of the universe as a whole implies the reduction of its varied content to a single consciousness that suspends space and time. In respect to what we have just said one can raise a question of the sense of that single consciousness which integrates all pieces and moments into the wholeness of the universe, and contemplates the universe as an event, a flash of memory. This question arises if one compares human history with cosmic history. One cannot achieve the presence of historical facts literally, but one can establish a kind of inference, reenactment, through the chain of witness in the continuity of collective human subjectivity. The reenactment of historical events assumes their invocation in the condition when the actual temporality of these events is suspended and integrated in the consciousness of the present. This suspension of time reveals itself as an inherent property of transcendental subjectivity, a property that one usually calls memory. Then one can enquire whether the suspension of time in the invocation of the whole universe represents a particular archetype of human memory that could be named as memory of "all in all." This memory would correspond exactly to that standing in front of the universe or communion with it which is implanted in the very fact of our conscious consubstantial embodiment in it. In fact, one conjectures that by invoking the image of the universe as a whole we effectively reenact this hidden memory through its progressive unfolding in numerous theories of the universe. However, this type of intentionality, associated with memory, with respect to the universe as a whole remains unfulfilled. The universe as a whole remains unknowable not in the sense that it cannot eventually become knowable, but because of human beings' finitude, and their non-attunement to it. Here, however, human beings function in a paradoxical condition, as was described by A. W. Moore: "Given their self-conscious awareness of their own finitude, humans neither want nor are able to represent themselves as infinite in a full and unbridled sense. While regarding themselves as standing in a God-like epistemic relation of creative intellectual determination to the world, they also, by that very same token, regard the world as limited by how they determine it, just one possible world among others. . . . While identifying themselves with the world as an infinite whole, they at the same time identify the world with themselves as a finite whole. . . . They aspire to be outside the world holding it all together; and for this the world has to be circumscribed."[41] The universe in its instantaneous synthesis is always given in excess. The contemporary explosion of cosmological theories and their precarious character demonstrate that the more we know about particular facts of the universe the less we understand its sense and the ground of its facticity.[42] Cosmological theories and their perception by the public mind represent an endless hermeneutics, and endless "exegesis" of the available experimental and theoretical texts about the universe.[43]

41. A. W. Moore, "Human Finitude, Ineffability, Idealism, Contingency," *NOÛS* 26, no. 4 (1992): 433.

42. By paraphrasing Meister Eckhart's famous citation of his favorite text (Isa. 45:15) concerning the ascension to the knowledge God ("The more one seeks you the less one finds you," Sermon 15) one can state, in cosmology, that "the more one seeks the universe the less one finds it." In a weak form the same thought was expressed by R. Penrose in the concluding section of his book *The Road to Reality*, titled "Deep Questions Answered, Deeper Question Posed." R. Penrose, *The Road to Reality* (London: Vintage, 2005), 1043–45.

43. This endless hermeneutics in cosmology manifests the endless commitment to a theoretical task that does not allow any procrastination, since each experimental and theoretical result in cosmology is temporary and must be kept aside while looking for the successive result, believed to be the limit of convergent theoretical approximations. This unceasing commitment in turn

One can talk about open-ended epistemology in articulating the universe as a whole. The term "the universe as a whole" does not constitute a name for some object that exists antecedently with respect to the human enquiry. One cannot define this "object" discursively before the actual process of its study begins. Since the definition of the universe as a single unique object would require an indeterminately large number of sentences,[44] not only related to distinct parts or structural units, but also related to the open-ended advance of knowledge in which description not only accumulates sentences but qualitatively changes in time due to the changing character of interaction between man and the universe, there is no preconceived and universal (in terms of historical time) methodology of cosmological research: this methodology as part of knowledge in general is open-ended. The methodology used in modern cosmology is historically contingent and, by definition, cannot pretend to have a universal status. In this sense the definition of the universe implies not only statements about facts and physical realities, but theories and models of the universe being intrinsically its own part. In other words, the expression "the universe as a whole" implies the need to explain what language of saying and what sense of the term are used in this case, that is, to indicate what theory or symbol of the universe is used. The historically contingent fullness of the notion of the universe is then achieved through the complex of theory and its constituted correlates. However, as long ago pointed out by Munitz, the "relativity" or fragmentary nature of such symbolic complexes does not remove the major unavoidable fact of the universe as

reveals an intrinsic teleology of cosmological research as where the *telos* of research, answering a perennial philosophical question "What is man's goal?" clarifies de facto that which can be called the nature of humanity. Correspondingly, the inherent teleology of cosmological research being related to the clarification of the question of the origin of the universe, clarifies not so much the question on the nature of the universe, but the nature of humanity. (See chapter 6 in this book, on *telos* of cosmological explanation as related to the question of humanity's origin.)

44. Munitz, "Approaches to Existence," 337.

communion, common to all theory makers.[45] In this sense another important aspect of the apophatic approach in cosmology emerges: namely as a logically consistent freedom of expressing experience of the universe subject to the condition that the content of this expression contains embodied agents disclosing the sense of the universe. In view of this the constant balance between the fantasizing tendencies of any theoretical cosmology and the inevitability of biological embodiment remains untouched and as such represents a stabilizing factor of any cosmologizing, making doubtful and unrealistic all theories of the universe that diminish or disregard the presence of humanity, which articulates the universe.[46]

As a result of our discussion, one can conjecture that the sense of identity of the universe through the instantaneous synthesis of communion can be defined as the enhypostasized mode of the universe's inward existence in human subjectivity, which does not need a spatial and temporal representation (another way of expressing the instantaneous synthesis). It is then clear that such an identity is not an abstract philosophical notion, not an impersonal substance or the totality of all convergent sense-impressions and their objective correlates, but a transferred feature of subjectivity, its response to the pressure of existence in the world, under which the identity of the "I" is being constituted. The contemplation of the identity of the universe is similar to the contemplation of the identity of

45. M. Munitz, *Space, Time and Creation* (Glencoe, IL: Free Press, 1957), 69-70.

46. The ultimate truth of the universe, which is experienced directly in the fact of life and affirmed in existential faith, and which is inaccessible to precise grasp by the discursive reason, leaves only a trace of its presence with no definite logical location "between" the affirmation and negation of the universe. Thus the *form* of antithetic propositions, as a pair of theses and antitheses, shapes constructively the operation of *open* epistemology in philosophical cosmology: all statements about the universe as a whole are always mysterious and "contradictory," leading human reason to incessant wonder between the poles of conviction and doubt. When reason is tired of this wonder it submits itself deliberately to the silence of faith, that is, to believing in being existent, as a true knowledge of the universe in the *negative certitude*, namely that the universe *is*, but one cannot give an accomplished answer to the question "What or why is the universe?"

one's "I" as a breaking through the anonymity of existence and its solitude,[47] its ecstatic rupture toward the unconcealment of being. This contemplation can be compared with experience of life as existence in solitude, that experience which does not dissolve in social tasks and objectives, but rather corresponds to a child's perception of being, given in its sheer facticity, as mystery with no beginning and no end. Life as communion creates the sense of its coinherence with the universe (*coaevus universo*), its fundamental attachment to the universe when the universe exists only insofar as a hypostatic communicant exists. This pre-predicative mode of perception obviously does not enquire about the grounds of the universe's facticity, for if it were to do so, it would be tantamount to enquiring into the facticity of one's "I." As a consequence, in this attitude, the question about the origin of the universe in an objectivistic sense does not arise, because the universe is not an object, but communion.

Now let us articulate the transition from the instantaneous synthesis of communion with the universe to its representation in the phenomenality of objects in cosmology. Communion implies that the universe exists for me only through my body, so that my body turns out to be the center of disclosure of the universe. But my conscious contemplation of my body is not a simple act of depositing and treating my body as *ad extra* to my subjectivity. My body and consciousness coinhere, so that any separation of my body from me as identifiable self, in thought, can only be an abstraction with no existential meaning. The immediacy of this coinherence, since it takes place only insofar as life continues, has no spatial and temporal

47. E. Levinas, *Time and the Other* (Pittsburgh: Duquesne University Press, 1987), 41. Similar to Levinas, this breakthrough toward one's identity and hence the universe's identity cannot be achieved either through knowledge or ecstatic transcendence toward the universe (which would imply one's disappearance as person). Thus when we speak about the universe as communion we mean that solitude can be exceeded while the identities of both a human being and the universe are preserved as distinct.

dimension. If, in thought, my body is abstracted as an external thing, as a physical object, in a similar way, the universe, as an overall context, including my body, is abstracted as a thing among other things and is treated as an object. One again recalls G. Marcel saying that the more one insists on the objectivity of things, cutting off the umbilical cord that links them with one's existence, the more the world would convert into an illusion, which in the long run self-annihilates by a simple fact that it ignores this one.[48] The "cutting of the umbilical cord between human subjectivity and the universe," exercised mentally, when the primary "contemplation of the fullness of life and its coeternity with all being" stops, leads one to enquire, along the lines of the natural attitude, into the origin of things: where they come from, whether they are finite, and where they go, as if this "where" would be an anterior or posterior something possessing the phenomenality of objects. Such a consciousness starts to enquire about the sense and origin of the object-universe, for to understand it as *an object* one should know its origin, where it comes from: what is that antecedent something which was the originary for the universe. In this sense the form of enquiry into the facticity of the universe, which presupposes the explication of its antecedents, represents another general transcendental condition for studying the universe as an object. Here one observes an entry of temporality into the discourse: the representation of the universe as an object is possible only if it can be expounded in terms of its consecutive stages of appearance, which in turn entail (as we know from cosmology) the presence of such a "moment" in its allegedly existing originary past where all had its ultimate origin. In different words, to know the universe is to know its history. This is a transcendental condition similar to that which is present in the field of the human sciences:

48. Cf. G. Marcel, *Du Refus à l'Invocation* (Paris: Gallimard, 1940), 32.

to know humanity means to know its history, sociology, sociology of scientific enquiry, and so forth. In this case the originally inherent "identity of the universe" as its unity (instantaneous synthesis) enters human subjectivity under the disguise of the universe's antecedent unity, its origin. The idea of the origin of the universe becomes an inevitable result of the naturalizing tendencies of consciousness to project the pre-predicative instantaneous synthesis outwardly, making it explicable in intersubjective form. The circle of interiority is transcended and manifests itself in the search for the remote origin of "all in all,"[49] in the Big Bang positioned in the past as the antecedent unity of the whole. The search for this origin explicates the hidden anxiety of the self of a cosmologist after the umbilical cord with the universe (seen only as an object) was cut off. Thus the working of the initial instantaneous synthesis hidden within existential anxieties promotes itself through the fact that it remains an ultimate transcendental delimiter of the constitution of the universe in its outward, object-like phenomenality.

Constitution of the Notion of the Universe: A General Analysis

In everyday life and in all aspects of scientific experience where classical physics is applicable, to deal with objects as they have been "constituted" may sound a bit odd, for to deal with the constitution of objects is the prerogative of a philosophical consideration that is no way an objective of physical cosmology. Indeed, if such a constitution had taken place, it was in the ontogenic past of human beings (in every individual person this constitution is transferred through social practices as a matter of adaptation to the social and

49. Cf. 1 Cor. 15:28.

physical conditions of living). In this sense the basic conditions of the constitution of objects have been at human disposal since those original times and they need not be questioned. Consequently the classical sense of objectivity as referring to material objects "out there" is accepted as already given.

But in cosmology things are different. The basic condition of the constitution of objects at large scales of space and time are not available and hence we are forced to think about cosmology not in terms of objects, but the ways its subject matter, that is, the universe as a whole is constituted. If one refers back to classical physics one understands the conditions of constitution in space and time in terms of two major requirements: the continuity and reversibility of the temporal sequences of phenomena. If they are implemented in a particular situation they give ground to the idea that there is something permanent and substantial preserving its *identity* across space-time, something that is endowed with properties and can be involved in a physically causal manner in sequences of events. But none of these conditions can be enforced at the scale of the universe as a whole.

Since by definition the universe encompasses all space and time, so that its "trajectory in space and time" cannot be constructed and the universe's identity cannot be tested "across space and time" (unless in a "superspace" of the multiverse), the universe is not an already-constituted object and its identity manifests in constantly renewing contingent events of knowing. One can speak about the identity of the observable cosmos as enduring through the ages, but this is not an identity of the object subjected to the repeated tests of classical physics. The universe is unique and cannot be rerun. This naturally entails that the criterion of reproducibility of phenomena across a large range of variations of perceptible and experimental histories does not stand in cosmology. One can admit that there are some given aspects of the visible universe that constitute a common already-given background for all historically consecutive, although contingent, observations (some aspects of our galaxy, for example). The important thing is that the universe as a whole is subject to advancing redefinition of its constituents through changing the transcendental conditions of its observability (related to the advance of embodiment through technology) and demand for its mathematical expressibility. In this sense the universe's constitution is effectively an open-ended mode of being of the universe itself, or, as Heidegger would say, the constitution of the universe is bringing it forth into unconcealment, into openness of disclosure.

The anomalies with the constitution of the universe can be elucidated through the appeal to the Kantian distinction between the principle of mechanism and the maxim of teleology which can be applied in the study of nature. According to the mechanistic methodology the scheme of constitution based on ordinary causality requires free substitution of well-defined antecedent conditions in order to check that a certain effect is determined by a certain antecedent. In cosmology this definition of causality cannot be applied to the universe as a whole unless theoretically, that is, on the level of *eidetic* variation of that which is given. Indeed, in order to describe the given contingent state of affairs in the observable universe, one can produce, depending on models, many versions of the initial conditions that are supposed to launch the evolution of the universe toward the present. The freedom of constructing these models in order to explain away the contingency of the present, points out, in fact, that the implied physical causality is a constructed causality, the causality of the reversed temporal order, where the antecedent conditions are in fact postcedent and it is the past of the universe that is reconstructed on the grounds of the present.[50] The efficacy of physical causality in the allegedly "from-past-to-present" temporal order is manifested in the movement of its reconstruction, a reconstruction that as such takes place in the opposite, that is, "from-present-to-future" order. In other words, the constitution of the physical causality in the past takes place through a *teleological* move of thought as directed to the future, where the *telos*, the goal of cosmological explanation is exactly that sought physical causality of the past. Thus the very possibility of applying the principle of mechanism to study of the universe in its frozen past is ordained by a hidden teleological requirement for the explication (expressibility) of the universe.

The transcendental sense of what is meant by constitution arrives from a Kantian stance on it: a constituted object is neither isomorphic to a real object existing in itself, nor reducible to a figment of imagination. An object appears as an intentional correlate of subjectivity subject to two essential things: first, there must be a structural framework that makes intention semantically significant and communicable, and second, the intentions must imply fulfillment, that is, they are not empty. Objects are by no means construed as part of external reality in the strongest sense (this would make the intentions unfulfillable, for the grasp of this reality through mathematical synthesis would be contradictory to the finitude of humanity related to embodiment); yet objects are independent of particular subjects: they donate themselves to subjects in such a way that not everything in this donation is controlled by the subject and thus not everything is subject to mathematical synthesis (one speaks here, using again the Kantian terminology, of "particulars"

50. The Big Bang, which is allegedly the past of the universe being reconstructed from the present, appears to be the *telos of cosmological explanation*, the *telos* not as *nexus finalis* in a Kantian sense, but as an intrinsic purposiveness of human action in cosmological research. See more details in A. Nesteruk, "Cosmology and Teleology: Purposiveness in the Study of the Universe through the Reading of Kant's Third Critique," *Journal of Siberian Federal University. Humanities & Social Sciences* 5, no. 9 (2012): 1304–35.

that escape exhaustion through discursive definitions; elsewhere these particular were named "recalcitrant"[51]).

The situation with the universe as an object of intention is aggravated by the fundamental inaccessibility of its alleged totality,[52] which makes all intentions to be fundamentally unfulfillable (not only on the grounds of impossibility of a mathematical synthesis). In this sense the constitution of the universe as an "object" is never accomplished and thus is being taken as in an ongoing present continuous tense. In fact, one cannot talk about constitution as an ever-accomplished act, but one can talk about constitution as never-ending fulfillment of the intention to have the sense of the universe. Correspondingly, the objectivity in cosmology no longer means a complete detachment of entities and properties of its constructs from the constructing cognitive faculties, but the coordination of phenomena associated with the visible universe (as well as theoretical models of universe as a whole) across a variety of instrumental and communal theoretical circumstances. The invariants introduced through this coordination exceed the manifest modalities of contingent objects and their theories and refer to some basic structural properties that are present in all cosmological theories and representations and that form the conditions of knowability. For example, the basic geometrical structure of space-time adopted in the standard cosmological model corresponds to the belief ("cosmological principle") in the spatial and substantial homogeneity of the universe. This belief comes forth as one of the signifiers of the universe's objectivity. For naïve physical realists or empiricists, this claim for the objectivity of the universe through the geometrical structure would

51. R. E. Butts, "Teleology and Scientific Method in Kant's Critique of Judgment," *NOÛS* 24, no. 1 (1990): 2.

52. The affirmation of the inaccessibility of the totality of the universe nevertheless implies a certain grasp of this totality. This creates a paradox: that the infinite is shown to us in the conditions of our finitude. See more discussion on grasping infinity in Moore, *The Infinite*, 218–33.

not be convincing since this allegedly existing structure cannot be verified because of a limited observational access to space of the universe along the past light-cone. They would demand an element of reality available to empirical verification.[53] To avoid the naïveté of any straightforward ontological commitment it is reasonable to treat the cosmological principle as a transcendental requirement for knowabilty and mathematical expressibility of the universe, and not as a matter of fact.

Thus transcendentalism naturally enters the discourse. First of all the stance of realists in their ambition to have reality available as "present in presence" is replaced by a neutralization, namely, that any conceptual *structure* representing the universe is not obliged to represent the universe as it is really in itself. Realism in a transcendental sense claims only that the universe is being constituted through employment of these structures and thus its reality is the reality of this constitution. The major requirement here is that the structure must avoid paradoxes, so that its elements form a *coherent* set; this structure is generic as much as possible and then it is able to unify the largest volume of cosmological knowledge at a given historical segment. In cosmology, the invocation of such a structure implies an appeal to metaphysical propositions about the wholeness of the universe, which resemble beliefs. In this sense the coherence of the elements of a structure follows from the coherence of beliefs. The sense of reality is thus based not in a straightforward correspondence with that which is available to the senses (which is not possible anyway in the case of the universe as a whole), but on the relative stability and coherence of the adopted representation of the universe related to communal beliefs (that is, to a community of cosmologists).

53. The uniformity of the universe can be a result of special initial conditions that cannot be empirically verified. See, for example, A. Albrecht, "Cosmic Inflation and the Arrow of Time," in *Science and Ultimate Reality: From Quantum to Cosmos*, ed. J. D. Barrow, P. C. W. Davies, and C. Harper (Cambridge: Cambridge University Press, 2003), 375.

The sense of reality of the universe is related to the historicity of its constitution and to the community of those who claim its objectivity. The individual subjectivity is replaced by intersubjectivity in a transcendental sense, that is, in the sense of general conditions of knowledge formulated on the level of the community of cosmologists.[54]

By invoking the idea of structures in cosmology one naturally implies the employment of mathematical physics, bearing in mind its historical advance in the last two centuries which extended the transcendental stance in physics beyond its initial Kantian scope. It is mathematical physics that implies the introduction of "mathematical constructs," that leads to the transcendental problematics,[55] and is concerned first of all with the general-relativistic extension of classical mechanics by means of mathematical representations of space and, in particular, their *symmetries*, which lead to conservation principles and corresponding observables. Through the idea of a global symmetry of space the transcendentalism enters into cosmology in an unexpected way, namely through postulating a principle of *nonobservability* of absolute kinematical magnitudes related to the large-scale structure of the universe. These quantities can be used for the explication of the behavior of the universe but they cannot be observed and measured.

By postulating the cosmological principle (that is, the uniformity of space) one reduces to non-sense and hence to nonobservability any particular physical location. This nonobservability reduces the

54. Coherentist epistemology is sometimes described as knowledge without a foundation of certainty. In this sense it stands in sharp contrast with the foundationalist approach, which demands that the knowledge of the actual, and even of the probable (so far intelligible), requires a foundation of certainty. The characteristic feature of this epistemology is that it is social, that is, emphasizing the fundamentally social nature of human knowledge. See, for example, N. Rescher, *The Coherence Theory of Truth* (New York: University Press of America, 1989), 316-33.

55. Cf. the "Introduction" in Bitbol et al., *Constituting Objectivity*, 3, where the authors associate the Kantian "mathematical construction" with that which in modern terms can be called "computational synthesis."

description of the universe to a minimal number of parameters, thus making possible its description in terms of variables related to the overall structure. The form-invariance of metric, as well as other cosmological quantities, demanded by the cosmological principle, entail the possibility of introducing such global parameters as the expansion scale factor, energy density, and pressure of the global cosmic "fluid" related to clusters of galaxies. These parameters are related to the totality of the universe in space and thus contain only dependence on cosmic time. We do not observe the structure of space: it is homogeneous (uniform) and hence effectively nonexistent, that is, the notion of "location in space" is deprived of any physical meaning.[56] At the expense of this we describe cosmological evolution through functions that satisfy some equations but whose values can be established only through observation from our location; they are nonobservable at the global scale, because of the constraints of physical causality.[57] In this sense the cosmological principle as a principle of nonobservability of space reveals itself as a transcendental principle: the *nonobservability* of absolute kinematical properties of objects in the universe entails their theoretical *explicability*.

This explicability can easily be illustrated by reference to Noether's theorem, which links space-time symmetries to conservation laws, that is, to existence of the integrals of motion and hence with observables, or, expressed differently, the observability of the conserved magnitudes at the expense of losing information about their absolute location in space and time. According to the proponents of the so-called formalized (scientific) epistemology, Noether's theorem made evident that the loss of some information

56. In fact, the information about this location is lost. Certainly this point is also strengthened by the fact that the space described by metric in relativistic cosmology loses its independent property of being a "background structure," acquiring the features of a dynamical variable.
57. The fact that we observe the universe along the surface of the past light-cone implies that we have an astronomical access to large-scale cosmological parameters only in their past values.

due to the symmetry properties "entails a gain of determination for the physical system itself by exhibiting the corresponding invariant quantities which contribute strongly to its identity."[58] While this is true for individual objects, in cosmology the symmetry of space implied by the cosmological principle does provide us with the global parameters that describe the universe, that is, constitute its identity, but these parameters are not conserved: they satisfy Einstein equations that describe them as functions of time.[59] Still one can claim that the implementation of the cosmological principle assigns some identity to the universe through these functions.

Here transcendentalism exercises itself in a mode that, being intrinsically Kantian, ultimately leads to an effect that is inverse with respect to that of Kant: the cosmological principle starts by abstracting from experience of the visible universe by subordinating it to the principle of unity expressed through the uniformity of the universe. However, it is this invocation of the unity that ultimately leads the way to the introduction of the analytical properties of this unity (scale factor, for example), which are subject to investigation in mathematical equations and calculations. One cannot say that this movement leads directly to the intuitable manifold since the cosmological functions, in spite of mathematical manipulations with them, are not observable and empirically sensible. Still, these referents contribute through computational operations with them to the constitution of the universe beyond the purely metaphysical assertions of its unity. Thus we see that the transcendental stance in cosmology does not imply an appeal to some abstract compendium

58. F. Bailly, "About the Emergence of Invariances in Physics: From 'Substantial' Conservation to Formal Invariance," in *Quantum Mechanics, Mathematics, Cognition and Action: Proposals for a Formalized Epistemology*, ed. M. M. Mugur-Schächter and A. van der Merwe (Dordrecht: Kluwer, 2002), 378.
59. The cosmological principle of uniformity of space is different from the so-called "perfect cosmological principle," according to which time is uniform and the universe is static.

of the cognitive, intellectual faculties of cosmologists which predetermine in advance the scope of their observations and theories. The transcendental dimension of the cosmological enquiry enters as a component of the philosophy of constitution, that is, as a natural mode of phenomenological appropriation of cosmology.

As we mentioned before, a theoretical explicability of a transcendental kind does not require that theoretical constructs correspond to empirical (sensible) reality. The constitution of objectivity is achieved through the generic nature of these constructs and their mutual coherence, accompanied by some metaphysical requirements, but with no simplistic ontological commitment. The latter thought becomes even clearer if one realizes that dropping off the ontological commitment, that is, the demand for observability, represents a form of "applied apophaticism," that is, a conviction that the signifiers of reality do not exhaust the reality of what is signified simply because some (contingent) aspects of this reality cannot be observed. In this sense, the enterprise of constitution based in theoretical explicability and its coherence is intrinsically apophatic (incomplete, unaccomplished) because it is related to the historically contingent communal agreement on the coherence of explicability.

The fundamental question one now faces is the status of mathematical constructs related to the theoretical explication and hence the constitution of objectivity in cosmology. The pretence of some adherents of a radical mathematical realism,[60] runs against the transcendental stance on the neutrality of the constituted nature with respect to ontological commitments, as well as against the claims of the justification schemes based on coherence of explanation. Making the argument more precise, the question is: Where, in what particular place, does the persistent ideal identity of the invoked mathematical

60. See in this respect M. Tegmark, *Our Mathematical Universe*, chs. 10, 12.

entities become insufficient in order to claim that a whole theoretical explication by means of these entities does not reach its ontological objective and remains ever incomplete? One can look for an answer by making a distinction between the operation of sheer abstraction from the manifold of intuition, which can be mathematical and whose ontological commitment remains uncertain, and a mathematical synthesis of that which is constituted through calculations and computations adjusted through the rules of correspondence with observations. This distinction suggests that physical objectivity cannot be tantamount to an ontology of some independent substantial reality simply because the computational synthesis is rather a prescription for how to understand the universe and correspondingly what can be known of it. Since mathematical physics is linked to the computational synthesis, its very possibility is restricted by the conditions of experimental accessibility and by the intellectual criteria of coherence of constructs. These conditions of observability and selection of mathematical constructs on the grounds of communal metaphysical requirements are therefore constitutive of the very concept of a physical object. Then, the concept of the universe, that is, of that which is donated empirically and in rubrics of noncontradicting thought, cannot imply the objectivity of an independent reality, that is, an ontology; rather one speaks only about a "weak" objectivity. A phenomenological sense of this term can be illustrated through reference to J. Ladrière: "The theoretical apparatus is thus not a description in the ordinary sense, as presentation of an entity, supposedly given, and of its properties, it is the characterisation of something which is not a thing, but a *structural path along which a thing comes, from the ultimate horizon of every givenness, to the actual presence in which it is effectively given to apprehension.*"[61] Seen in this way, the mathematical synthesis in cosmology is an open-ended enterprise, an asymptotic approximation to reality. The assumption of the possibility of a complete mathematical reconstruction of such an ontological reality would ascribe the human mind with excessive intellectual capacities that transcend its finitude related to the limits of embodiment.

However, in spite of the fact that physical objectivity established through mathematical physics is not ontological, it is not entirely subjective either: it is not descriptive, but prescriptive, because it uses the conditions of accessibility to the realities that are beyond sensible experience, as well as to certain metaphysical ideas and ways of their mathematical expressibility. The way such a prescription is done encodes a general paradox that is intrinsic to how the sense of physical reality is constituted. Indeed, by taking into account the conditions of empirical accessibility and mathematical expressibility of universal metaphysical ideas, this prescriptive procedure does not include the "theory" of structures of subjectivity (which makes possible access to both empirical and intelligible reality, as well as to metaphysical ideas) in the theory of the physical universe. In other words, a predication of the universe does not reflect explicitly the underlying conditions of this predication (not only physical, but mental). One can suggest that this paradox in a way explicates an essential feature in the constitution of human subjectivity, namely that this subjectivity is *meaningfully defined*, although apophatically, through the extent that it cannot be included in its own product, namely cosmological theory.[62] The working of subjectivity in this

61. J. Ladrière, "Physical Reality: A Phenomenological Approach," *Dialectica* 43, no. 1-2 (1989): 138 (emphasis added). Similarly, P. Heelan approached the concept of reality as being naturalistic and evolutionary, and it is "people and praxis that provide it with the categories of the real." P. Heelan, "Experiment and Theory: Constitution and Reality," *The Journal of Philosophy* 85, no. 10 (1988): 524. The term "weak objectivity" was also used in Bitbol et al., *Constituting Objectivity*, 4.

62. In this sense the assertion of B. Carr on the lack of description of consciousness in physics can be considered as a *negative* but useful definition of how subjectivity works in cosmology. B. Carr, "On the Origin, Evolution and Purpose of the Physical Universe," in *Modern Cosmology and Philosophy*, ed. J. Leslie (New York: Prometheus, 1998), 152 (cf. footnote 33 of the previous

ambivalent way, when its contents as manifestations of its working do not contain explanations of the very possibility of this working, represents a fundamental movement of disclosure, which is nothing other than the pure emergence of the meaning of this very subjectivity, that is, its constitution.

Then, when we come to categories and principles of physical objectivity in cosmology interpreted mathematically, the conditions of accessibility to metaphysical ideas and corresponding references to empirical reality are incorporated in them. For example, when cosmology invokes the notion of the universe in terms of "totality," "unity," and "wholeness," notions that are metaphysical rather than physical, this invocation includes the conditions of their semantic explicability and mathematical expressibility. The cosmological principle as a product of a reasoned metaphysical and coherent convention among cosmologists generates the sense of a weak objectivity of what it attempts to describe, that is, the universe as a whole. Combined with a theoretical model of the universe's expansion, the cosmological principle makes possible a pictorial representation of the universe as a whole in spite of an a priori

chapter). M. Bitbol characterizes the absence of mind from our scientific picture of the world in terms of progressive "self-dismissal." M. Bitbol, "From the Anthropic Principle to the Subject Principle," in *The Anthropic Principle: Proceedings of the Second Venice Conference on Cosmology and Philosophy*, ed. F. Bertola and U. Curi (Cambridge: Cambridge University Press, 1993), 92. However, this progressive self-dismissal is inherent in the very logic of the scientific view of the world, because this picture itself is exactly what we call mind. The definition of mind, its content, is explicated by the scientific view of the world. In this sense one cannot define mind prior to this picture, because its very definition would require mind to exist prior to it. As E. Schrödinger expressed similarly: "The reason why our sentient, percipient and thinking ego is met nowhere within our scientific world picture can easily be indicated in seven words: because it is itself that world picture." E. Schrödinger, *What Is Life? The Physical Aspect of the Living Cell* with *Mind and Matter & Autobiographical Sketches* (Cambridge: Cambridge University Press, 1992), 128. In a phenomenological context the problem discussed here is related to the fact that in predicating objects science ignores the paths that led to the structure of relations pertaining to objects. And phenomenology since Husserl argues that these paths that manifest the human activity lie precisely in the ontological ground of objects. See, for example, P. Kerszberg, "From the Lifeworld to the Exact Sciences and Back," in *Husserl and the Sciences: Selected Perspectives*, ed. R. Feist (Ottawa: University of Ottawa Press, 2003), 189-212.

philosophical conviction that any such representation is precarious and in the best case iconic.[63] This confirms further the transcendental nature of this principle: it makes possible the graphical symbolism of the universe as means of its manifestation. Certainly this symbolism, being only a form of signification, does not exhaust the sense of that which is signified. Nevertheless it generates a certain sense of identity of the universe.

The transcendental move toward explication of the universe in the context of the cosmological principle gives an example of how reason proceeds from complexity (and contingency) of the given (empirical particulars) to the simplicity (and necessity) of representation of groups of objects subjected to high symmetries. To explain something means to overcome the facticity of the contingently given by reducing its phenomenality (or appearing) to a minimum and by replacing it by an intellectual intuition of that which is allegedly necessary. The structure of the global space in cosmology cannot appear, so that its phenomenality is impossible. What appears is predetermined by physical causality, which by means of the past light-cone corresponding to the home-place, selects a particular fragment of the universe (its particular phenomenality), which is seen as cosmic display. The transcendental nature of the cosmological principle is strengthened by the fact that the homogeneity of the universe in space cannot be extrapolated from observations. What one sees in the universe as uniform (or, more precisely, isotropic), de facto, belongs to different temporal eras because of the causality based in the finitude of the speed of light (light-cone). That piece of the universe which we see from our home-place contains objects belonging not only to different locations with respect to us but also

63. The graphs that attempt to express the wholeness of the universe related to its infinitely small observable segment can be found, for example, in E. Harrison, *Cosmology: The Science of the Universe* (Cambridge: Cambridge University Press, 2000), 375–86.

to different times. However, physical cosmology teaches us that the universe is not homogeneous in time: it evolves, and observations point toward this. In this sense the universe manifests contingency not only in terms of space (our space as related to home-place) but also in terms of time (the era when human beings exist).[64] If the spatial contingency is removed through an appeal to the cosmological principle (and as we have said, this principle guarantees the explicability of the universe), the contingencies related to time are tackled differently by constructing the dynamics of properties related to the subspaces of space-time corresponding to fixed moments of time. In other words, unlike the cosmological principle which acts as a metaphysical postulate (having in itself some teleological connotations related to expressibility and hence to the purposiveness of research[65]), the requirement for the existence of a universal temporal dynamics of global cosmological parameters represents another type of transcendental requirement for the knowability of the universe as a whole, which contributes to the universe's constitution and its "weak" objectivity. As it was expressed by Albrecht: "The arrow of time, as it is currently understood, simply has to be used as an 'input' to any theory of the universe."[66] If such a temporal dynamics exists, then the cosmological principle is implanted in it not only as the possibility of mathematical expressibility of this dynamics (equations for the scale factor $a(t)$ and energy density $\rho(t)$), but in the initial conditions of those dynamical principles that drive the universe. The dynamical inhomogeneity in time through the laws of evolution of cosmological parameters does not remove the problem

64. It is interesting to point out that this contingency is however a special one, for it corresponds to our presence in the universe and the very possibility for the universe to be observed as it is. See Primack and Abrams, *The View from the Centre of the Universe*, 117-18; also L. M. Krauss, *A Universe from Nothing* (New York: Free Press, 2012), 105-19.

65. See chapter 6.

66. Albrecht, "Cosmic Inflation and the Arrow of Time," 384.

of contingent facticity of the observable universe: it just transfers this problem to the allegedly existent remote past, that is, to the contingent initial conditions.[67] This move reveals itself as another transcendental constraint on expressibility of the universe in terms of the mathematical (related to the initial conditions) and thus as a condition of its constitution. This transcendental constraint pertains to all cosmic mythologies that retrace the generative steps that have led from the primordial past to the present.

The schema of representation is the movement from the undifferentiated unity (the initial condition, the Big Bang) to a qualitatively differentiated multiplicity in a way such that initial origin is saturated in that it precontains the entire process of formation of the diverse structure of the cosmos. Thus all particular cosmologies that follow these steps represent theories of constitution of the universe. Constitutive thought reenacts the production of the visible things, their manifest being, from the origin. The intelligibility achieved in constitutive thought of the universe reproduces in its own development the original unfolding that posits the universe as the universe and in which (that is, in unfolding) the universe itself consists. In other words, when in thought one enquires into the beginning of the universe as its origin, this thought ultimately enquires into its own beginning. Understood in this key, the conditions of accessibility to and explicability of the universe are ultimately implanted in the embodied subjectivity; in other words the universe, by allowing human beings to come into existence, predetermines the possibility of its own explicability and constitution. This last point brings us to assert with a new force that cosmological

67. It is interesting that theoretical cosmology always wants to get rid of the contingency of the initial conditions by developing theories of origin with the most generic initial conditions. However, as it is understood nowadays, if cosmology wants to give account of the arrow of time the initial conditions must be nongeneric.

discourse as constitution and explication of the universe is inseparable from the problem of explication of human subjectivity.

Indeed, humanity intrinsically enters all cosmological propositions because the universe, being a subject matter of cosmology, implies the presence of this humanity as cognizing subject. The universe cannot be considered as an object in the sense of classical physics, that is, a corporeal entity invariant with respect to space and time. In this sense the famous Copernican "turn" in description of the astronomical universe (through denying the selectiveness in a particular view of the universe) represents a certain attempt to ascribe bodily character to the universe by transferring this quality from the immediately accessible Earth to the remote cosmos. Humanity is earth-centered, so that the universe is weighed not only against humanity as such, but as corporeal and earthly existence. In the words of E. Husserl, the transcendental argument in constituting the universe is: "[I]f the earth is constituted with animate organisms and corporeality, then the 'sky' is also necessarily constituted as the field of what is outermost, yet which can be spatially experienced for me and all of us—with respect to the earth basis."[68] It is this type of transcendental reasoning that gives rise to a kind of "identity" of the universe, although of a nonclassical type, whose spatio-temporal objectivity is replaced by the weak objectivity of a transcendental type, which does not imply detachment of properties of the universe from subjectivity, but instead implies the coordination of new phenomena and theoretical essences into patterns of epistemological invariants that hold across historical and technological circumstances. These circumstances are linked to the conditions of embodiment (and consubstantiality). Human beings carry within themselves all

68. E. Husserl, "Foundational Investigations of the Phenomenological Origin of the Spatiality of Nature," in *Husserl Shorter Works*, ed. P. McCormick and F. A. Elliston (Notre Dame: University of Notre Dame Press, 1981), 227.

those properties of the universe that allow life to emerge and exist. The actual knowledge of the universe is then seen as an ongoing incarnation of humanity in the universe. Correspondingly, the sense of the wording "the universe as a whole" cannot be abstracted from the transcendental delimiters of the historically advancing incarnate humanity. The very word "universe" is of human origin, so that cosmology even if it pretends to describe the universe as some "physical out there," devoid of humanity, is still imbued with the human presence, although implicitly and in an inarticulate form.

The ongoing incarnation of humanity in the universe through which cosmology unfolds and thus the universe unfolds enters the definition of human nature but at the same time manifests itself as sheer contingency. In this sense the universe, as a phenomenon of consciousness, appears of itself: the gaze at the night sky delivers to a human being the sense of a picturesque visual display of varieties of patterns and shapes that bedazzle observers in the inexpressible mystery of their givenness. As a pre-predicative phenomenon, the universe is here and now, so that its totality is connected with the totality of the personal perceptive and intellectual experience, rather than with the totality, as a result of mental accomplishment, invoking the sense of extension in time and space. The thus-experienced universe corresponds to the ancient "cosmos" as the category of beauty and personal relationship. Physical cosmology does not treat the universe as an aesthetical (or even ethical) cosmos of the human soul. In cosmology the objects in the universe are not considered anymore as phenomena, or perceptible data, but rather as possible places for habitation of potential observers (based on the *hypothesis* of a formal interchange of home-places having a counterpart in physical reality[69]); however, this representation entailing spatial extension

69. Cf. P. Kerszberg, "The Phenomenological Analysis of the Earth's Motion," *Philosophy and Phenomenological Research* 48, no. 2 (1987): 206.

remains hypothetical. In this representation the universe is seen as global space and time with some corporeal connotations reminiscent of the ancient receptacle idea.[70] As result, a body-like image of the universe, as a certain substance contained by this receptacle, emerges. It is only in the vision of the universe as extended that it comes to mind that humanity occupies one particular place in it and that the meaning of this place was articulated differently during the history of astronomy and cosmology. However, the awareness of our position in the universe forms a transcendental condition of knowing the universe in the natural attitude: the universe must be extended in space and time in order to be explicable.

Here the transition in attitude takes place from a pre-predicative instantaneous synthesis of the universe toward its representation as a spatially uniform extension. Thus the cosmological principle acts as that epistemological tool which allows one to express outwardly communion with the whole universe.

The reification of the transcendental stance on cosmology comes through a strong affirmation that physical objectivity generally originates in contexts of historicity and instrumental situations. The ability to see in the universe what we see is selected not only by our particular location in space, but also by some general epistemological delimiters, such as our primary cognitive faculties and capacities of intellectual synthesis.[71] Something similar is asserted in the so-called

70. See a very nice review of ancient concepts of space, including that of a receptacle, in T. Torrance, "The Greek Conception of Space in the Background of Early Christian Theology," in T. F. Torrance, *Divine Meaning: Studies in Patristic Hermeneutics* (Edinburgh: T. & T. Clark, 1995), 289-342.

71. This is a well-known Kantian position on the possibility of knowledge: we can know only those things that can be adequated with our capacities of sense and understanding. This general transcendental stance was mimicked in some works related to astronomy and cosmology whose authors argued that human cognitive faculties and instrumental arrangements influence the way the universe appears to us. See, for example, M. Harwit, *Cosmic Discovery* (Brighton, UK: Harvester, 1981), and W. I. McLaughlin, "Kantian Epistemology as an Alternative to Heroic Astronomy," *Vistas in Astronomy* 28 (1985): 611-39); as well as J. D. Barrow and S. P. Bhavsar, "Filaments: What the Astronomer's Eye Tells the Astronomer's Brain," *Quarterly*

Weak Anthropic Principle (inference), namely that what we observe in the universe is related to the conditions of observability linked to embodiment.[72] It is important to understand, however, that these conditions of observability, as general epistemological delimiters, do not restrict the methodology of research, so that the limitations of research are rather related to the overall path of knowledge, rather than to a particular state of affairs with the fragment of this knowledge here and now.[73] The overall view of the visible universe is limited by our capabilities to receive signals from the cosmos via particular physical equipment, so that what we can observe is linked to the particular historical conditions and advance of technology. This means that when books on astronomy and cosmology present colorful pictures of stars, galaxies, and other objects in the sky, as well as produce generalizations toward the invisible universe, they deliver to the reader the cumulative result in a vision of the universe as related to a particular historical path of the constitution of the universe. Indeed, the history of cosmology is not very long: on the scale of a few dozen thousand years of existence of self-conscious humanity, humanity, for the first time in history, within a very short period of, let us say, a hundred years, is capable of making complex cosmological observations relying on the advances of technology. Here we have an interesting "methodological" situation: a science that effectively developed within the last hundred years attempts to predicate the universe in the wholeness of its temporal span. Hence, it

Journal of Royal Astronomical Society 28 (1987): 109-28; J. D. Barrow, "Patterns of Explanation in Cosmology," in *The Anthropic Principle*, ed. F. Bertola and V. Curi, 1-15; R. L. Bishop, "What We Bring We Find," *Journal of the Royal Astronomical Society of Canada* 80, no. 5 (1986): 224-29.

72. As asserted by Bitbol, the Weak Anthropic Principle "establishes a retrospective link between two sets of phenomena: the universe and the human body." Bitbol, "From the Anthropic Principle to the Subject Principle," 100.

73. S. Roush, "Copernicus, Kant, and the Anthropic Cosmological Principles," *Studies in History and Philosophy of Modern Physics* 34 (2003): 33-35.

is interesting to raise a question as to what philosophical convictions could justify such a "scientific" methodology, which attempts to make claims about the totality of all from within an infinitesimal part of it?[74] This question has two dimensions: the first one can be called historical, that is, whether it is really possible to reconstruct the historical past from within the present and what will be the status of this past. The second dimension raises a most general epistemological problem: how and why the physically and biologically local position of humanity in the universe makes it at the same time fundamentally nonlocal in its all-encompassing grasp of the universe?

The addressing of the first dimension of the question above would require from us an attempt based in an analogy between cosmology and all other historically based natural sciences such as geology, paleontology, or biology (the issue of history as a human science represents a slightly different matter, because of the nature of disclosed facts). Cosmology exhibits a similarity with historical research proper, which deals with artifacts and testimonies (but not repeated experiments). In cosmology, all artifacts are *images* (in a widely understood sense, be it optical, radio, or other type of material carriers of signatures of the cosmos) that are collected, classified, and interpreted. This collection, classification, and interpretation is humanly made from a vantage point of here and now, according to principles of reason. It is because of this that reason imposes locally disclosed physical laws (by way of transcendental argument) on the whole universe, thus transferring to it some properties of embodied subjectivity (here the *hypothesis* of a formal interchange of home-

74. H. Alfven as far back as 1978 expressed his doubts about the scientific status of cosmology, which dares to predicate realities of the early universe: ". . . it must be absolutely clear that as a scientist makes a guess about the state of the universe some billion years ago, the chance that this guess is realistic is negligible. If he takes this guess as a starting point of his theory, this is unlikely to be a scientific theory but very likely will be a myth." H. Alfven, "Cosmology: Myth or Science?" in *Cosmology, History and Theology*, ed. W. Yourgrau and A. D. Breck (New York: Plenum, 1977): 13.

places is implemented once again) and thus making the universe realistic, not in a naïve sense of a thing-like objectivity, but as an invariant of unfolding quests and contexts appearing through the ongoing embodiment of humanity in the universe. The view of the universe as being an objective invariant with respect to observational methods, and the affirmation of its intended identity being established by these methods, is nevertheless fundamentally historical, depending on the whole path of the history of science. Thus the alleged objectivity and neutrality of physical cosmology with respect to epistemic claims is of a conditional kind: cosmology fulfills itself only in the framework of one particular "trajectory" of humanity's history, which is linked to the technological advance. Scientific cosmology is based on *observations* that imply the extension of humanity's means of perception through technology and thus humanity's further embodiment in the universe; whereas the universe as unknowable whole defines and constitutes the form of this embodiment. Along this path, cosmology represents an endless hermeneutics of humanity's interaction with the universe, thus manifesting cosmology's intrinsic connection with anthropology in a profound philosophical sense, namely that it is through cosmology that humanity comes forth as a measure of unconcealment, that is, truth of being.[75]

Addressing the second dimension of the question above leads us back to the general philosophical problem of the ambivalence of the human position in the universe, which is encoded in the paradox of human subjectivity. To address this last dimension is to attempt to shed light on the nature of the paradox of subjectivity. This in turn brings us to the perennial question of human consciousness and the possibility of knowing the universe in general. In naturalistic

75. Cf. Heidegger's treatment of the famous thesis of Protagoras on "man as a measure of all things." M. Heidegger, *Nietzsche*, vol. IV: *Nihilism* (San Francisco: Harper, 1991), 91–95.

terms human consciousness transcends the realm of its embodiment, becoming nonlocal in space and time. Cosmology has to rely on this property of consciousness, and it represents the universe through mental images and constructs by this consciousness. The ability to create constructs and to associate them with physical reality is deeply based in the same *hypothesis* of a formal interchange of home-places that we have mentioned before, but now having a counterpart not in physical reality but in the intelligible (*cosmos noetos*). Thus the constitution of the universe naturally requires the extension of the cosmological quest to the intelligible realm, thus extending the very body of physical cosmology beyond the physical as such. The physical itself becomes a matter of constitution appealing to some general rules of the understanding and reason, that is, to the principle of rationality. And this is the reason why our next move will be to look closely at how the intelligible image of the universe, the image of the physical cosmos in the *cosmos noetos*, is constituted.

4

The Universe as a Construct

Epistemic Beliefs and Coherence of Explanation

The real world is not a thing founded in itself, that can in a significant manner be established as an independent existence. Recognition of the world as it comes from God cannot . . . be achieved by cognitions crystallising into separate judgments that have an independent meaning and assert definite facts. It can be gained only by symbolic construction.

–Hermann Weyl, Mind and Nature, p. 50

From the Cosmological Principle to the Origin of the Universe

Our intention now is to provide a concise and symbolic (graphical) description of the universe as a whole in order to explicate an epistemological meaning of such a description, in particular its dependence upon some irreducible beliefs that make this description possible. To do this we need to give a brief overview of the major methodological presumption in cosmology, namely the cosmological principle.[1] Since we cannot empirically verify the statistics of distribution of matter from other locations in space we have to speculate on the overall distribution of matter in the universe, appealing to philosophical and hence physically untestable assumptions. The universe seems to be isotropic on the scales corresponding to clusters of galaxies. This local isotropy, being an empirical observation, is a contingent fact and it does not entail that the universe should look isotropic from every possible location in it. However, the idea of our indifferent position in space was a prevalent trend after the scientific revolutions of the seventeenth century, so that modern cosmologists felt obliged to apply the observable isotropy to all locations in the universe. This principle of indifference, the cosmological principle, postulates the uniform distribution of matter in the universe and the uniformity of space.[2] Only under this assumption is any scientific methodology of studying the universe as a whole in cosmology possible. Indeed, the uniformity of the universe is needed in order to predicate its properties in terms of the same physical laws in locations that are inaccessible to our reach. The integrity of our intelligence must correspond to the integrity of the cosmos, and this integrity is best expressed in terms of its uniformity. The global picture of the universe would not be possible if, in every corner of it, physical laws would be different and objects and phenomena would be infinitely varied. The cosmological principle reduces the description of "cosmic matter" (with the

1. See chapter 3. See also A. Nesteruk, "Cosmology and Teleology: Purposiveness in the Study of the Universe through the Reading of Kant's Third Critique," *Journal of Siberian Federal University. Humanities & Social Sciences* 5, no. 9 (2012): 1304–35.

2. This principle, inaugurated by E. A. Milne (after Einstein's suggestion) as "cosmological principle," that is, "all places in the universe are alike," has many motivations. One of them, which is closer to the spirit of physics, was to extend the observable isotropy of the universe at the scale of clusters of galaxies to every possible location. Here we are dealing with a refined version of the hypothesis of a formal interchange of home-places (as having a counterpart in physical reality) related to the phenomenon of isotropy. Thus the uniformity of matter distribution as well as space itself can be understood as the postulate of an isotropic view of the universe from every possible location.

constituting element of this matter to be a cluster of galaxies) to two macroscopic parameters—density and pressure.[3] Indeed, theory cannot deal with the variety of specific and concrete objects that are astronomically observed. The contingent facticity of these objects is transferred to the contingent facticity of the generic symmetry, that is, to the uniformity of their distribution in space. Thus the introduction of the cosmological principle does not explain away the contingent facticity of this principle itself: as such it does not explain *why* the universe is uniform. The postulate of cosmic spatial uniformity introduces a fundamental *construct of cosmic matter* that makes it possible to talk reasonably about the universe as a whole, including its spatial structure.

The "cosmological principle" changes our perception of the contingency of our spatial position in the universe. If the universe is uniform, a potential observer traveling across the universe would observe statistically one and the same picture of the universe. This evidently diminishes any drama related to the spatial contingency of humanity in the universe: we could be anywhere and would observe the same. This means that that fragment of the universe which is visible to us gives a fair representation of all that is possible.

The situation is more complex in terms of that particular temporal era in which we are present in the universe, if the universe is subject to change (evolution). If, hypothetically, the universe as a whole were to be static, that is, without any large-scale change in time, its description would be very simple and would correspond to a couple of contingent numbers (its density and pressure) that characterize the

3. The concept of matter of the universe in relativistic cosmology is similar to the concept of an ideal gas, in which real physical particles are treated as point-like objects, material points. Any set of material points can be taken then as prototype, which can be filled in with different meanings. For the ideal gas one can take a prototype of chaotic motion of material points and substitute molecules for these points. In cosmology, by substituting material points by clusters of galaxies one can obtain the notion of "matter of the Universe." C. Misner, K. Thorn, and J. A. Wheeler, *Gravitation* (New York: Freeman, 1973), 711–13.

large-scale structure, and any change and becoming would amount to the evolution of specific objects, such as galaxies and stars. The novelty in the universe would emerge from local physical processes, whereas the overall large-scale universe would remain the same. On the large scale the displayed uniformity would remain, and no hope for its "explanation" would exist. This is the reason why the idea of the evolution of the universe as a whole becomes so important: it gives us a chance to transfer the problem of its contingent facticity as observed here and now to the problem of its origin from some unknown initial state that would exhibit, either in terms of known physical laws or some mathematical argument, apodictic features, so that the annoying question about the facticity of its display could be removed if the initial conditions leading to this display are "explained."

Cosmology received this chance when the experimentally observed red shift in spectra of galaxies was interpreted as their mutual recession due to a universal kinematic expansion of the universe.[4] The link between expansion of matter and expansion of space is asserted by general relativity theory. This expansion, observed through luminous objects such as galaxies, seems to be isotropic. In combination with the "cosmological principle" the observed expansion in the universe is presented as taking place uniformly in all locations in the universe. This means that our image of the expanding universe can be transferred to any other location in which the picture of expansion would be the same. In this sense there is no geometrical center of this expansion: one can talk of

4. The red shift in astronomical objects is varied with their distance from our galaxy, which was established by E. Hubble as far back as 1929. The velocity of recession of a galaxy is proportional to its distance with the coefficient known as the "Hubble constant" H0, which is treated as a fundamental cosmological parameter characterizing the rate of expansion of the universe as a whole. Here, for the sake of our objectives, we disregard possible objections to this interpretation of red shifts in galaxies' spectra that raise doubts regarding the idea of the expanding universe.

the expansion of the universe about every particular point. This is a counterintuitive result that refines the cosmological principle as related not only to spatial locations, but also to how the evolution of the universe looks from these locations, namely that it looks the same. In combination with the fact that the received signals come from remote parts of the universe, one can state that the past of the universe looks the same from different locations. And this leads to another counterintuitive result: that the universe is seen as the frozen past.[5] The past of the universe is made manifest through its present image. Thus the distinction between the universe's past and present, which is usually made by analogy with human history (some events in the past of the human history are not definitely in the present), is problematic: one cannot understand and interpret the universe as we see it here and now without referring to its past: when one pronounces the word "universe" one affirms the totality of its temporal spans. In this sense the subject matter of cosmology, that is, the universe in the entirety of its spatial and temporal spans, represents the unity of all its locations and eras, unifying all differentiated happenings in the

5. See a representation of this statement by the graph of "The Cosmic Spheres of Time" in J. Primack and N. E. Abrams, *The View from the Centre of the Universe* (London: Fourth Estate, 2006), 135. See also the same authors, *The New Universe and the Human Future* (New Haven and London: Yale University Press, 2011), 74. The seeing of the universe as the frozen past connotes with the notion of the "block universe" according to which all points of space-time have an equal ontological status, so that no fundamental meaning can be ascribed to the distinction between "past," "present," and "future." See more details, for example, in C. J. Isham and J. C. Polkinghorne, "The Debate over the Block Universe," in *Quantum Cosmology and the Laws of Nature*, ed. R. J. Russell, N. Murphy, and C. J. Isham (Berkeley: CTNS, 1996), 139–47.

universe in a kind of *event*.[6] As we will see later, this is related not only to the visible universe, but to the universe as a whole.

The cosmological principle applied to geometry predicts three possible models of the spatial structure of the universe (closed, flat, and open), all of which now must be linked to the universal expansion of space corresponding observationally to the mutual recession of galaxies and their clusters. The three models of space can easily be generalized in order to accommodate this expansion through introducing a notion of the universal scale factor $a(t)$ as a growing function of "cosmic time," which stretches the spatial metric corresponding to all three models. There are two points that must be noted about this new cosmological *construct*: first of all, "cosmic time" is a construction obtained through imagining that one can place clocks everywhere in the universe, for example in all galaxies, so that they move along their world-lines together with the overall kinematic expansion, and are somehow synchronized at its beginning understood here simply as a reversal of expansion to its initial point. After introducing the construct of time it becomes possible to describe the overall geometrical becoming in the universe, its evolution in terms of the scale factor $a(t)$ as a function of time t. The dynamics of this scale factor as well as the dynamics of cosmic perfect fluid is subjected to Einstein's equations. These are differential equations for $a(t)$, whose solution is reversible in time, that is, the

6. The given description of the universe corresponds to its disclosure by human beings in the course of their history. Because of the finitude of this history, which itself can be treated as an *event, humankind too may be seen an event.* This notion was introduced in my *Light from the East* (Minneapolis: Fortress Press, 2003), 194–214; the disclosure of the universe as a particular action of the human spirit can be treated as an event within the humankind event. The notion of an event entails the presence of fundamental contingency. It is because of this contingency that cosmology attempts at all costs to get rid of this notion by grounding the counterintuitive content of this event in something apodictic and undeniable, inferred through a kind of an extralogical argument. In this aspiration cosmology exercises a leap of faith, making it similar to theology, which predicates humanity's existence and the presence of the world in this particular condition as an event of their communion with God, who is transcendent indeed.

equations can be used either to predict the solution in the "future" given the conditions posed now, or, alternatively to formulate the solution in the "past."[7] In this sense they describe the reversible transition of the universe from one state to another: they do not describe the real becoming as emergence of novelty in the universe. This then implies that the contingent values of the cosmological functions, for example $a(t_0)$ at present time t_0, do not stop being contingent through the reversed (backward in time) cosmological expansion: their values in the past are just recalculated through the solutions of the Einstein equation values at present. It is in this sense that one must be aware that the cosmological expansion cannot be associated with the arrow of time that is observed empirically,[8] so it seems that the "t" variable of cosmic time has little to do with the time of human experience, which is intrinsically irreversible. The dynamics of the universe at large scales is reversible and thus does not account for the second law of thermodynamics which, as it is believed, is linked to the special initial conditions.[9] The geometrical model of the evolving universe provides simply a description of certain changes that happened at large scales, but real becoming as creation and transformation of new forms of matter at smaller scales require for their description an appeal to physics that has a rather "local," earthly origin.

The solving of the Einstein equations for $a(t)$ requires one to start with a particular equation of state for matter. If the point of departure is the present universe in which free propagation of light

7. Assuming that the reversal of initial velocities is possible. See P. Davies, *The Physics of Time Asymmetry* (London: Surrey University Press, 1974), 22–27.
8. This point was made clear, with reference to R. Tolman, in R. Penrose, "Singularities and Time-Asymmetry," in *General Relativity: An Einstein Centenary Survey*, ed. S. Hawking and W. Israel (Cambridge: Cambridge University Press, 1979), 581–638.
9. This point was emphatically defended by Penrose in many of his writings. See, for example, *The Emperor's New Mind* (Oxford: Oxford University Press, 1989), 440–47; *The Road to Reality* (London: Vintage, 2005), 726–32, 765-69.

takes place, and its matter content can be treated as "dust" with no pressure. In this case the density of matter decreases as the universe expands in inverse proportion to the so-called commoving volume, which grows in time as $a^3(t)$, so that $\rho(t) \sim a^{-3}(t)$. If time is reversed, so that the expansion becomes contraction, $a(t)$ decreases and, as result, the density of matter grows. Theory predicts that it grows to such an extent that the evolution of the universe divides roughly into two completely different stages: the present stage in which matter is decoupled from radiation; and the early stage when the spatial size of the universe was a thousand times less than it is at present, and the universe was opaque. Before the transition to the dust-dominated era, according to theory, the universe represented a mixture of radiation and hot matter.

From the very inception of this nonstationary model it was a great temptation among cosmologists to extrapolate the contraction of the universe in the reversed time order to a limit, when the spatial size of the universe tends to zero. This mathematical limit represented a problem of interpretation, since all physical parameters such as density, pressure, curvature, and so forth acquired nonphysical infinite values. The limit $t=0$ together with $a(t)=0$ was treated as the temporal beginning of the visible universe. The construct of this beginning seems to be contentious because one can hardly understand whether $t=0$ indicates the beginning of the visible universe only (and all other physical parts of the universe admitted by the cosmological principle), or the beginning of the overall spatial structure if it is taken as preexistent (similar to the Newtonian absolute space). In any case all contingent facticity of the present-day visible universe at large scales is transferred to the cosmological singularity, which becomes the initial condition for cosmological equations and which exhibits some idiosyncratic properties because all physical parameters become infinite in it. It was understood long before this that cosmological singularities represent strange "initial conditions" of the universe, whose facticity can hardly to be explained within available physics; classical physics collapses at the singularity[10] and since no credible quantum description of the initial state of the universe exists so far, we deal here with a problem that *de facto* exceeds the scope of physics and tests its limits. For example, it is difficult to understand what the beginning of the universe could mean in terms of space and time. Did time or space exist before the expansion started; in other words: did the expansion begin in preexisting space and time, or were space and time brought into existence at the origin and thus their actual presence explicates the act of their beginning? It is possible to use a simple diagram to illustrate the problems arising with the interpretation of the origin of space and time of the universe.[11]

10. J. A. Wheeler articulated this point in order to assert the intrinsic mutability of physics, including its conceptual ingredients, such as space and time. See, for example, "From Relativity to Mutability," in *The Physicist's Conception of Nature*, ed. J. Mehra (Dordrecht: Reidel, 1973), 202–47, as well as his book *At Home in the Universe* (New York: American Institute of Physics, 1994).
11. See more details on cosmological diagrams used by us in E. R. Harrison, *Cosmology: The Science of the Universe* (Cambridge: Cambridge University Press, 1986), 215, 375–87.

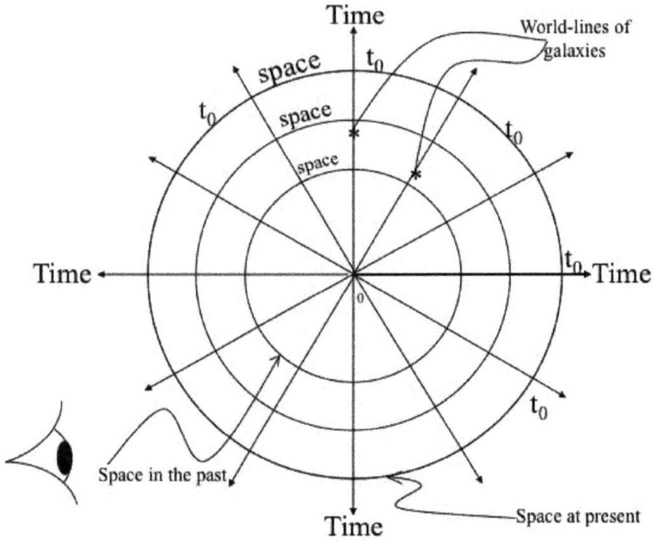

Figure 1

This diagram attempts to express the unity of space and time of the universe as being generated from their "origination event" depicted by a point at the center of the diagram. The diagram consists of a series of expanding concentric circles that aim to represent spatial sections of space-time. The circles expand from the initial zero point, which symbolizes the origin of the universe. The radii correspond to the world-lines of particular objects (clusters of galaxies, for example) that somehow originate at the singularity (corresponding to zero linear scale) and diverge in all directions. It is along these lines that time, associated with the imagined clocks of all galaxies experiencing the overall kinematic expansion, is measured (this time is somehow synchronized at the zero point, that is, the beginning). The fact that the spatial sections (that is, concentric circles) in this diagram are compact must not be interpreted as if we are dealing with a topologically closed universe. If these imagined circles are associated with some structural units of the universe (galaxies or their clusters), their expansion reflects only the process of the mutual recession of galaxies. The major conceptual difficulty with the interpretation of this diagram is to conceive the meaning of the point of origin of the world-lines. One must not treat this diagram as if it depicts the actual process of origination and expansion in preexistent space or time. Actually this origin is not *in* space and in time, so that its depiction as a point in the plane of the page is a metaphor. However, the diagram as a whole can be treated as representing the global structure of space and time in the natural attitude, that is, as if they existed objectively and independently of the human observer who appeared in the universe at its late stage. The distinction between past, present, and future has a purely symbolic nature (associated with the radius of a circle, or progression of the world-line) as divisions in abstract objective time.

It is worth noticing that the enquiring intellect is implicitly present in this picture: this picture is a mental creation, that is, its content is the product of human subjectivity which is present in all articulations of the universe; it places the singularity, which is supposed to initiate all being, including this same incarnate subjectivity, outside itself. This is the reason why this subjectivity experiences difficulties with explicating the sense of the cosmological singularity: it is impossible to explicate within the natural attitude that which is supposed to be in the foundation of everything with which this subjectivity deals, including the facticity of subjectivity itself. When the enquiring intellect draws such a diagram it positions itself outside the universe, as if it could look at it from some vantage point. However, this mental operation is possible only in abstraction, because one cannot get out of this universe, which would be tantamount to breaking the conditions of conscious embodiment in it.

The construct of the initial singularity brings into play a certain correction in perception of causality in the universe and of temporality. Logically, the ideas of the initial singularity appeared as the result of theoretical reversal in time of the presently observed expansion of the astronomical cosmos. This reversal corresponds to the reversal in the solutions of cosmological equations, which are extrapolated to their nexus at the point with time equal zero. The thus-obtained state represents the initial state of the universe in the past. However, since the Einstein equations are time-reversible they transfer (according to efficient causality) the information encoded in the present state of the universe backward in time. In this sense, technically, the state of the universe at the singularity (related to its large-scale structure) is isomorphic to the state of the universe at present, that is, to that which is observed here and now (in spite of the fact that the numerical values of some cosmological parameters become unbounded at the singularity). One can say that the distinction between present and past is a matter of commonsense convention, for, as we have mentioned above, the cosmological expansion does not account for the growth of entropy, that is, for the irreversibility of time.

If the cosmological singularity is only considered as a construct that outlines the limits of physics in exploring the nature of facticity of things, then cosmology does not run into a problem of justifying the physical status of this singularity. Singularity, in this view, becomes no more than a parable of the mystery of the facticity of the universe whose appearance in cosmology is inevitable, simply because the efficacious *telos* of cosmological research requires one to introduce a notion of the overall unity of the universe whose role is played by the singularity. However, if the singularity is considered as the point of initiation of all contingent display in the universe in a physical sense (as a *nexus finalis* in the reversed temporal order), so that the distinction between the singularity and the present state of the universe is associated with the asymmetry between past and future in existential, irreversible time, there arises a problem because the singularity as such is fundamentally nonobservable (apart from, as it is believed, its remote consequences in the present), so that any claim for the viability and truth of its concept must follow a different pattern of justification in comparison with that for experimental science. This can be simply understood if one realizes that, by placing the singularity in the past as the foundation of the explanation of the present, and assuming temporal causality between that past and the present, one exercises an act of belief in which a characteristic transition from an intelligible entity (construct of singularity) to the empirically real (present-day universe) takes place. This kind of transition does not follow the pattern of explanation based on the principle of correspondence because there is no independent empirical verification of the idea of the singularity. It fits theory on the grounds of epistemic coherence and aesthetical criteria. Correspondingly, the cosmological scenario of the origin of the universe (not its physically visible past) cannot be verified because the universe cannot be subjected to physical experimentation,[12] so that all inferences in cosmology are grounded in the realm of extrapolation and analogy with other "historical" natural sciences.

One may further clarify the spatio-temporal representation of the universe in the standard cosmological model and, in particular, the sense of the cosmological singularity, or the Big Bang, by taking into account the special position of the human observer in the universe.

12. One cannot rerun the universe with the same or altered initial conditions to see what would happen if they were different. See G. F. R. Ellis, "Issues in the Philosophy of Cosmology," in *Handbook of the Philosophy of Science, Philosophy of Physics*, Part B, ed. J. Butterfield and J. Earman (Amsterdam: Elsevier, 2007), 1183–1283 (Thesis A1, 1216).

In spite of the fact that the cosmological principle claims that the location of human observers is mediocre in order to create an average statistical picture of the universe, it turns out to be that "what" human agents can actually observe is subject to general causal limitations following from the physical laws established in the terrestrial domain. The universe, considered as luminous objects, invisible radiations, and cosmic particles (which are consubstantial with us), is perceived from the given space-time location through the so-called past light-cone along which electromagnetic signals travel with the speed of light.

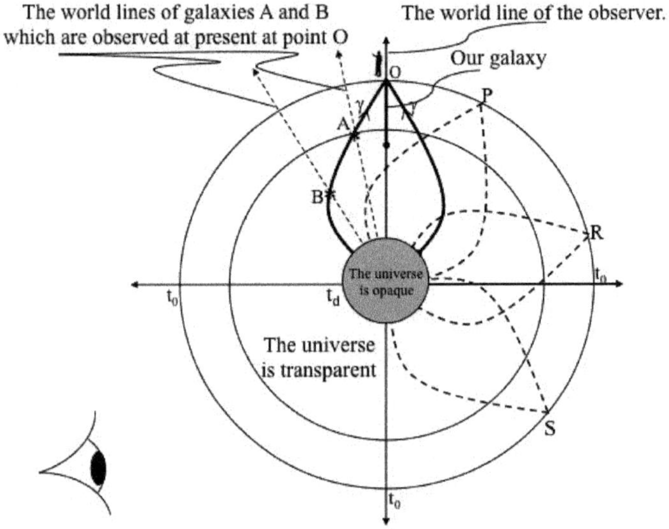

Figure 2

In Figure 2 this situation is depicted through photons (γ), traveling on the surface to the curved light-cone and reaching the observer at present. In this case the geometrical manifold of the observable universe represents two symmetric curves (onion-like shapes) that originate at the cosmological singularity and reach the observer's location. The universe is observed along the past light-cone and the maximal spatial distance of objects from our location whose radiation could be detected is determined approximately by the separation between a galaxy B in Figure 2 and the radius corresponding to observer's location.[13] It is not necessary to be a qualified mathematician in order to realize that if the wholeness of the universe as space-time is anticipated in this diagram as the two-dimensional interior of the circle with the radius corresponding to the present time t_0, then the observable universe represents a one-dimensional manifold whose quantitative measure with respect to the whole universe is zero. In other words, the observable universe in this view is an infinitely small contingent piece of the allegedly existing whole. In Figure 2 this is indicated by a random choice of the point (with a tiny human figure) on the circumference symbolizing the whole space at present. Since there are infinitely many potential locations on this circumference, the contingent choice of a particular one (linked to the contingent choice of the past light-cone) makes this contingency acutely felt through the incommensurability in measure of the one-dimensional circumference and a chosen point which, strictly speaking, has no dimension at all.

One could raise an issue here as to whether it is legitimate to attempt to infer the universe as a whole from an infinitely small part of it. One must remember, however, that Figure 1 and Figure 2 are both drawn in the natural attitude, that is, both treat the universe within the phenomenality pertaining to objects; correspondingly, in order to make sense of these objects one has to appeal to the cosmological principle, which allows one to speculate about the universe as a whole while being chained to one single point of it.

13. The numerical estimate for this distance l_{max} is given in A. Rothman and G. Ellis, "Lost Horizons," *American Journal of Physics* 61 (1993): 886.

Indeed without this principle any representation of the universe as a whole would not be possible: we could responsibly speculate on the origin and evolution of the visible astronomical universe but not about the universe as a whole originating from the same Big Bang. The cosmological principle makes irrelevant the question about an observer's location on the circumference, because, according to it, all its points corresponding to all possible spatial locations are equivalent (for example, the density of cosmological matter is the same at different points such as O, P, R, S). This means that we, as observers, could be anywhere. Correspondingly, the past light-cone could be anywhere, giving, in accordance with the same principle, a similar picture of all possible observable universes. In Figure 2 this situation would correspond to rotating the light-cone, keeping its apex on the same circumference and thus covering entirely the interior of the circle. It is because of the possibility of this rotation that the difference in measure between the interior of the circle and the past light-cone epistemologically disappears: the infinite multitude of light-cones corresponding to all possible contingent locations of observers covers entirely the interior of the universe, thus giving statistically the same picture of its structure from any imaginable location.

The result of this simple analysis is that the cosmological singularity whose existence is inferred from the expansion of the visible universe is unique because it would be predicted in all possible cosmological scenarios based on observations from all possible locations. This implies that the *construct* of the cosmological singularity as related to the *whole* universe is ultimately based on the cosmological principle. It follows then that similar to the certainty of belief from within which originates the cosmological principle, the validity and truth of the construct of the cosmological singularity is also situated in the certainty of belief. Together they form a *coherent* framework of interrelated beliefs that form the basis of cosmological methodology.

One can also add that the cosmological singularity (or the Big Bang), as a direct consequence of the cosmological principle, reveals itself as a transcendental principle of explicability of the universe as a whole with corresponding teleological connotations related to the methodology of research: to provide a coherent view of the universe as a whole one has to introduce a unification principle of all, which is imitated by the concept of the Big Bang.

The cosmological principle removes the ambiguity of the human observer encoded in Figure 2 (this ambiguity is another explication of the paradox of human subjectivity in the world discussed in chapter 2). Indeed, on the one hand human subjectivity is present outside the schemata of the universe, because the universe as a whole (including its observable part) is the result of a theoretical hypothesizing: it is reflected at Figure 2 through an eye looking over the universe. On the other hand, human observers, being embodied creatures, establish their insights about the physical universe from a particular location in space, thus selecting that part of reality which is linked through physical causation to the place of embodiment. The cosmological principle allows one to balance these two approaches to the universe by telling us that our particular location in the universe and its constructed image can be reproduced from all possible physical locations. The universe becomes not only intelligible, which is implied epistemologically by the cosmological principle, but also intelligent: the universe is represented as a continuum of potentially possible human-like observers. In this case it effectively reproduces an old idea of Plato that the universe is alive in a very nontrivial sense: the universe allows the presence of human intelligence in it not only in an anthropic physical sense, but in the sense that all predications about the universe contain the deposit of the human hypostatic subjectivity.

It is clear from Figure 2 that human observers, in their contingent and indifferent location in the universe, see only the past of the universe: while gazing at the celestial sphere they see the images of that which was emitted by cosmic objects long ago. On the level of perceived phenomena we deal with the image of *the past in the present*. However, this past is not a fixed past referred to a particular historical stage of the universe, but it is the accumulated image of different objects at different distances, and hence different eras. Theoretically we are dealing not with a particular past that could be marked in terms of a fixed figure of years passed after the Big Bang, but an integrated past bearing images of different galaxies and their remote ancestors since the times of their formation. This can be illustrated with the help of Figure 2, which shows that we receive simultaneously signals from galaxies A and B, which are at different distances from the world-line of the observer. This simple geometrical fact implies that these images correspond to different times at which signals from the galaxies have been emitted: the images of galaxies A and B come from the same location in the celestial sphere, but they correspond to objects that have crossed the surface of the past light-cone at different moments of time. It is not difficult to see from Figure 2 that both these galaxies are beyond our reach at present (their world-lines at present are separated from us by a space-like interval) so that we receive only the optical signatures of their *past* existence. When we observe the images of these galaxies we must, on strictly philosophical grounds, assert their existence only as phenomena. Any hypothetical affirmation of their physical existence can only be made by using the language of past tense.

Some elucidation of the assertion of the accumulation of the past in the universe's display as a phenomenon can be achieved by changing the perspective and not considering the Big Bang as "out there," but contemplating it as being encapsulated in the display of the universe here and now. This is in correspondence with a simple truth that the point of the beginning of the universe has no location in space because space appears together with this point, so that while looking at all possible directions in the sky we, strictly speaking, look toward a nowhere that appears to us in the guise of everywhere: the singularity is out there but, at the same time, it is nowhere.[14]

In view of this, a comment must be made related to the cosmological principle. Let us recall that this principle, as affirming the uniformity of distribution of matter in space, is based on the observed isotropy of this distribution from our vantage point. One may ask what the meaning of this isotropy is; from Figure 2 it is clear that the observed isotropic distribution of matter in space corresponds not to what "happens" in the universe at present, but is related to the past, because whatever is observed as a phenomenon corresponds physically to the accumulated past. In this sense the alleged isotropy of "space" is not isotropy *in* the present, because we cannot observe spatial locations of distant objects at present apart from our galaxy (which is depicted in Figure 2 by the bold line going down from point O and which is not isotropic at all). In other words, one can assert that the distribution of matter is isotropic only in the sense of the accumulated past. It is from this fact that an inference is usually made, by means of extrapolation, to assert the isotropy and then uniformity of matter not in terms of the integrated past but at every particular era of the past and in the present. Formally this is expressed, for example, by assigning to the density of matter a constant value for all possible locations at a given moment of cosmic time, that is, at all locations on the circumference in Figures 1–2. This observation strengthens the conviction that the cosmological principle (that is, the universe is uniform at any space-life surface corresponding to

14. Compare with the graph of "The Cosmic Spheres of Time," in Primack and Abrams, *The View from the Centre of the Universe*, 135;

a moment of cosmic time) is based on the idea that the integrated image of the past of the universe can be decomposed in terms of consequent stages of the universe's evolution. Said philosophically, in order to deal with the contingent image of the frozen past of the universe in scientific terms, one should decompose this image into consequent layers of reality corresponding to different cosmic eras. Thus it is from this decomposition that the idea of the universal cosmic time receives its epistemological justification.

Generalizing what has been said so far, in observational cosmology we deal with the phenomena that, according to theory based in the laws of relativity, are living images of the past or its remote consequences. Unlike the events of human history whose reenactment requires the appeal to the witnesses of the past through documents and archives, through the communion with the minds of those who were present behind the artifacts, in cosmology we have a different situation in which a certain past of the universe is constantly present: the artifact of the universe's past is always given to us and is being constantly gazed at. Since this is not a fixed past, but the accumulated past, the past stretched through time but frozen in its image given to us, human beings live in the presence of the extended and never-ending *event* of the past whose contingent facticity remains a mystery. Thus the contingent facticity of the observable universe represents the contingency of the accumulated image of the past: the universe is contingent, being "an event" with extended spatio-temporal characteristics. This event, theoretically explicated through the past light-cone of Figure 2, is linked to the human observer which itself represents an extended event of the overall human history. This is a different way of stating a Christian conviction that it is cosmic history that is treated in theology as part of the human history of salvation, and not vice versa. It is here that cosmology effectively explicates its hidden theological commitment linked to the Divine image in humanity which articulates the universe from within its history.

A scientific attempt to unfold the phenomenon of the observable universe as existing in space and time presupposes an idea of the continuum of space and time as a preexisting entity. However, all objects corresponding to their observable images "exist" in different moments of preexistent time. In this sense our intuition of existence of these objects manifests itself as fundamentally nonlocal in time, that is, we affirm existence not in proportion and connection with the fact of our existence, which is local in space and in time, but as supra-temporal or trans-temporal, when all moments of allegedly preexistent time are reduced to the facticity of their observability at present.[15] The universe as a phenomenon is here and now: it is that which we see in the sky and perceive as an immediate medium of our embodiment. There is no distinction between past, present, and future in this phenomenon. In this sense the phenomenon of the universe as a whole is contingent because it is given. Scientific cosmology, in its instinctive desire to overcome this contingency, appeals to the idea of the originary origin of the universe (for example, its temporal origin), in order to objectify the contingency of its givenness by shifting it into the remote past under the guise of an ill-articulated apodicticity. However, the idea of time and the "past" that stands behind such an appeal ultimately originates in human historicity, whose facticity cannot be scientifically accounted for. This point reasserts that which was before expressed theologically, namely that it is cosmic history that unfolds from within human history and not vice versa, because the primacy of existence (understood as communion) belongs to hypostatic human beings,

15. This assertion can be considered as a certain variation of the idea of a block-universe, where all points related to the past light-cone—related to our vantage location in the cosmos—are considered as having an equal ontological status.

and not inorganic rocks in the cosmos, in spite of the fact that the conditions of embodiment originate in the stardust.

The Universe as a Construct: Its Rationality in Rubrics of Faith

Let us reflect upon the links of the notion of the universe as a whole to the life-world as an attuned medium of all contexts and thematizations of the universe associated with the conditions of corporeity. The reader will have noticed that we have used the term "construct" several times when introducing some basic mathematical statements about the universe; for example Figure 1 depicting the expanded space-time can be considered as a construct. "Construct" in this case represents a certain departure from the immediately given and a combination of the empirically given, which has already been accumulated in knowledge of the universe (for example astronomical observations) with an intelligible image of reality as a whole (global space-time manifold). The validity and efficacy of constructs in theory is determined through the rules of correspondence with the reality of the empirically given (what could be called reification), the rules that also include other forms of approval and reference to truth relying on the experience of communities in the life-world (this could be called objectification).[16]

For example, the construct of the spatially uniform cosmological fluid is obtained from further extension of that ideation which is used in earthly physics to construct the notion of the ideal gas. The ideal gas is used as a *gestalt* for a cosmological fluid where its elements (atoms) are replaced by clusters of galaxies. The notion of a cosmological fluid has its origin in empirical physics, but in its

16. Cf. H. Margenau, *The Nature of Physical Reality: A Philosophy of Modern Physics* (Woodbridge, CT: Ox Bow, 1977), 70.

content it exceeds the realm of the empirically observed and relies on an intuition of the global space, which represents another construct. As mentioned before, the constituents of the cosmological fluid have a precarious ontological status related to their nonlocality in time, so that it is the human mind that brings together different elements of the cosmic display in order to construct a model of this "fluid" as if related to a particular moment of time. It is important to realize that constructs, as mental accomplishments, do not presume any strong commitment to realism. In this sense if one asks a question about the correspondence of cosmological constructs to empirical reality, one must admit that this correspondence exists for sure only in the sense that the constructs are produced by us who are part of this reality. Certainly the concept of the universe as a whole depicted in Figure 1 must contain a place for the reality of the life-world. But, as has been already demonstrated through Figure 2, the quantitative measure of this reality is infinitely small with respect to the universe as a whole, so that in physical terms the diagram in Figure 1 has a link to the empirically given only at one point corresponding to the place of humanity in space and time.

In order to relate the construct of global space of the universe to the life-world, one must understand how the global space is related to the spatiality of the earthly world and the corporeal spatiality of human beings. In other words, there must be some legitimization in the transition from the centrality of human beings in their attuned space of the life-world (or the centrality of humanity on earth as its home-place) to the periphery locations in a thematized and actually infinite space. This transition is effectively connected with mental procedures that can be called ideation and abstraction. Their essence requires some verbal reification.

The attuned space of the lived space associated with corporeity has a character of self-givenness "in flesh," that is, presence "in person," in the sense that space and objects in it are given in confrontation with functioning corporeity. Obviously this can be said about the space of the planet Earth and all objects in it. In a certain sense one can say a similar thing about the images in the celestial sphere, which enter the reality of the life-world as a certain horizon. In cosmology the extension in cosmic space loses the character of presence "in flesh" simply because the mathematical components of this extension have no direct relationship to corporeity. However, in similarity with that intellectual procedure which led us to the construct of the perfect cosmological fluid, one exercises here ideation, as a special case of abstraction, which disregards all particularities of spatiality present in the individual subjects or objects. This ideation does not bring simply something common in a given multitude of objects, rather on the basis of perception "in flesh" it intuits the universality of the essence. This ideation is crucial for introducing constructs with respect to which a *new type of objectivity* is constituted as a new type of intentionality in which the founding acts of corporeal intuition are not included in this objectivity. This new intentionality contains those determinations with respect to the universe that are not found in the objects initiating this intuition. But if the construct of the global space of the universe is introduced along the lines of this new intentionality, its components, as essential universals and ideal objects, are indifferent not only to the number of empirical cases that could represent them but to the possibility of their empirical realization and hence verification at all. One can say that the global space of the universe appears as immediately self-given in the overall intuition of the universe as communion in the life-world, but this self-givenness is not directly related to the aspect of corporeity and has rather a categorical nature. However, at this stage of constitution this global space is not yet mathematized and hence one cannot say that it is purely nonsensible. For example, it receives a pictorial representation in Figure 1 that is not nonsensible in spite of the fact that it already contains the elements of a mathematical (geometrical) ideation. But what is important, and this was the achievement of the phenomenological analysis of mathematical knowledge, is that in the foundation of such a constitution lies the intuition of mathematical continuum.[17] This intuition correlates with the sense of unity of experience that pertains to every subjectivity: the unity and continuity of the field of consciousness cascades toward the unity of reality which includes the concreteness of corporeity. Corporeity as existence-in-situation entails the sense of belonging, and this is intuited as a continuum whose presentation is achieved by using mathematical ideas. Now it is not difficult to realize that the pictorial presentation of the universe in Figure 1 is based on the assumption that the underlying mathematical continuum (corresponding to the existential unity of experience) is geometrized (as space) in the class of pseudo-Euclidian metrics under the condition of uniformity of space. Once again, this construct contains both sensible elements (its image as such) as well as categorical elements. In this sense by asserting the idea of global space and its representation through Figure 1 we avoid either the commitment to realism or to idealism. The fact that this kind of representation is a construction, but not a purely mathematical concept, originates in its being obtained through the accumulation of facts in the field of astronomy and astrophysics, so that it is a historical intellectual achievement and, unlike pure mathematical ideations, does not possess a trans-temporal nature. The construct is not part of the physical reality of the universe, but, at the

17. See, for example, on the phenomenological treatment of the intuition of continuum, two papers of G. Longo: "The Mathematical Continuum: From Intuition to Logic," in *Naturalising Phenomenology: Issues in Contemporary Phenomenology and Cognitive Science*, ed. J. Petitot, F. J. Varela, B. Pachoud, and J.-M. Roy (Stanford: Stanford University Press, 1999), 401–25; "The Constructed Objectivity of Mathematics and the Cognitive Subject," in *Quantum Mechanics, Mathematics, Cognition and Action: Proposals for a Formalized Epistemology*, ed. M. Mugur-Schachter and A. van der Merwe (Dordrecht: Kluwer, 2002), 433–62.

same time, being an element of a cosmological narrative it is a part of the reality of cosmology as a cultural phenomenon. In this sense the positing of constructs in cosmological theories represents the work of a different type of intentionality which, being related to the natural attitude through an attempt of an inferential causation from the empirical, yet points toward a fundamentally different intentionality relating to the questioning on the facticity of the unity of experience in the conditions of corporeity; but this, as we mentioned before, brings us to the intentionality of existential faith in its particular realization as faith in the unity of experience (and hence as consistency of nature). Hence one understands that all basic notions of the standard cosmological model, including the cosmological principle, constructs of global metrics and perfect fluid, the dynamics of the scale factor, and the prediction of the Big Bang are all constructs related to the realm of the immediately given only indirectly in the sense described in previous paragraphs. It becomes evident that the existential belief in the unity of reality corresponding to the unity of conscious experience permeates the whole standard cosmological model, including all its constructs.

The question now is: What makes this standard model so convincing for scientific communities and a wider audience? Why, in spite of the fact that cosmology is characterized by uncertainty and untestability,[18] does it remain appealing to the scientific and common sense and its constructs are treated in realistic terms? Here we come to the question of logical and philosophical requirements on constructs and their epistemic function in the particular case of cosmology.[19]

First of all, according to the very definition of a scientific enterprise it is assumed that it is to disclose some aspects of reality through an empirical contact with things, thus phenomenalizing them and

18. Ellis, "Issues in the Philosophy of Cosmology," 1259, 1274.
19. Cf. Margenau, *The Nature of Physical Reality*, 75.

making them immanent to the enquiring subjectivity. In that case, when cosmology attempts to predicate the wholeness of the universe, it is obvious that immanent phenomalization is not possible or a priori incomplete because there is not and never will be direct empirical contact with that which is implied by the notion of the universe as a whole. This situation is well known in many parts of theoretical physics, where the predicated realities and objects are represented by abstract mathematical forms and their physical existence can be verified only in a mediated way through special experimental equipment. Contemporary critical realists assign to these realities physical existence.[20] In the case of critical realism, one believes that if the constitution of phenomena takes place at the theoretical level, there must be rules of *correspondence* between what is theoretically introduced and what is observed empirically. In this sense the difference between objects (empirical versus theoretical) and the extent of their immanent phenomenality is not of an ontological kind: in both cases they have a similar ontic status. In a way, if some empirical phenomena (in a physical sense) are modeled mathematically, both their empirical evidence and mathematical representation exhaust what they are aimed at (intentionality conditions that which appears). The correspondence principle, as a constituent of the intentionality pertaining to the natural attitude, guarantees the link between visible and observable on the one hand, and that which is logical, mathematical, and nonobservable on the other. But, since the mathematical and nonobservable enter as constituents of the correspondence rules, it is clear that these rules

20. See, for example, P. Allen, *Ernan McMullin and Critical Realism in the Science-Theology Dialogue* (Farnham, UK: Ashgate, 2006). According to W. H. Newton-Smith's terminology, "One who gives a realist construal of all scientific sentences will be called a global realist" (*The Rationality of Science* (London: Routledge, 1996), 29. A claim for a radical mathematical realism identifying mathematical constructs with physical realities can be found in M. Tegmark, *Our Mathematical Universe: My Quest for the Ultimate Nature of Reality* (London: Allen Lane, 2014), chs. 10, 12.

are not only enforced by data, but in many ways by the internal consistency of the facts and constructs based in these rules.

The situation becomes crucially different if the correspondence principle does not work. This happens in cosmology where theory attempts to predicate (on the basis of extrapolation) something about the long distant past of the universe with no hope of verifying theoretical constructs related to this past by means of direct observations in this past. In this case, even if mathematics is applied for modeling some aspects of the universe, there is always a possibility that this mathematics is incomplete and historically contingent; that is, it will be eventually replaced by something new, so that the theoretical vision of reality (its construct) will change. Indeed, if one speculates about the past of the universe on the grounds of a simple physical causation, one can assert that the varied display of the cosmos (which reaches us through light and other channels of physical information) is the remote consequence of some originary event that took place in the past (and cosmology attempts to predicate in theoretical terms) and is beyond the conditions of observability. The frozen image of the past of the universe is given to us through its display here and now, but the sense of this past can only be conceived in certain limits, because this very past is "present in absence." Cosmology attempts to phenomenalize this past through its theories, which are incomplete and constantly corrigible. In other words, any attempt to reduce our knowledge of the past of the universe to the limits of the constituting subjectivity of a cosmologist is a reasonable but never-ending, inexhaustible enterprise. Then the question arises: What is the sense of a theoretical modeling of the past in terms of constructs, with respect to which the correspondence principle (as related to a fixed temporality of objects) does not work? The answer comes from an intuition that theoretical cosmology (not observational astronomy and astrophysics), *de facto*, bases its methodology not on the correspondence principle, but in the *coherence of epistemic justification*. E. McMullin argues that cosmology, as well as other disciplines that attempt to reconstruct the past (such as geology, paleontology, or biology) rely on *retrodiction* as that foundational principle which can bring into focus the past of the universe. The acceptance of this principle "is due to the cumulative success of the historical sciences, of geology, of paleontology, and of evolutionary biology. Success is not measured here as it might be in physics and chemistry but is a matter of *coherence* rather than of novel prediction. The *coherence* lies not just in the particular historical reconstruction of a long-past geological or biological episode but in the ways in which one reconstruction supports another, and the scope of the concepts and explanatory concepts on which the reconstruction is based gradually widens."[21] Here, however, we have to face an ontological question as to what extent the implied *coherence* of justification entails truth. For McMullin, who associates retrodiction with a realistic methodology, this entailment is paramount because "when reconstructions of quite different sorts of evidence drawn from geology and evolutionary biology, say, begin to 'jump together,' as it were, begin to blend fairly harmoniously into a single story, then our conviction grows that the story is not just coherent but is also close to truth."[22] Certainly there remains a question whether the experience of dealing with geology and biology is so easily transferable toward cosmology, in particular toward what is concerned with the origin of the universe as a whole, but not only of that part of it where we find ourselves.

The legitimacy of invoking coherence theories of justification in cosmology proceeds from the fact that cosmology starts its discourse

21. E. McMullin, "Long Ago and Far Away: Cosmology and Extrapolation," in *Bang: The Evolving Cosmos*, ed. R. Fuller (Saint Peter, MN: Gustavus Adolphus College, 1994), 120.
22. Ibid.

with a set of propositions, which can be qualified as beliefs, rather than verified assertions. We have discussed above that physical cosmology is only possible under the assumption that there is a fundamental uniformity in the universe in space (as well as in time, in what concerns physical laws). As we have mentioned above, the "cosmological principle" cannot be empirically verified. Let us now analyze carefully how this basic belief enters all cosmological constructs and makes the whole theory coherent. In order to avoid repetition and simplify visual perception we introduce in Figure 3 a simplified graphical representation of the epistemic structure of the construct of the universe in classical cosmology, produced in analogy with the analysis of constructs in physics by H. Margenau.[23]

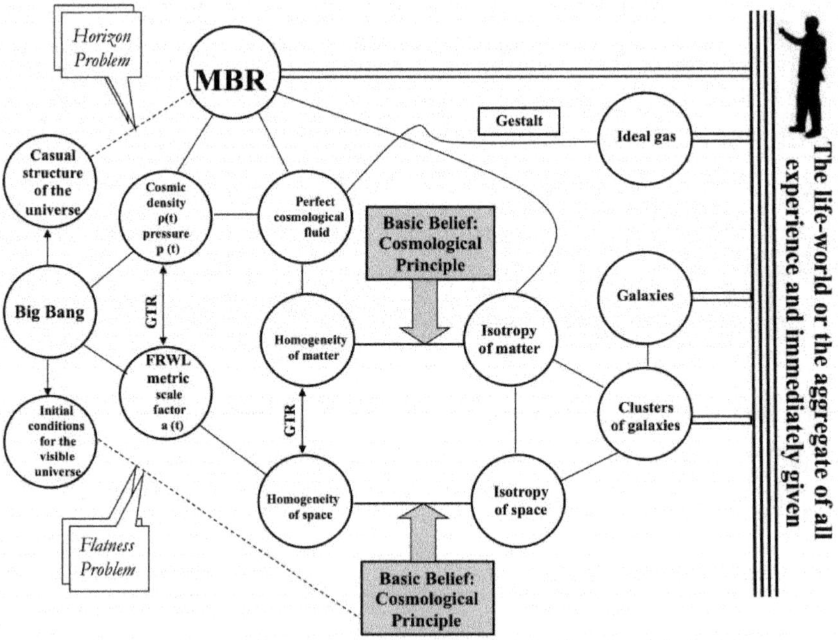

Figure 3

23. Margenau, *The Nature of Physical Reality*, 84–88.

At the right-hand side of this graph one finds a representation of a sensible world, the world in which human embodiment takes place and which determines in physical and social terms that aggregate of experience which, by borrowing phenomenological terminology, is called the life-world. To the left from the life-world the reader finds a space of knowledge that is populated by major epistemic elements. The circles represent constructs. The shaded rectangle illustrates a basic belief in cosmology. All constructs are linked by single lines that illustrate formal connections between them. Some constructs are linked to the life-world through double lines that represent epistemic connections. One sees that many cosmological constructs, in accordance with what we discussed before, do not have epistemic connections with the life-world, that is, immediate observations and measurements. However, all constructs are connected, creating a coherent volume of theoretical knowledge. Their coherence and the very possibility of this knowledge depends on the basic belief in the uniformity of the universe, whose efficacious presence is depicted through a box in the center of the diagram. This belief makes it possible to proceed beyond the contingencies of observations from a particular location in space to space as a whole, which itself represents a construct because, in spite of its obvious presence in the conditions of the life-world, is not present in its entirety and is thus sensibly unavailable, being supplemented by the mental construction.[24] To give this intuition of space a physical content, one has to postulate that one can shift our home-place and potentially experience a similar structure of space everywhere. The most natural attribute of this shift is a simple spatial translation that presupposes the uniformity of the overall structure. This presupposition forms a basic belief, that is, the cosmological principle, which allows one to apprehend the

24. E. Ströker, *Investigations in Philosophy of Space* (Athens: Ohio University Press, 1965), 176–224.

totality of space physically. This belief is exercised in the natural attitude, thus positioning all shifted home-places as physically real.[25] A similar belief asserts that the distribution of the material content of the universe that is observed astronomically from our home-place is not only isotropic for us but for all possible shifted home-places. This entails the overall uniformity of matter across the visible and invisible universe. The link between space and matter is established through General Relativity Theory's assertion of the relationality of space and matter. Correspondingly, both constructs, the global isotropic and uniform metric space, as well as the perfect cosmological fluid consisting of clusters of galaxies, are based in the certainty of belief. The fundamental role of the cosmological principle is that being implemented, it allows one to use the formal connection between space-time structure of the universe and its material content across the global structure of the universe. The choice of the physically motivated equation of state (for example, the equation for dust in present-era cosmology) in the cosmological perfect fluid allows one to develop a formal connection between the constructs of the energy density of the cosmological fluid and the universal scale factor $a(t)$, which in turn introduces new constructs. These formal connections follow from the Einstein equations and lead to the conclusion that since the scale factor grows in terms of the metric time that the universe expands. It is this last connection which leads to the introduction of the notion of the hot (radiation-dominated) universe if the expansion above is reversed. Through a limiting procedure when the cosmic time tends to zero, another construct of the beginning of the visible universe (the Big Bang) is introduced. This construct as such represents a limiting reference point with respect to all other possible constructs. For while the construct of the Big Bang

25. See P. Kerszberg, "The Phenomenological Analysis of the Earth's Motion," *Philosophy and Phenomenological Research* 48, no. 2 (1987): 177–208.

is the highest possible term of cosmological theory to which ultimate aspirations are addressed; physically it is supposed to be treated as that initial point in the state of the universe which is responsible for all other physical effects. On the one hand the Big Bang becomes the goal of the explanatory process; on the other hand, physically it corresponds to the original foundation from which everything unfolds. One must remember here that in order to draw a conclusion as to the Big Bang being the all-encompassing "beginning" of the universe as a whole, one needs to have a basic belief that the universe is uniform. In this sense all constructs presented at the very left-hand side of Figure 3 are imbued with this belief and hence represent the formal constructions in rubrics of belief.

If we generalize the latter observation one can state that the depiction of the universe as a whole through the diagram in Figure 2 (which contains in encapsulated form all formal connections among the cosmological constructs from Figure 3) represents a generalizing construct that is deeply dependent upon the basic belief in the uniformity of the universe. It is not difficult to realize that the epistemic connection between the construct of the universe as a whole and the life-world exists only through the top point in Figure 2, describing the observer and, correspondingly, along the past light-cone (that is, the visible universe), whereas all other parts of the allegedly existing spatial structure of the universe are only in formal connection with the construct of the visible universe. These formal connections are possible only through the basic belief that the structure of the universe outside the visible realm can be potentially comprehended by some hypothetical observers similar to that one of the visible universe. However, this potentiality as an eidetic variation of home-places does not actualize its physicality because the similar *necessary* conditions for the observers to exist outside the visible universe do not guarantee their actual existence, that is, the fulfillment of the *sufficient* conditions. In other words, the hypothesis of the similarity of the necessary conditions for embodiment in other places of the universe (related to similar physico-biological conditions) does not entail automatically the fulfillment of the sufficient conditions for such an embodiment, that is, for the actual existence of observers, thus remaining no more than an eidetic intuition, concerning realm of the intelligible, that is, a belief.

One must however admit that the resulting picture of the universe (recapitulated differently in Figures 1–3) is in a high degree epistemically coherent because its constructs align in an aesthetically attractive theory through multiple connections that can be easily seen from Figure 3. Apart from the construct of the Many Worlds (multiverse), which is usually invoked for explaining away the problem of contingent facticity of the initial conditions in the visible universe, all other constructs are connected with each other and have some epistemic references related to the life-world. The construct of the multiverse, which, by using terminology of Margenau, can be called peninsular,[26] is linked to the construct of the visible universe through mental causation, that is, on the level of intentionality and not physical causality. In this sense its status is crucially different from all other constructs that in one way or another have some epistemic connections. The construct of the multiverse in this sense requires another sort of belief in the possibility of shifting of home-places, this time not in terms of space and time but in terms of different types of worlds (universes) or types of being. Here one can detect an idea similar to that of the cosmological principle, that is, the idea of "democracy" among the worlds as a generalized principle of indifference that removes all particular specificity of our universe. The major problem with the construct of multiverse is that it does

26. Margenau, *The Nature of Physical Reality*, 86.

not stand in any realistic sense of causality with other constructs in the visible universe. Even if the visible universe is thought as one out of many members of the multiverse, it is completely unclear as to how to describe in terms of real physical processes the phenomenal actualization of this particular universe, that is, the phenomenal facticity of our universe. Since any imagined mechanism would, by default, transcend this universe, its verification is impossible, leaving the whole conjecture to the field of beliefs. In this sense the construct of the multiverse does not fall under the rubric of epistemically justified beliefs at all. It lacks the coherence in agreement among members of the cosmological community and in this sense it remains hypothetical and problematic.[27]

Unlike peninsular constructs such as multiverse, which by themselves cannot have any direct relation to the life-world, the construct of the visible universe possesses a heuristic quality of predicting some new properties of the universe that are subject to empirical testing. On the one hand we have an epistemic coherence among different cosmological constructs that follow from their mutually dependent nature under the assumption of cosmic uniformity. On the other hand there is a certain percolation of this epistemic coherence toward coherence of truth by predicting new epistemic links with the life-world. An interesting historical example of this is the prediction of the cosmic microwave background radiation (MBR) as a remaining matter ingredient from the early hot stage in the universe's evolution which was detected in 1965 (in Figure 3 the construct of the MBR has an epistemic connection with the life-world). According to the theory, the MBR represents a newly predicted construct that turned out to have (through technology) epistemic connection with the world of experience. However, even

27. See a strong critique of the concept of multiverse by G. Ellis, "Does the Multiverse Really Exist?," *Scientific American*, August 2011, 38–43.

in this case one must be cautious in asserting the correspondence between the theory which predicts the Big Bang and the observable MBR, because the latter is interpreted (through constructs) as the remote consequence of that which is asserted as physically existent in the past. We are unable to verify all details of the cosmological scenario by making direct experiments that reproduce in any feasible physical sense that long-gone past.[28] In this sense the predication of the past takes place on the basis of coherence of constructs-beliefs, coherence that is supported by the communal convention in established cosmology. One can argue that the very sense of the past is established from the present, so that one cannot affirm this past as physically existent on the grounds of correspondence with the present (in spite of an obvious temptation to use analogy with other historical sciences, such as geology or paleontology). The correspondence between the observed phenomena and their preexistent past takes place on the level of intentionality, but not that of physical causality when the past is assumed to exist in a sense different from what the universe displays as being de facto the image of the past. In this sense the discovery of the MBR becomes a signifier of that which allegedly took place in the past of the universe but as such does not exhaust the whole content of what is signified. The discovery of MBR does not change the status of cosmological theory of the past whose truth is asserted through epistemic coherence; it just strengthens this coherence by referring one of its signifiers to the reality of the life-world.

The important requirement for constructs is their extensibility: indeed any working theory cannot be static and involves growth of new elements. But this extensibility as a fact of scientific process

28. As expressed by G. Ellis in one of his theses on philosophy of cosmology: "The universe itself cannot be subjected to physical experimentation. We cannot re-run the universe with the same or altered conditions to see what would happen if they were different, so we cannot carry out scientific experiments on the universe itself" ("Issues in the Philosophy of Cosmology," 1216).

can or cannot depend on the formal connections among constructs. In some cases it arises on a so-to-speak meta-empirical level when theories bring forward some puzzles and paradoxes that do not contradict observations but disturb the consciousness of physicists who are not satisfied by their sheer presence and want to get rid of them. A famous example of such an extensibility is the extension of the hot Big Bang cosmology toward the so-called inflationary cosmology as a reaction toward three cosmological problems: of flatness, horizon, and monopole. Two of these problems are mentioned in Figure 3, and formal links with the results of their theoretical resolution in inflationary cosmology concerning the causal structure and initial conditions in the universe are also mentioned.

Coherence of Epistemic Justification in Cosmology

Now we would like to articulate with more precision what is meant by epistemic coherence in general and why it seems plausible to conjecture that cosmology follows this route in justification of its theories. If one proceeds in cosmological study of the early universe beyond the observable limit toward the universe before the decoupling of matter from radiation, one has to hypothesize about entities and corresponding physical mechanisms that are not directly observable and sometimes not related to any known forms of matter.[29] These hypotheses, being abstractions from experience, in many ways function as intentional objects that by their function in theory can have no direct relation to that which is observed (they can also be described as metaphysical assumptions). There is an element of irreducible belief present in their invocation, which

29. One means here the inflaton field in inflationary cosmology, dark matter, dark energy, etc. See more on this in Ellis, "Issues in the Philosophy of Cosmology," 1208–11.

reflects the fact that cosmology is driven not only by strong logical connections following from established physical causality but from the intentionality of cosmologists who are inspired by intuitions about the unity of the universe encoded in its past. Since the initial conditions of the universe cannot be tested, any predication of this beginning must entail a certain epistemic justification that cannot be based on correspondence with the empirical reality. Since this predication does take place, it implies belief in the realities of what is predicated. The presence of such beliefs makes sense of the success of modern cosmology, its popularity and ability to preach about the universe, as if cosmology's truth would be the truth of really existing things. Indeed it is because the principle of correspondence cannot be employed directly in the cosmology of the early universe, that it implicitly bases assurance in its hypotheses and models in the *coherence* and mathematical rigor of its theories.[30] In other words, the justification of cosmological theories comes not from their direct reference to the observable facts, but through coherence of explanation, which is achieved by applying a set of mutually consistent and connected beliefs that aim ultimately to codify in terms of mathematics the fundamentally contingent display of the large-scale universe. The coherence theory of justification holds that a belief is justified to the extent to which the belief-set of which it is a member is coherent.[31] "According to the coherence theory, to say that a statement is true or false is to say that it coheres or fails to cohere with a system of other statements; that it is a member of a system whose elements are related to each other by ties of logical

30. See on the coherence theories of justification, for example, R. Audi, *Epistemology* (London: Routledge, 1998), 187–204. See also J. Bowker, *The Sacred Neuron* (London: I. B. Tauris, 2005), 118–48, in which the author persuasively argues on the importance of coherence considerations in science and religion as a different form of justification in comparison with the correspondence principle.

31. J. Dancy, *Introduction to Contemporary Epistemology* (Oxford: Basil Blackwell, 1989), 116.

implication."[32] In different words, what is at issue in a coherence theory is a matter of a proposition's relation to other propositions and not its coherence with *reality* or with the *facts of matter*. This is similar to Margenau's requirement for constructs to possess *logical fertility* and to obey logical laws: "It asserts little more than that they have relational meaning. But in no sense does the present requirement make it necessary for the proposition involving constructs to be materially true, to have an existential counterpart,"[33] that is, that they cohere with the facts of matter. Coherence theories of justification operate with propositions-beliefs, or constructs-beliefs. As we have already seen, all major cosmological constructs contain the presence of a basic belief in the uniformity of the universe. In inflationary cosmology the belief that there exists a material field Φ (inflaton) that drives the evolution of the universe during the very early period is invoked in order to construct a theoretical model of the exponentially growing period of expansion, which in turn solves some problems of the radiation-dominated cosmology, thus making the cosmological model even more coherent.[34]

However, the major problem here is that epistemic coherence does not guarantee that knowledge progresses toward truth. Justification can grow, but there is no criterion that it delivers truth: cosmological models can become more sophisticated and expanded, but there will still be a problem whether their advance guarantees any convergence toward that alleged reality which they aim to describe. In technical philosophical terms, this situation sounds as if there is no conduction from the coherence of epistemic justification to coherence of truth. It

32. A. R. White, "Coherence Theory of Truth," in *The Encyclopaedia of Philosophy*, vol. 2, ed. P. Edwards (New York: Macmillan, 1967), 130.
33. Margenau, *The Nature of Physical Reality*, 82.
34. See more details in A. Nesteruk, "The Universe as a Construct: Epistemic Beliefs and Coherence of Justification in Modern Cosmology," *Journal of Siberian Federal University. Humanities & Social Sciences* 5, no. 9 (2012): 1304–35.

is in this sense that a coherentist epistemology can be characterized as knowledge without a foundation of certainty. Since the coherence of epistemic justification in cosmology has to abandon the principle of correspondence with empirical reality and a foundation of certainty, it has to appeal to different criteria in asserting the truth of these theories. Cosmology, in what relates to radical mathematization, in similarity with the coherence approach, maintains that truth is accessible in the extralogical realm where all criteria of reasonability as its foundation do not work. For example, by insisting that there are many disjoint universes which comprise a totality, "cosmology of the multiverse" enters a certain contradiction with the main stance of existential philosophy: it predicates universes where no condition of embodiment is possible. In spite of the existential futility of such predications, which can have sense as no more than an eidetic variation of the *possible* in order to affirm the *actual*, cosmology finds a kind of "extralogical" justification for the existence of such universes. This extralogicality follows exactly from the fact that the *discursive entailment* is replaced by *beliefs*. But for beliefs to sustain the challenge of skepticism one needs a *communal*, that is, conventional argument.[35] This implies that epistemic justification in theoretical cosmology where the correspondence principle cannot be applied relies on the acceptance of certain ideas about the universe by a community.[36] The community of cosmologists then establishes the sense of truth of that which is inferred from a theory. The coherent system of beliefs in cosmology determines as justified all sorts of statements about the remote past of the universe. It is typical for the coherence theorist

35. In this sense the coherence of justification in cosmology works similarly to theology, where the catholicity ("sobornost") of the church acts as the guarantor of collective wisdom, which opposes any sort of ethical individualism in religion. See more details in A. Nesteruk, *The Universe as Communion* (London: T. & T. Clark, 2008), 211–19.

36. N. Rescher, *The Coherence Theory of Truth* (New York: University Press of America, 1989), 331–33.

not to be constrained to only that which one will someday be able to verify. The validity of cosmology's propositions about the past of the universe is thus not under obligation to be tested in any direct observations because the very reality of this past is established on the grounds of *coherence* of a certain set of *beliefs* about this past.

If cosmology relies on the coherence of its own statements it is enclosed in itself and cannot be assessed from an outside system of thought. Since there is no direct link between coherence of justification and coherence of truth, which naturally requires breaking out of the system of coherent suppositions, cosmology can afford to create as many theories allegedly explaining the origin of the universe as it wishes, without even a slight idea whether these theories correspond to truth. In fact, the question of truth is inappropriate in this context because everybody, philosophically honest, understands in advance that the fullness of truth of what concerns the foundations of the universe cannot be grasped through some fragmented theories. All references to correspondence with the available empirical material do not reach their aim, because the process of adjusting theories of the early universe in order to fit observable data is in a state of permanent advance, so that all theories, seen through the prism of the philosophically understood "certitudes négatives" with respect to any knowledge of the universe as a whole, seem to be metaphors and aberrations of the incessant human desire to know the universe. They also manifest a fundamental human incapacity to achieve this goal on the grounds of discursive thinking. In this case, the whole pattern of coherentist epistemic inference in cosmology functions under rubrics of the belief in the possibility of knowledge of the universe, *attempting* to express it through reduction to an object-like phenomenality. This phenomenality is limited and hence does not exhaust the sense of the universe that has been signified by it. Thus cosmological knowledge acquires the features of an apophatic enterprise retaining the ongoing possibility for further explication of the universe.

Now it is not difficult to conjecture along the lines of phenomenological reasoning that the communal nature of knowledge established on the grounds of epistemic coherence leads to the view that physical reality (and the universe) is a mental accomplishment[37] ("hypostasis of mental creations"[38]). Here a distinction is implied between nature as it appears in primary perceptual experience and nature-for-physicists, as an *ideal* limit of the allegedly convergent sequences of "images of nature" that are constructed in the course of history. Any particular articulation of what is called nature can be assigned the character of a historical event. The articulation of the past is thus an event within the life-world of a particular community, loaded with a sense of the community's lived past and of decisions to be made in the future. As P. Heelan points out, "It is not the case that every historical event is also an event of a scientific kind . . . , but when the local community is

37. See, for example, R. Walker, *Coherence Theory of Truth, Realism, Anti-Realism, Realism* (London: Routledge, 1988), 19. As the coherence theorist would say, the nature of objective reality is determined by the coherent set of beliefs about it. Independently of this M. Munitz, discussing whether the universe as whole can be discovered, suggests that it would be better "to say that the concept of the universe as a whole is a creative, constructive achievement, an invention, not a discovery." M. Munitz, *The Question of Reality* (Princeton: Princeton University Press, 1990), 141. (A similar idea of the universe as being invented is advocated in the book of L. Brisson and F. W. Meyerstein. *Inventing the Universe. Plato's Timaeus, The Big Bang, And the Problem of Scientific Knowledge*, Albany: State University of New York Press, 1995.) The fact that the universe as a whole is a construction of our thought can be inferred from a counterintuitive sense of what can be called its "existence." The term "existence" cannot be applied to the universe in a sense pertaining to ordinary objects available as their unity at hand through their pieces and moments. The existence of cosmic objects, such as clusters of galaxies, for example, is problematic because each galaxy in a cluster is seen by us at different time in its history: thus the cluster we observe is a mental construction. This mental construction, being referred to the embodied subjectivity, thus reveals all signs of its historical contingency. In this sense the whole construct of the "universe as a whole," being an ideal accomplishment still bears in itself some features of contingent formation.

38. A. Gurwitsch, *Phenomenology and the Theory of Science* (Evanston, IL: Northwestern University Press, 1974), 44. Elements of nature as "mental creation" also appeared in the terminology of A. Einstein. See his "Physics and Reality," in *Ideas and Opinions* (London: Souvenir, 1973), 291.

one of the expert witnesses, then the scientific data produced by that community are also historical events in relation to that community."[39] In his classical paper on phenomenology and physics, H. Margenau argued along the same lines that "physical reality" is best defined as the totality of all valid constructs and rules of correspondence. In this approach the universe is defined not as a static, but as a dynamic formation: ". . . the universe grows as valid constructs are being discovered. Physical entities do not exist in a stagnant and immutable sense but are constantly coming into being."[40] The reality changes with the flux of experience.[41] However, for Margenau the belief of many scientists in the convergence of the system of the entire set of physical explanations that would deliver them an ideal of their aspirations, that is, a unique and ultimate set of constructs for which we would reserve the name "nature" or "reality," is problematic because it is not capable of scientific proof.[42] Since this convergence assumes a sort of historical process, it implies the postulate of history which is not physics.[43] Historicity, according to Margenau, involves knowing what "arises through a union of a knower and his object of knowledge."[44] Thus the very ideal of "reality" independent of the process of knowledge seems to be dependent on factors linked to human existence, which develops the sense of history and defines

39. P. Heelan, "An Anti-Epistemological or Ontological Interpretation of the Quantum Theory and Theories Like It," in *Continental and Postmodern Perspectives in the Philosophy of Science*, ed. B. E. Babich, D. B. Bergoffen, and S. V. Glynn (Aldershot, UK: Avebury, 1992), 66.
40. H. Margenau, "Phenomenology and Physics," *Philosophy and Phenomenological Research* 5, no. 2 (1944): 278.
41. Margenau, *The Nature of Physical Reality*, 295.
42. Margenau, *The Nature of Physical Reality*, 76.
43. H. Margenau, "Physics and Ontology," *Philosophy of Science* 19, no. 4 (1952): 343.
44. H. Margenau, "Physical versus Historical Reality," *Philosophy of Science* 19, no. 3 (1952): 343. The fact that the ideal of science searching for the ultimate "reality" is historical by its constitution—that the concept of "objective nature" can only be a mental accomplishment—makes doubtful recent attempts to advocate for the radical mathematization of nature, in which the "final theory" would be free of human baggage (see, for example, Tegmark, *Our Mathematical Universe*, chs. 10, 12).

its goals. In this case the above-mentioned convergence of "images of reality" can have its source rather in a philosophical argument asserting the existence of a certain *telos* in the human research activity that drives this convergence to its fulfillment, but this argument exceeds the scope of scientific justification and is grounded in beliefs about humanity as transcending the certainty of nature itself (that is, a certain commitment to seeing humanity as made in the image of God). The situation in modern cosmology, where the ever-increasing set of theoretical constructs reveals the components of matter content of the universe that escapes physical description (dark matter and dark energy, for example), points exactly to the danger of idealization of a scientific description of the universe as ultimate and accomplished in an ahistorical sense: the more details we know the less we understand the whole. In this sense the ideal of convergence of constructs in cosmology remains no more than wishful thinking.

The point of view on the historical contingency of scientific research and thus the fundamental conditionality of its results and views of reality, which we exemplified above, raises the conviction that the statements of cosmology (with respect to realities inaccessible to any empirical verification) established on the grounds of coherence and logical fertility of its constructs cannot have truth-values independently of our verification and, because it is *our* verification, it can never be conclusive. In spite of an explicit belief of physical cosmologists in the possible convergence of these statements to a kind of truth that lies beyond our reach, at every particular stage of research the truth of what these statements deliver turns out to be contingent and incomplete, open to further exhaustion through research. To say that the verification of this or that statement in cosmology is never conclusive[45] is to say that although our assertion

45. Cf. Ellis's Thesis of Uncertainty: "Ultimate uncertainty is a key aspect of cosmology. Scientific exploration can tell us much about the universe but not about its ultimate nature, or even much

of this statement may well be warranted in the circumstances, our warrant for it is always defeasible: new elements of theory or insertion of new indirect data could always make the assertion in question unjustifiable at all. In the case of the lack of empirical verification the cosmological statement has no truth-condition independent of the capacity of the scientific community to recognize it as true. Thus the claim of cosmology for objectivity and neutrality does not hold. One can only speak about a weak objectivity that includes the transcendental conditions of establishing truth.[46]

There are two philosophical qualifications that can be made with respect to this last conclusion. For philosophers working in the natural attitude, such a position would raise some suspicion of being anti-realistic. If one denies the verification-transcendent truth (even in a weak form as an ideal of a convergent set of self-correcting explanations based upon the ultimate rules of correspondence) one effectively adopts an anti-realistic view that truth is not independent of our capacity to find out about it, or, in other words, to have beliefs about it in a particular context. The meaning of this "context" can be very different according to the field of research, starting from a simple sensual perception in an experimental science and finishing by a more sophisticated scheme, let us say in theology. Anti-realism emerges naturally in that particular modification of a coherence theory that does not think of the set of truths as a determinate totality; it is the case we discuss here: what we can recognize as true in cosmology is indeterminate and open-ended. Scientific truth in this approach is not that hypothetic uniqueness which transcends the conditions of knowledge, but is determined by the fundamental plurality of that which we are able to discover and recognize as that kind of truth.

about some of its major geometrical and physical characteristics. Some of this uncertainty may be resolved, but much will remain. Cosmological theory should acknowledge this uncertainty" ("Issues in the Philosophy of Cosmology," 1274).

46. See more on this in previous chapter.

If, however, one adopts a phenomenological stance in which any knowledge is possible only within the noetico-noematic correlation, the suspicion of anti-realism falls away, simply because the certainty of knowledge is immanent to the constituting consciousness, so that, by definition, knowledge of the universe cannot escape the conditions of its origin in a particular realization of consciousness (be it personal or collectively historical). The issue of the verification-transcendent in this case becomes a question on the possibility of retaining transcendence in knowledge that is immanent. The stance of the coherence theories of verification and truth, in particular in their anti-realistic versions, points toward the possibility of such a transcendence simply because it claims that the process of knowledge is intrinsically incomplete and open-ended, leaving the immanent discursive consciousness with an image of reality and some statements of its truth without any exhaustion of that subject matter which it aims at. In this sense the knowledge of the universe as a specific contingently historical process, based in many ways on the conventional agreements of the community of scientists, never exhausts the sense of the universe, or, the excess of intuition of the universe through communion over its knowledge through discursive reason. Indeed one can attempt to express the experience of admiration by the forces of the universe through very complicated mathematical theories, but all of them will remain no more than symbolic and metaphoric images of that anticipated unity and infinity of the universe which is present to the incarnate human subjectivity through belonging to it, through a partial consubstantiality with it. For example, since there is no empirical access to the alleged happening of the Big Bang, all that we express about it by using cosmological theories can be characterized as metaphors and esoteric symbolism based in mathematical formalism. The beauty of this symbolism, its coherence, give us some assurance to believe in the possibility of the Big Bang as a principle of explanation and justification. However, the "truth" of the Big Bang in an ontological sense remains unclear and, what is more important, fundamentally inaccessible. In other words, all cosmological theories give us some symbolic representation of that toward which they aspire (the universe as a whole or its encapsulated image in the Big Bang), but it is that which will never be known and reached in a sense of certain truth. The apophaticism in cosmological research is thus present as the limitation of thought: it wanders around the idea of the Big Bang, but it will never reach it as ultimate origin of the universe. In this case all competing theories are epistemologically and axiologically equal, but no one can pretend to claim the fullness of truth and the knowability of the Big Bang as that intended ideal which is implied in a hidden teleology of cosmological explanation. Thus all cosmological knowledge is apophatic in the sense of its limited validity determined by the boundaries of the physical, because of the open-endedness of the intended horizon and a fundamental inexhaustibility of truth about the universe by means of discursive thinking. However, in order to realize this fact, one should shift cosmologists' consciousness toward a phenomenological attitude, which is capable of bracketing all theoretical statements about reality and to conceive them as varieties of expression of the human intuition about the entirety and identity of the universe. But this attitude is simply not available to cosmologists themselves. They will never agree with the verdict of philosophy that all eidetic imagination in cosmology, incarnate in complicated formulae, is only a wandering around truth, but not truth itself. At the same time it is exactly the limited nature of our knowledge of the universe, its apophatic character, which makes it possible to render the *belief* in the transcendent other of all that we see in the universe, not as an ideal of convergent rules of correspondence with something that is out there waiting for our grasp, but, on the contrary, as that unobjectifiable givenness whose gaze upon us constitutes our subjectivity through the never-ending enquiry about the universe.

The last question we need to briefly discuss in the context of coherence of justification is the issue of mathematization. Namely, the question is: If cosmology predicates reality in mathematical terms following the rules of logic, does it leave any chance to claim that cosmology is still based on the coherence of epistemic justification with an inevitably following open-endedness of its justification? Mathematical truths are not beliefs but logical rules invariant to all situations and, as some claim, existing independently of the "human baggage." However, there is still a slot for a proponent of coherentism to claim that the usage of this or that particular mathematics in this or that particular physical theory is a matter of a human choice and thus of belief in the efficacy of mathematics. The choice of particular mathematical models in cosmology is dictated by basic beliefs. For example, the cosmological principle as philosophical statement entails the choice of particular geometries to describe the universe. The "initial methodological condition" for using these geometries is not in mathematics itself but beyond, in the belief of uniformity of space. As soon as a mathematical model is chosen, all further computations follow the "iron" logic of the algorithm, so that their realized "apodicticity" is still initiated by basic beliefs. In another example, when the scalar field in inflationary cosmologies (inflaton) is chosen as that hypothetical ingredient which is coupled to geometry and drives its evolution, the game of construction of a suitable potential and equations for this field is predetermined by this very choice. But this choice is not an empirically driven or correspondence-based rule; it is a conjecture in a certainty of belief that turns out to be very fertile in solving paradoxes of the standard cosmological model.[47]

To generalize, one can say that the use of a particular mathematical model and logic is determined by the human choice. If some aspects of the physical universe are mathematized and expressed in simple logic, the intuitive content of the notions in question is reduced to zero. However, it is the very choice of what logic and mathematics to employ that is driven by intuition whose excess, by definition, makes the advance of all knowledge possible. This implies that even a coherent mathematical description of some global aspects of the universe does not entail the exhaustion of the "phenomenon of the universe" through mathematics (mathematical signifiers do not exhaust the content of the universe as physical or existential outcomes of the physical laws). Mathematical simplicity and logical coherence are related only to those aspects of the universe which allow in principle their representation in the phenomenality of objects. Some proponents of extreme mathematization of the universe believe that if mathematical structure exists, it reflects the existence of a corresponding physical reality.[48] Even if this were to be true, the problem is that the existence of mathematical structure as articulated by human consciousness does not account for its own contingent facticity: it cannot transcend its own givenness. The self-explanatory justification of mathematics is not even possible if one gets rid of

47. See on the argument of coherence of epistemic justification in inflationary cosmology my papers: A. Nesteruk, "The Universe as a Construct: Epistemic Beliefs and Coherence of Justification in Modern Cosmology," as well as "Coherence of Epistemic Justification versus the Principle of Correspondence in Modern Cosmology," *Izvestiya: Herzen University Journal of Humanities and Sciences* (St. Petersburg) no. 156 (2013): 59–67.

48. A popular trend in philosophy of mathematics, which effectively advocates such a view, is so-called structural realism. It is enough to give a couple of references with respect to it: H. Lyre, "Structural Realism and Abductive-Transcendental Arguments," in *Constituting Objectivity: Transcendental Perspectives on Modern Physics*, ed. M. Bitbol, P. Kerszberg, and J. Petitot (Dordrecht: Springer, 2009), 491–501; a popularized version of structural realism can be found in S. Shapiro, *Thinking about Mathematics: The Philosophy of Mathematics* (Oxford: Oxford University Press, 2000), 257–89.

the human baggage present in mathematical articulations and thus postulates mathematics as noncontingent apodicticity. However, this suggestion is tantamount to the claim that mathematics is related to impersonal, anonymous, disembodied consciousness, whose very facticity remains an utter mystery.

Classical phenomenology could not pose the question of consciousness's facticity, for any transcendence of the field of consciousness was ultimately prohibited by the phenomenological reduction. In a new phenomenological development, mostly related to its merger with theology and other aspects of the human sciences, the question about the facticity of discursive thinking is being posed as the question about the possibility of retaining the intuition of the initial whole behind the discursive mode of thinking that dissects this whole. It is the question of retaining transcendence as that donating intuition which is not subject to immanent constitution and which is always behind the facticity of thinking. It is because of this intuition that the presence of the universe is never fully disclosed through logic and mathematics. The very contingent facticity of mathematics is seen thus as a part of the fundamental incomprehensible facticity of the universe. The universe manifests itself here: it exists, and it is given to us in its particular mode of phenomenalization through mathematics linked to the conditions of our embodiment.

Thus we have seen that coherence of epistemic justification in cosmology works at the level of intuition by ordaining all mathematical models that are employed by cosmologists for achieving their computational synthesis of the universe. In this sense the principle of epistemic justification becomes, in a way, a *maxim* of reason in a Kantian sense, rather than an analytical prescription. The further justification of this thesis will be given in chapter 6.

5

The Origin of the Universe and Event of Birth

Phenomenological Parallels

I cannot really stand aside from the universe, even in thought. Only by a meaningless pretence can I place myself at some vague point outside it, and from thence reproduce on a small scale the successive stages of its genesis. Nor can I place myself outside myself . . . and question myself upon my own genesis. I mean of course the genesis of my non-empirical, or metaphysical reality. The problem of the genesis of the I and of the genesis of the universe are just one and the same problem, or, more exactly, one and the same insoluble, the insolubility being bound up with my very position, my existence, and the radical metaphysical fact of that existence.

–Marcel, Being and Having, p. 24

Introduction: The Origin of the Universe in Modern Cosmology

It is a matter of common understanding nowadays that with all respect to the grandiose achievements of modern science and cosmology in particular, scientific discourse in general, being efficient in interpreting the most nonhuman aspects of reality, feels helpless in advancing the understanding of the mystery of human persons who are those agencies that make the universe palpable and act as its voice. This hidden predicament lies in the fact that cosmology is being constructed from within the *natural attitude* that alienates its subject matter, by making its object (the universe) devoid of any references to human subjectivity, thus removing insights into the conditions of that phenomenality which it deals with. The aim of this chapter is to suggest how this predicament could be overcome in a particular problem of the origin of the universe. In other words, how to interpret the Big Bang idea in the perspective of the interior life of a human person. Here we implement that possible strategy, proposed in the introduction, of dealing with cosmology as a mirror of anthropology.

It is known that the so-called anthropic inference in cosmology attempts to link the natural conditions of human embodiment with the fundamental physical parameters that are responsible for the stability and actual display of the physical universe and make it possible for conscious life to exist. However, what is articulated in the anthropic inference is physico-biological conditions of human existence: it makes a nearly trivial observation that, indeed, there is an underlying consubstantiality between human observers and some stuff of the universe expressed in some technical terms, such as, for example, the so-called fine-tuning of physical constants (whose precise numerical values are crucial for sustenance of earthly-based life-forms).[1] However, even in this case, the very facticity of this consubstantiality, its particular contingent noetico-noematic

1. See a now-classical book of J. Barrow and F. Tipler, *The Cosmological Anthropic Principle* (Oxford: Oxford University Press, 1986), as well as a recent collection of articles on fine-tuning: J. Barrow, S. Morris, S. Freeland, and C. Harper, eds., *Fitness of the Cosmos for Life: Biochemistry and Fine-Tuning* (Cambridge: Cambridge University Press, 2008).

givenness (fine-tuning), is not accounted for from within scientific method and, probably, cannot be accounted for at all. But it is this very facticity (that is, the facticity of the universe suitable for life as well as the facticity of life itself) which is ontologically responsible for the necessary conditions for multihypostatic embodiments of humanity, that is, the plurality of conscious subjects, a *plurality* that cannot be simply explained in terms of the natural. Embodied persons, as particular contingent realizations of anonymous and impersonal physical laws, disconcert cosmologists, who are eager to dismiss any trace of them as not falling under the rubrics of scientific objectivity and explanation: scientific objectivity does not deal with contingencies in spite of the fact that it is itself a mental creation of a contingent formation in the universe. Scientific objectivity cannot come to terms with the fact that the world, disclosed and constituted, is contingent in its immanence with this contingent formation—human consciousness.[2] The whole edifice of cosmology manifests the universe but conceals intelligence as its noetic pole. *The universe possesses in physical cosmology such a phenomenality as to show that the author of the scientific artwork of cosmology cannot be shown.* This view points to a philosophical link between the problem of the facticity of multihypostatic human consciousness and the facticity of the universe. Indeed, since all consciousness is intentional, the primary object of this intention in cosmology is the universe in which this consciousness embodies. Consciousness gazes at the universe as that premise which sustains the very possibility of consciousness. Hence, in the same way that scientific anthropology

2. Here one can observe the presence of two tendencies: on the one hand, physics and cosmology represent an example of eidetic reduction of the historical plurality to ahistorical physical laws (driven by differential equations with no intrinsic time); on the other hand, the internal temporality of consciousness, realized uniquely in different persons and thus making an ontological difference between them (as well as subjective time), are "contra-reduced," so that the cosmological time is considered as existing objectively and independently of the human subjectivity, as if it could be grasped in the natural attitude.

and psychology cannot tackle the problem of any singular hypostatic existence, cosmology cannot adequately approach the problem of the contingent facticity of the observed universe, that is, its particular phenomenality linked to human agents. The constructive and nontrivial content of the anthropic inference is exactly the demonstration, by using physical propositions, of the interplay between a fundamental contingency of the universe (*noematic* pole) and that embodied consciousness (*noetic* pole) which grasps the universe. However, this contingency as such remains an ultimate unexplainable mystery.

Still, physical cosmology persistently attempts to escape a philosophical verdict on the unknowability of the facticity of the universe and to *explain it away* by referring to the initial conditions in the universe as if they would contain all "information" about the present and future of the universe. Let us reflect upon this motivation. The objective of physical cosmology is to "explain" the specificity of the state of affairs in the empirical, observable cosmos. We have a display of different objects in the sky, different structural and physical phenomena, observed by various physical means. There are two aspects in this variety of the natural phenomena: on the one hand we have some species of similar objects (let us say galaxies or their clusters) that can be classified into groups; on the other hand there is a particular givenness of objects from these groups that display themselves in the cosmos in an absolutely contingent way (for example, there are some particularly famous galaxies associated with special directions in the sky: an obvious example is the Andromeda nebula, which is seen by us through the constellation of Andromeda). On a smaller scale there are particular patterns of stars in our galaxy, which since ancient times have been grouped into constellations, whose display in the sky is contingent and given to us as it is. Another example of astronomical contingency is our planetary system, which contains nine planets of particular sizes and parameters of their orbits. Since Kepler and later Newton the movement of planets is explained in terms of periods of revolution around their orbits in dependence on their spatial distance from the sun using physical laws, but astronomy cannot explain the fact as to why there are nine planets (instead of twenty-nine, for example) and why the concrete parameters of their revolution around the sun were set in the order that is observable now. In other words, neither astronomy nor physics can account for the initial conditions of the movements of various contingent objects in the universe in order to explain the specificity of their display in the sky (certainly a strong anthropic explanation from the fact of life can always be invoked: there must be necessary conditions in the very beginning of the universe that would allow life to develop). Said philosophically, the universe displays the contingent state of affairs on a huge spatial scale whose facticity and origin cosmology attempts to interpret, including the very physical laws that are employed in it.

In prescientific eras "explanations" of the variety in the cosmos were given in mythological terms.[3] In modern terms, the idea of mythological explanations amounts to an attempt to reduce the variety of contingent objects in the sky and peculiarities in their motions to a minimum, by representing the universal qualities of the cosmic display through symmetries and harmonies and by making all contingent happenings as derivatives and spontaneous outcomes from the rule.[4] A characteristic feature of such an attempt is a

3. A concise expression of what is meant here can be found in J. Ladrière, *Language and Belief* (Dublin: Gill & Macmillan, 1972), 153: "The cosmological myth is an account of cosmogenesis. It tells how the world was made, how the contemporary world that stands before our eyes developed from what went before, from the nonworld, the formless. . . . The schema of the representation is a successive unfolding in which there is a movement from homogeneous unity to a qualitatively differentiated multiplicity . . ."

4. One can refer, as an example, to the cosmological principle as a principle of nonobservability of contingent deviations from uniformity. All observable nonuniformity such as our solar system

conviction that the underlying "world" of highly symmetric (Platonic) forms is actually coexisting with the empirical display, but is not perceived through the senses. This underlying world was treated as an ever-existing reality, with no change and hence with some stable universal patterns so that the stability of the universe as its identity in the background of the flux of mundane things was secured through reference to these intelligible patterns (Platonic forms).

What makes contemporary scientific cosmology similar to and at the same time different from mythological cosmologies is that, as a scientific discipline, cosmology can function only on the condition that its "object" (the universe as a whole) sustains *identity* in time. Indeed, according to Kant experience is possible only if time is involved as an element of unity and synthesis. However, in cosmology, this is not an innocent requirement: for example, in the case of a static and infinite universe (that is, by assumption, existing forever) the identity of the universe in time, is a tautology, for there is, *de facto*, no objective physical time in such a universe—it is static. Correspondingly, the observable universe (because of the postulated finitude of the speed of light) would be an infinitesimal part of an incommensurably bigger whole that is principally beyond empirical grasp. In this case only could one legitimately talk about the identity of the visible part of the universe and infer (in a leap of faith) to the identity of the whole static and infinite universe. This in turn entails that the facticity of the observable universe would be the first and last principle of its explanation, for it would not be possible to refer to any other state of the universe considered as an originary cause or an origin of the universe as we see it.

The situation changes drastically in a standard cosmological model based on the idea of the universal expansion. The universe is

of the Milky Way exhibit the breakdown of that symmetry on a much lesser scale than clusters of galaxies.

expanding and the idea of the evolution of the universe is taken exactly as that time-synthesis which is needed in order to preserve the unity of the universe as a whole (it provides us with images of the universe at different moments of time as a consequence of the finitude of the speed of light). The fact that the term "past" can be assigned to the universe as different from the everlasting present of the static universe (in a static universe the past is associated with aging of the light signals but not with the evolving nature of its space related to the material stuff) is exactly related to the principle that there was an ultimate *temporal* beginning of the universe from which all its stages emerged, and because of that one can observe the universe backward in time and in the entirety of its temporal span (excluding its dark ages before the separation of matter and radiation took place).

The ancient idea of the underlying eternal world becomes replaced here by another, but similar idea. This time cosmology finds refuge in saying that the present state of the observable universe (with its annoying contingent facticity) is the frozen instantaneous display of the temporal evolution of this universe from some remote initial state which in its seeds (through the laws of physics implanted there) contained the potentiality of all various features in the cosmic display. This kind of thinking naturally invokes the idea of the originary "past" of the universe whose partial consequences are displayed, a "past" as such that is not observable but "existed" physically long before that display which humanity looks at and reflects upon. What is also believed (and this is in some way supported by theory) is that this initial (early) state of the universe, in terms of physics and its logical presentation to human minds, is much "simpler" than the grandiose variety of different spatially and temporally disjoint objects displayed in the present universe, because the initial condition is an idiosyncratic state of matter with either infinite or uncertain parameters. In spite of this uncertainty the motivation of cosmology remains the same: to explain away the contingent variety of objects and their species in the universe (as seen in the perspective of their "present") by "sweeping these contingencies under the rug" of their undifferentiated unity in the "past" of the universe.

The hope that such an attempt will be successful suffers from two logical flaws that can be elucidated through philosophical analogies. First of all, a theoretical attempt to reduce all variety of forms and structures in the universe to rubrics of the allegedly existing undifferentiated unity (which could play a role of a fundamental *substance*) does not make this unity visible. One infers to this unity from a premise that it is accessible to us only remotely through its differentiation and particularization in the course of time. It is not difficult to see that such an attempt to pronounce on the underlying unity of the universe is reminiscent of the ancient ambitions to claim that the substance of the world, that is, the universals of the world, are in *water*, *fire*, *apeiron*, and so on.[5] Cosmology silently follows the same route of thought by postulating such a primordial substance at the beginning of the universe that is ultimately responsible for the variety of objects available in the cosmic display. However, here, there is a seeming difference with an ancient view: *water* of Thales or *apeiron* of Heraclitus were abstract notions but allegedly coexisting with the empirical display of the world. In this sense one could say that there were no temporal extensions within these substances. A rather contemporary way of saying *water* and *apeiron* could correspond to a microscopic level of reality, but *here* and *now* so that there is no "evolutional extension between universals and their empirical incarnations." Unlike this, physical cosmology places the undifferentiated substance of the Big Bang in the past universe,

5. On the analogy of the ancient Greek quest for the underlying matter and contemporary physics see G. Feinberg, "Physics and the Thales Problem," *The Journal of Philosophy* 63, no. 1 (1966): 5-17.

assuming that there is a time-like extension in this substance which ultimately leads to the present display of the universe. (There is a reversed assertion: whatever direction we observe deeper and deeper in the sky, we ultimately "observe" the Big Bang, that is, its predicted outcome—the microwave background radiation.) Thus the difference between mythological and contemporary views of the world in terms of universals is related to the treatment of extension inherent in the underlying substance: spatial (in ancient cosmology) versus temporal (in a modern one). However, such a perception of the difference comes from a scientifically based conviction that when one looks at the sky of the universe one looks at its remote spatially and temporally extended past: this is implied by causality based in the finitude of the speed of propagation of light in the universe. But, and this is important, since the size of the visible universe is decreasing if one looks backward in time (due to its reversed expansion), one can also say that by "looking" (not optically, for light could not travel before the universe cooled down and split from matter) at the Big Bang we are "looking" (that is, are making insights) deeply in space (that is, into the microscopic scales of space as we do through microscopes or in experiments in nuclear physics). In other words, one can say that a frozen display of the universe represents a geometrized passage of time. In this sense one can further conjecture that what is called the evolution of the universe in time from the initial Big Bang is effectively the unrolling of the universe through different scales in space. (One must pay attention to the fact that this unrolling is driven by dynamical laws that do not contain any intrinsic historicity, so that this unrolling is rather to be called not evolution but dynamic development determined by the initial conditions.)

If the idea of the undifferentiated substance is laid in the foundation of the visible universe, then the problem becomes of how to explain in general the process of differentiation in the universe, that is, generation of basic elements and structural units that form varieties of things in the universe. Still this would not help too much in terms of individual contingent things. By understanding that individual things are not subject to scientific descriptions, cosmology intends to get rid of this individuality at all by saying that what is really interesting is to understand how a class of impersonal objects emerged from *something* undifferentiated and homogeneous. One speaks here not of contingent individual exemplar objects/beings, but of species of objects unified through realized physical forces (for example, not of this or that galaxy, but of galaxies in general; not of this or that planet, but planets in general; not of this or that human person, but of human beings in general).

The idea of the Big Bang, as the idiosyncratic state of the universe in its asymptotically distinct past in which nothing was similar to that with which we deal in physics, is considered as the prototype, as the seed of all possible things in the universe that develop from this Big Bang. The problem, however, is that even if we describe the Big Bang in terms of a physical state that has been mathematically constructed as a backward-in-time extrapolation of what we have in the universe now, the Big Bang turns out to be merely a very clever *disguise* of that which we experience as the now of the universe. In the hypothesis of the Big Bang, the contingent facticity of the universe as it is observed here and now is reinterpreted in terms of a certain temporal origin in which the undifferentiated "substance" was "set up" in such a state as to evolve into the visible universe. The procedure of "naming" this initial state is supposed to play the role of disclosing the universe's identity and hence acts in thought as a disguised name of its present-day displayed facticity. It is not difficult to comprehend, however, that the problem of contingent facticity of the universe can only be *explained away* by this type of reasoning through referring to the initial conditions, because there is no way to *explain* the contingent facticity of these initial conditions themselves.

Indeed, since every cosmologist works under the assumption that there is continuous physical causation in the universe, so that there is a chain of causal explanations of what happened in the universe if we extrapolate its behavior backward in time, it is not difficult to realize that the equations that drive the universe backward in time, in fact, effectively encode the variety of existing objects in the spatial display in the astronomical universe into the same variety extrapolated backward in time, which is now treated as the "initial" condition for those physical states from which it has been backwardly constructed. However, any hypotheses of the facticity of these initial conditions in the universe remain no more than hypotheses with no chance of their instatement to the status of laws. Being hypothetical, these initial physical conditions of the universe acquire some particular imagery in different models.

Let us discuss, as a purely illustrative example, an imagery presented in a famous idea of S. Hawking: that the universe was in a quantum state in the so-called past and did not have any point of origination. The universe was in a space-like state where all temporality, associated with the flux of time and irreversibility, was suspended.[6] This idea was a counterreaction to the representation of the space-time structure of the early universe as a curvilinear cone, with an apex that is usually associated with the Big Bang.[7] In this

6. According to Hawking, temporal flux is a "figment of imagination," so that the "real" underlying world is atemporal, that is, either trans-temporal or simply ever-existing. S. Hawking, *A Brief History of Time: From the Big Bang to Black Holes* (London: Bantam, 1988), 139. Here we have an example of dismissal not only of the internal time consciousness, subjective time, historical time, but also the objective physical time. Temporality as a basic category of the world of living beings is eliminated.
7. The detailed explanation of this can be found in C. J. Isham, "Creation of the Universe as a Quantum Process," in *Physics, Philosophy and Theology: A Common Quest for Understanding*, ed. R. J. Russell et al. (Vatican City State: Vatican Observatory, 1988), 375-408. See also A. Nesteruk, *Light from the East* (Minneapolis: Fortress Press, 2003), 134-41.

representation the universe has a boundary as its temporal origin whose nature cannot be accounted by the physics that is suitable for the interior of the cone. In other words, the initial conditions at the Big Bang are not subject to the laws of physics within the universe and therefore some metaphysical assumptions must be made on the nature of their contingent facticity. This "incompleteness" of physics which requires philosophical inputs has been recognized by physicists themselves and was the leading motif for Hawking to remove it by suggesting sophisticated theories of how to avoid temporal origin in cosmology at all. This has been done by Hawking through a mathematical trick of replacing the temporal variable in equations for the wave function of the universe (which is usually associated with the empirical flowing time), by an abstract imaginary quantity that behaves like space ("imaginary" refers here to a special type of complex numbers).[8] The aim of this quantum cosmology is to announce that there was no time in the early universe; the universe did not have a point of temporal origination: it just exists endlessly, from eternity. It exists as compact four-dimensional space, which is, however, contemplated by human observers (time is a figment of imagination, according to Hawking) in its "remote consequences" as a three-dimensional spatial structure evolving in physical time. The temporal original boundary of the universe is removed: thus there is nothing beyond the universe either in terms of space or time; correspondingly, there is no need to appeal to any transworldly factors responsible for the facticity of the universe so that the universe is fully graspable by the human intellect, which in this case is just isomorphic to it.[9] Certainly this type of scenario is fundamentally untestable. One could raise a question as to whether the mathematical

8. Hawking, *A Brief History of Time*, 139. The physical meaning of imaginary time is completely unclear. For Hawking, however, it was not a problem at all, for, according to his suggestion, "the so-called imaginary time is really the real time, and that what we call real time is just a figment of our imaginations."

beauty of this theory and its intrinsic coherence, as an explanatory device, could provide a justification for this model to reflect "truth." However, scientists (as well as philosophers and theologians) were very much influenced by this model of the universe, because, in a way, the issue of contingency of its *temporal creation* (not creation in a sense of *ex nihilo*) was "explained away" whereas the facticity of the very "initial conditions" of the universe described through the "nonboundary" definition, or in terms of preexisting Euclidian space, was not addressed and cannot be addressed at all.[10]

Assessing Hawking's model philosophically one can suggest only two interpretations: either his scenario is purely Platonic, reproducing an ancient model of frozen time or eternity, and, correspondingly, ordinary time as an image of this eternity,[11] so that the quantum universe represents an intelligible "substance" with no relation to physical reality whatsoever, or alternatively, it provides us with a new incarnation of the old Greek idea of substance (the wave function of the universe is this substance), which lies in the foundation of all possible differentiations among objects. In this latter case the appeal to the "past" of the universe corresponding to its quantum stage does not have any sense: as having no boundaries and no temporal flux it cannot have any "past": past arises only as a figment of imagination. The question then is not about the *temporal origin* or *origination* of the universe, but about its *originary foundation*,

9. The apology for such a cosmology led Hawking to the dismissal of the idea of God as creator, in his famous phrase "What place, then, for a creator?" (ibid., 141). This is in amazing correspondence with that which was predicted, and criticized, by E. Husserl in his *Crisis of the European Sciences*, namely that the extreme mathematization of nature makes human beings to believe that they rule their own affairs in the universe by believing that the universe as it is described mathematically is ultimately true, so that they replace God. E. Husserl, *The Crisis of European Sciences and Transcendental Phenomenology* (Evanston, IL: Northwestern University Press, 1970), 66.

10. See a detailed philosophical and theological analysis of Hawking's ideas in Nesteruk, *Light from the East*, 141–59.

11. On a neo-Platonic interpretation of Hawking's model see Nesteruk, ibid., 145–52.

that is, the foundation of its facticity as having nothing to do with its alleged origination in physical time. The origin is meant here not in a temporal sense, but as an ultimate cause of this universe, similar to the absolutely necessary being that was discussed by Kant in his fourth antinomy. In this sense any referral to the origin of the universe in terms of its "past" again loses its sense. This so-called past turns out to be no more than a fiction. Correspondingly, the evolution of the universe has an absolutely different sense: it reflects not so much physical processes as if they had taken place in the past, but a temporal representation of the underlying structure of the universe by means of intersubjective temporality, which in turn originates in internal time-consciousness. According to Hawking this temporal representation is physically initiated by a transition from a quantum state of the universe to its present empirically contemplated condition.

The Originary Foundation of the Universe's Temporality and Consciousness

One can make some generalizations on the meaning of the origin of temporality in the universe which is invoked in above-mentioned models of the initial conditions. Cosmology makes an assumption that time can be asserted through conscious acts as an attribute of the world constructed by physics, that is, as that type of reality which has been in place prior to the human embodied intelligence in the universe (the time of the universe is asserted as nonlived, abstract time). Then the alleged "origin" (as its originary foundation) of physical time is also thought in similar naturalistic terms. However, this origin is obscure not only in physical terms, for physics can deal only with the already temporal; it is also obscure in the perspective of the intersubjective temporality of consciousness itself. Indeed, if consciousness is embodied in the elements of the universe, any thinking of the origin of the universe must implicitly contain hints on how to think of the "origin" (originary foundation) of consciousness itself and vice versa, that is, the thinking of the "origin" of consciousness must implicitly contain hints on how to think of the origin of the universe.

And here we have to face a simple philosophical fact that consciousness (either related to the whole of humanity or to an individual) cannot deal with its own "origin," for it cannot stop the flow of intentionality and make an introspection upon itself from a perspective of non/preconsciousness. The temporal flow of consciousness is characteristic of human life, so that to exit it in order to "find" its pre/a/trans-temporal origin is not possible. In phenomenological terms, one has no access to the phenomenality of one's own conception or birth, that is, to one's coming into existence (as internal time-consciousness) from that "nonexistent" which is one's originary foundation. Detached from our originary foundation, the characteristic feature of its partial, a posteriori phenomenality is exactly that it does not show itself: it is present only as an efficacious origin of all states of life. Correspondingly, any constitution of such an originary state is an act in the already-existent consciousness, that is, in the present, in the context of the present. Thus the phenomenology of the originary foundation can only be established through the constitution of this "origin" through acts and insights taking place in the present. This raises a question on the sense of the reality of a thus constituted "origin": Is this "origin" indeed in the temporal antecedents belonging to the sphere of the already unconcealed (and thus it is indeed the origin *par excellence*), or, vice versa, is this "origin" the hidden name of the backwardly extrapolated present, that is, the name of the "past" as an open-ended constitution of its content taking place as a process directed to the future from the present?

It is then not difficult to realize that cosmology is doomed to deal with the same difficulty of not being capable of phenomenalizing the originary foundation of the universe and the source of its temporality. One possible way out of this phenomenological difficulty is to commit to a form of Platonic realism. If the thus-asserted "origin" of the universe is treated as no more than a construction, that is, as belonging to the realm of intelligible realities, this would not pose any problem, for the "temporal" status of such an intelligible entity would take the form of the immanent temporality of conscious acts that are directed from the present to future. The past of the universe as its origin would become an *ideal* whose content would be constantly filled in through the movement to an uncertain future. In this case one could avoid the antinomy-like difficulties of a Kantian kind because the past or the origin of the universe would be regarded as a material of an indefinite development of thought, in particular cosmological theory. Correspondingly, this development as an indefinite advance would be devoid of contradictions since no definite concept of the past or the originary foundation is envisaged. But this treatment of the past or the origin of the universe in a Platonic-like manner would be considered by physical cosmology as unsatisfactory, since the latter attempts to build its concept of the origin in rubrics of scientific objectivity and to treat it as "object."

Cosmology affirms that the time of the early universe *is* not an intelligible entity, but physical time, so that there is evolution of the universe in this time and what we observe in the universe here and now is the remote result of what had been in the universe long ago when the corporeality of subjects was not possible. Indeed, that which is remotely observed by the senses extended through technology as the frozen memory of the processes in the universe, cannot be participated in, or lived through, in principle. We see a frozen temporal span of the universe along the past light-cone, so that the more distant the object we see the more we see the past of the universe. However, there is a limit to this seeing because the universe was not transparent to light before the so-called decoupling of matter from radiation took place. Whatever properties of physical matter of the universe prior to this temporal limit that cosmology predicates, including the very cosmic time, they are not observable in principle, so the nature of realistic commitment with respect to its theories is rather uncertain. Temporality and the constituted reality of these so-called dark ages of the universe have a very limited and formal mathematical character to which no intuition corresponds: time is introduced in equations as a fictitious parameter incapable of any direct physical verification and thus has a relation to the physical nonlived time only within a belief-based ontological commitment.

As to the "origin" of the universe, what is happening in this type of reasoning is that human consciousness projects instinctively the intuition about the finite origin of this same consciousness (as posited in the immanent time of this consciousness) onto the outer world, the universe as if this universe existed long before the human presence was possible. G. Marcel argued that "we should never view the world's *past* under the aspect of datum—for under this aspect it is inevitably contradictory and unthinkable. We should only regard it as the material of an infinite rational development (a development conceived as potential and future and hence not contradictory)."[12]

Indeed, when cosmology predicates things of the universe, or the universe as a whole as an "object" being out there independently of that subjectivity which articulates the universe, it exercises itself in the *natural attitude* of consciousness. In this case the observed present

12. G. Marcel, *Metaphysical Journal* (London: Rockliff, 1952), 8 (emphasis added).

state of the universe, as well as its nonobservable but theoretically constituted past, are treated as equally objective, but with distinct physical references. If the attitude changes toward the philosophical one, that is, if past and present are seen from the point of view of generating consciousness, there is a certain equivalence between "the past" and "the present" based not on an objectivistic commitment with respect to them, but on the fact that they are both being constituted. While within the natural attitude it is supposed that there is a temporal evolution of the universe and hence there must be an asymmetry between the universe's past and its present, from a phenomenological point of view the situation is not so clear, because the intuition of "the past" (as having existed prior to any consciousness) is exercised in the present of consciousness of a cosmologist. In other words, "the past" of the universe represents an intentional correlate of the multiplicity of conscious acts of cosmologists at their immediately given "present."[13] In this case the posited causality between different states of the universe appears to be a projection of that immediately experienced temporality of internal states of consciousness or thematized historicity which enters the rubrics of consciousness. "The past" that forms an implicit component of any perception of the given present, as a pre-predicative experience of the present, enters the horizon of meanings in all thematizations of the world and, in particular, the physical universe. But in this case as pre-predicative experience, open to an indefinite constitution the "past" ceases to function under the aspect of *datum*. The past reveals itself as an originary intuition of the hidden

13. As was advocated by J. A. Wheeler, the past is theory since it has no existence except as it is recorded in the present. The very term "big bang," according to him, is merely a shorthand way to describe the cumulative consequence of billions upon billions of elementary acts of observer-participancy reaching back into the past. See more on Wheeler's ideas in his *At Home in the Universe* (New York: American Institute of Physics, 1996), in particular pp. 299–302.

antecedent of that state of affairs which humanity experiences in its developed stage.

Phenomenology of Birth

In order to make an analogy between the origin of the universe and the event of birth more convincing, let us discuss carefully a phenomenological approach to birth. The problem is that in analogy with the Big Bang, which cannot be seen but only experienced in its delayed consequences, I did not see my birth and I must rely on the account of my parents or other witnesses in order to attempt to grasp my birth as that occurrence which affects me through all my life, but I will never be able to reconstitute this event as a phenomenon. The phenomenon of birth gives itself without showing itself because it comes to pass as an event, that is, something without foundation, ground, as origin, but as one that is nonoriginary.[14] The exceptional and unique status of this event follows from the fact that birth gives itself together with the fact that it gives *me to myself*. This is the mechanism of how my birth phenomenalizes itself, for without this giving me to myself I would not be able to realize that it is I who am affected by birth. The phenomenon of birth thus exemplifies the condition for any phenomenon: the possibility of phenomenalization of something lies in the extent by which it gives itself. In this sense the phenomenon of birth is the first phenomenon that initiates the possibility of receiving all other phenomena. The phenomenon of

14. See on phenomenology of birth J.-L. Marion, "The Event, the Phenomenon and the Revealed," in *Transcendence in Philosophy and Religion*, ed. J. F. Faulconer (Bloomington and Indianapolis: Indiana University Press, 2003), 87–105, as well as his *In Excess: Studies of Saturated Phenomena* (New York: Fordham University Press, 2002), 41–44. See also C. Romano, *L'événement et le monde* (Paris: Presses Universitaires de France, 1998), 95–112, and M. Henry, *De la phénoménologie. Tome I. Phénoménologie de la vie* (Paris: Presses Universitaires de France, 2003), 123–42.

birth as a phenomenon par excellence, not being reducible to any preceding causes and being incommunicable and indemonstrable, forms that excess in human perception of life which always allows for an unpredictable and undescribable future, for an indefinite series of commentaries and insights on the sense of this birth which extends forward in time while being interpreted retrospectively. Not being a phenomenon given to myself, I always experience an intention to look at birth as a phenomenon that initiated me, my identity, my spiritual growth, ultimately my hypostatic uniqueness. Birth as an existential premise is always silently encoded in all my actions, which attempt to reconstitute it in order to come to terms with the fact that I was born without my consent and can do nothing about it. In a way, my birth can be seen as the never-ending continuation of my experience of life, but it is still inaccessible as a phenomenon to my direct gaze. My appropriation of birth is an unceasing endeavor, which is always delayed. It is delayed because any retrospective reflection contains as its basic element a condition of a delay: delay between the occurrence of my birth and innumerable intuitions of its meaning. This inability to comprehend my own birth can be presented as a paradox: birth shows me exactly that my own origin cannot be *shown*. In this sense, I as an original being, do not have an *originary origin*, that is, a ground to which I can refer in order to deduce the occurrence of my birth from a chain of worldly events. In fact the very idea of the possibility of grounding my birth in the chain of worldly events signifies a fundamental reduction or deprivation of the phenomenality of birth, of its excessive primordiality (not deficiency but excess). It is exactly because my birth is in the foundation of all derivative intentions to construct a chain of historical or cosmological transformations, which as antecedents would conclude in my birth, that all articulations are overwhelmed initially and irreducibly by the intuition of this incomprehensible and indemonstrable event of birth.

How then can my birth as a phenomenon, while not showing itself, affect me radically in the sense that it produces my unique existence as my contingent hypostatic incarnation? How can the origin of myself, which is present in all following events of my life, show itself in such a way that, effectively, it is indemonstrable? The answer to these questions comes from the realization that this showing has an "eschatological" character because the past of my birth is being shown to me only through its *anticipation* (as the intention to understand "who I am") as directed to the future. My birth has sense only as an "event" that phenomenalizes itself by endowing me with my indefinite, potential future. Being an indemonstrable phenomenon, birth reveals itself as an "event" that was never present to me in orders of "presence in presence" and always already imbued with the qualities of having passed, but never irrelevant for the present and outdated. But even in this "eschatological phenomenalization" my birth does not allow any demonstrability in a sense of communication: my birth for me is an event that cannot be grasped as a fact and correspondingly described in rubrics of thought and demonstrated, being irreproducible and surpassing any expectation and prediction. By rephrasing a passage from G. Marcel: I am my birth; I am the more my birth the less I treat it simply as a collection of events jotted in a notebook as possible answers to eventual questions. Does that not mean that between my birth and my actual experience there is a relation of sympathy, but that this relation is closely bound up with the function of my body? "Is not this global experience which is me, but which far from being capable of being objectified is the condition of any possible objectification, the *mediating element* which alone allows the attention to bear on itself, that is to say, allows it to be? And the impossibility of defining this *past-subject* [*birth-subject*, AN] which makes memory possible is only another way of expressing the impossibility of treating the *mediating element* as an object and of forming an idea of it."[15] Here Marcel establishes a link between the event of birth and that which he calls incarnation, or, in different usage, embodiment. This link is important because apart from a teleological sense of an event of birth it entails a purely material, physical sense related to the very conditions, that is, a possibility of this birth.

Indeed the event of birth (if we regard it as coming-into-being of unique persons) as an event is not accountable on the level of *sufficient* conditions of its happening; its outcome is unpredictable and unforeseeable: given the normal physical conditions, birth (conception) might not happen at all. However, the necessary conditions for this event to happen lie in the sphere of what preceded it, the physical plan. In this sense in spite of its sporadic and unique character, an event of birth as physical incarnation contains in itself that something which made the happening of this event possible. And when we say that birth gives itself in an unmediated and indemonstrable way, that is not to say it does not contain in itself and thus manifests the hidden conditions for it to take place. These conditions come with birth and follow birth in the same unmediated and indemonstrable way. This means that in no way can I treat myself as an absolute beginning. I can oversee the limits of my origin and look objectively at it, that is, to formulate for myself the necessary conditions that made it possible. In a simple case of a personal family history I must admit that my parents were that necessary condition which made my appearance possible. I did not choose my parents, and they are mine in an absolute sense. My embodiment links me ontologically to them, and any refusal of this

15. G. Marcel, *Journal Métaphysique* (Paris: Gallimard, 1935), 243. ET: *Metaphysical Journal* (London: Rockliff, 1952), 249, translation corrected.

fact could imply an existential suicide. In this sense it is I who am ontologically responsible for what they have been or what they are. My parents are such existents who participate in the very possibility of my existence. Thus, with regard to all possible outcomes of my relations with them what always remains between me and them is piety, irreducible unity and unbreakable communion of being. But my personal story can easily be extended to that "before" which lies in the foundation of my incarnation not only on the level of my parents as a biological species, but that "before and out there" which make it possible for life to exist at all. One means here physical conditions and ultimately the universe. Thus my act of birth entails not only an unbreakable communion with my parents but an unbreakable communion with the universe where I was born and an implicit premise of the very possibility of my articulations with regard to both my birth and the universe as a whole. Repeating a previous thought, I did not choose the universe where I was to be born. The universe then is mine in an absolute sense. I cannot disregard the universe in my life because its presence is implanted in my birth: I am in communion with the universe from the very moment of inception of my body and consciousness. In an incomprehensible way, by being incarnate in the universe through my birth, I am "carrying the universe" (as it was, is, and will be) with me. The conditions of my birth point to that fact that the universe is not entirely alien to me and that it sustained the fact of my coming into being, and in this very fact the universe manifests itself as a gift, as that which we receive together with a gift of life.[16]

16. One can make a parallel with studies of Kuiper who made explicit links between mythologies of creation and experience of conception (prenatal states) in ancient Indian cosmosgonies. Kuiper's stress was on the possibility of anamnesis of one's prenatal state in order to explicate the cosmogonic representation of creation of the world in any human life: "reexperiencing one's own conception might be of the greatest importance for our interpretation of religious phenomena, and particularly of cosmological myths" (F. B. J. Kuiper, "Cosmogony and Conception: A Query," in his *Ancient Indian Cosmology*, (New Delhi: Vikas Publ. House

To summarize, on the one hand the event of birth endows us with a future, so that its explication goes on continuously as a process directed to the future: I am becoming more and more my birth through the articulations of its efficacious *telos*; on the one hand the event of birth manifests to myself the hidden conditions of its very possibility as related to the physical incarnation, and on the other hand the physical conditions of my birth themselves receive an articulation from within the same efficacious *telos* as related to the sufficient conditions of existence of life which remain a hidden and indemonstrable mystery.

An Example from Christian Literature

In order to give a literary illustration of some counterintuitive observations made in the phenomenological analysis of birth, let us analyze the sense of birth assigned in the history of Christian thought and its theology to Jesus Christ, considered as an ultimate archetype of all humanity and whose incarnation in flesh recapitulated all creation. The very fact that we still remember the birth-nativity of Christ and that this event has ongoing efficaciousness in the present history of humanity, amounts to its unmediated and indemonstrable character accompanied by hermeneutics reflected not only in exegetical literature, but also in visual art, fiction, and poetry. In this sense the event of the birth of Christ remains a mystery, phenomenologically concealed from us, but endowing Christians with the future from the movement to which the sense of the nativity is unfolded.

Pvt Ltd, 1983), p. 113). However this intuition is not exactly what is advocated by us. The phenomenological concelament of conception (birth) goes beyond the prenatal states and its "anamnesis" can be thought only as the immemorial that consecrates life to the future, without us being able to return to the origin of our being.

Indeed the apostles and the church affirmed that the Nativity of Christ, apart from its occurrence in earthly history, contained the hidden message about the everlasting Kingdom which Christ opens to humanity and which nobody can close after him (Matt. 2:2; Luke 1:32-33; 2:11-12). The incarnation of the Logos of God in flesh was a manifestation of the end of the old age, and the beginning of the new, the age that is driven by the "logic" of the Kingdom of God, the age that is eschatological *per se*.

There is a hidden dualism in the Christ-event. On the one hand, in its outward appearance as *an* event in the conditions of created nature, it begins with the Nativity of Christ, the birth of a human baby Jesus (like many other babies), and extends through his life and teaching toward his [human] death on the Cross. On the other hand, in the Christ-event understood as the Incarnation of God in human flesh, Christ recapitulates the long history of mankind and in that history it rearticulates the hope for salvation, the promise of God to man and his Kingdom.[17] The hidden message of the Nativity, which later was set forth by Christ through his parable of the Kingdom, was a mystery, not accessible to everyone, but opened by Christ himself to his followers. And when his disciples asked him what this parable meant, he said, "To you it has been given to know the secrets of the Kingdom of God; but for others they are in parables, so that seeing they may not see, and hearing they may not understand" (Luke 8:9, 10). The message of the Kingdom, which is manifested through the birth of Christ in Bethlehem, points toward a mystery of the union of man and God in Christ, man who was born in Palestine and God who, being in flesh, did not cease to be present hypostatically in the entire universe, being its ruler and provider of its order and harmony. Being in Palestine, he still was in the Kingdom, about

17. See, for example, Irenaeus of Lyons, *Against the Heresies*, III 13, 1.

which he taught his disciples. It is because of this that when theology asserts that the incarnation recapitulates the whole creation, it also asserts that the whole creation (from its beginning to the very end) is recapitulated from the perspective of the Kingdom of God. It is in Christ's incarnation that the sense of the temporal span of the universe as having already passed but not outdated is revealed, so that the universe was created and fashioned to receive God in flesh in order for the inauguration of the future of the Kingdom to take place.

By pointing toward the age to come in his parables Christ encourages his disciples not to be preoccupied with questions about the facticity of his birth (and correspondingly to their own birth and familial attachments), not to pose questions such as "Why?" and "How?" Since the whole span of history is recapitulated by him, the contingent facticity of his arrival in the world in terms of space and time had no importance, whereas it had importance as the inauguration of the Kingdom whose realized presence endows with the future that which is here and now, in the past and present. This fragment of Christian literature can be considered as typology of anyone's birth: the preoccupation with the contingent facticity of one's embodiment, that is, birth, is in vain since it is phenomenologically concealed from one's grasp. One cannot understand the meaning of the birth of a new person apart from formulating some necessary conditions for its happening. However, the event of birth endows a person with freedom and a future; correspondingly, the process of disclosure of the sense of birth as well as the necessary conditions related to its embodiment aspects are directed to their future and endowed with a certain *telos* (closely related to the sufficient conditions of one's birth). Thus, in a Christian perspective, the sense of the event of one's birth comes not only from some familial or historically contingent circumstances, but from an existential fact that a newborn person is endowed with a future and a potential of salvation and that this endowment cannot be traced back as having antecedents. At the same time the hope for salvation makes it still possible not to forget of birth as a God-given event: the acceptance of birth is thus based not in memory (which is not possible) but in faith.

The question that naturally arises is whether, in its insistence on the unimportance of the historically contingent conditions of one's birth in terms of potential salvation of a person, Christian teaching neglects cosmological aspects of the possibility of life and, in particular, the very possibility of the incarnation of God in flesh of Jesus Christ? The answer is obviously "not." For the incarnation to take place on the earth, this universe must possess some features such that the generation of a human body capable of receiving God would be possible. This links the creation of the universe, its evolution and structure, to the phenomenon of man, its natural origin. The teaching on incarnation in theology articulates this link, making sense of the human presence in the universe as grounded in the will and love of the personal God, who transfers the image of his personality to human beings. This observation allows us to conjecture that the development of the universe before the incarnation of the Son of God in flesh on the earth, and after has, theologically speaking, a drastically different meaning. It was necessary for the universe to have such an origin and to be in a constructive development in order to sustain life on the earth and to allow God to condescend to us and to assume human flesh in order to initiate the new stage of salvation history. What will happen in the universe afterward is difficult to say, for according to the cosmological predictions, there is a natural limit to the extension of human life in the cosmos, so that the transferral of the saving *telos* to the universe will have a limited temporal span.

Apart from this trivial link between the natural conditions for persons' appearance in the universe and thus its physical structure and the possibility of the incarnation as the birth of Jesus, there is another more subtle connection between the incarnation and the origin of the universe as its creation. According to the Christian Creed, the incarnation of the Son of God was present in God's plan of salvation before the world was created. This implies that the actual event of the incarnation in rubrics of space and time took place having in its remote antecedents the creation of the world, which provides the necessary condition for the possibility of the incarnation. This implies in turn that since it was the incarnation of the Logos of God, by whom and through whom the world was created, the incarnation as a contingent happening in human history carries in itself some traces of the purposiveness of the universe toward this incarnation, and the explication of this purposiveness initiated in the event of the incarnation is exactly the ongoing articulation of the Kingdom of God which was announced in Christ's coming in this world.[18] Thus the very articulation of the Kingdom in this universe presupposes the universe's actual existence, that is, its history initiated in its creation. Thus the eschatological aspiration for the Kingdom of God implies the elucidation of the meaning of creation of the universe: to understand the sense of creation one must proceed along the lines of the saving *telos* in the humanity of Christ. In this sense the understanding of the incarnation of God cannot be detached from the disclosure of the sense of creation. Humanity, being endowed with

18. See a detailed discussion in A. Nesteruk, "The Cosmos of the World and the Cosmos of the Church: St. Maximus the Confessor, Modern Cosmology and the Sense of the Universe," in *Knowing the Purpose of Creation Through the Resurrection*, ed. Bishop Maxim (Vasiljevic) (Alhambra, CA: Sebastian, 2013), 319-28.

the Divine image, thus contains a hidden "memory"[19] of that creation when "all was in all."

The final theological analogy that somehow unites archetypically human birth and the origin of the universe proceeds from a nontrivial spatial paradox that inheres in the doctrine of the incarnation. Its essence, briefly, is: on the one hand, Jesus Christ, being in his nature fully human, lived being located in body in a particular place and time of earthly history; on the other hand, Jesus Christ was fully God, who did not leave his "divine place" and who, being God, not only was present in Palestine two thousand years ago but was always present in all places and times of the universe created by him and through him. One observes here a nontrivial, historico-topological relation between the finite track of Jesus in empirical space and time and its extraordinary link to the whole history of the visible universe.[20] The human nature in Christ was operating within the reality of empirical space and historical time, whereas Christ's divine nature was always beyond the empirical and intelligible aeons, in the uncreated realm of the Kingdom of God, which can be expressed symbolically in terms of the "boundaries" of the created if these "boundaries" are seen from the divine dimension. It is from this "outside" that Christ the Logos of God coordinated the empirical

19. The word "memory" is used here metaphorically as a constituting aspect of the human soul or its divine image. Correspondingly the "remembrance" of "all in all" of the creation comes through faith in God. In similarity with what Augustine wrote, that human soul remembers God "not because it knew him in Adam, or anywhere else before the life of this body, or when it was first made in order to be inserted into this body," for everything "has been erased by oblivion" (Augustine, *The Trinity*, XIV, xiv [21]), the "memory" of "all in all" originates in God himself. In his *Confessions*, X, xxvi (37), he asks and responds: "Where did I find you [God] to be able to learn of you . . . if not in the fact that you transcend me [above me]?" Indeed to have a memory of "all in all" one is to be in communion with God either in faith or in unbelief.

20. On the problem of paradoxes linked to the dogma of the incarnation and spatial structure of the universe see T. Torrance, *Space, Time and Incarnation* (Edinburgh: T. & T. Clark, 1997), as well as his paper with reference to patristic sources, "The Relation of the Incarnation to Space in Nicene Theology," in *The Ecumenical World of Orthodox Civilization, Russia and Orthodoxy*, vol. 3, *Essays in Honour of Georges Florovsky*, ed. A. Blane and T. E. Bird (The Hague: Mouton, 1974), 43–70.

space where he indwelled in the body with the rest of the created universe, including its *temporal origination* which is detected by cosmology. In this sense the incarnate Christ, being fully human, knew everything about the beginning of the universe, because remaining God, he "saw" this beginning from the "other" side of it. He was present in all locations and eras in the universe hypostatically and hence through knowledge of it. Thus he looked at the spatial and temporal span of the universe from a perspective of an act of creation, which as such could have no spatial and temporal distinctions. Thus the whole universe was "seen" by Christ-Son of God as a single point, as a flash of the unconditional love of God with respect to the world. Here the archetypical "memory" of the "all in all" in human beings receives its justification from an observation that for Christ the spatial and temporal extension in the world did not exist because he, while being fully human, limited by the rubrics of space and time, still saw the world through the eyes of God. In this seeing the temporal origin of the world was a particular, fully human projection of the unity of all creation experienced by the Divinity of Christ. Correspondingly, humanity, by being endowed with the divine image, incessantly searches for this origin as the lost communion with "all in all."

From Phenomenology of Birth to the Big Bang as a *Telos* of Cosmological Explanation

Now we are in a position to summarize our discussion by bringing together all observations concerned with both the idea of the origin of the universe and inexplicable nature of one's birth. From what we have already said one can easily grasp that the search for the origin of the universe is rooted in a psychological desire to understand one's own origin, that is, the sense of one's own biological birth, understood in a philosophical sense as the mystery of hypostatic incarnation. In the same way as an event of one's birth is unavailable to phenomenalization in consciousness, whereas its phenomenality unfolds through its future-directed constitution, one can say that in cosmology the origin of the universe is present only in its actual absence, so that all attempts to articulate this origin (as intentionality of consciousness directed to the future) are doomed to deal with the unfolding facticity of the universe without any hope of achieving the "presence of the past in presence."[21] It is in this sense that the very advance of cosmology toward understanding the past of the universe constantly deals not only with its unknowable *essence* but also with an unavoidable *absence*. In spite of the metaphysical fact that the origin of the universe, as its foundation, as well as the origin of one's person are radically unavailable to humanity, cosmology still intends toward the principally unknowable and absent past as if it might become knowable and present at some distant future. Here cosmology exercises not so much its intrinsic creativity of constructing a variety of mathematical cosmological models, but rather that intrinsic *teleology*, that is, the purposiveness of enquiry into the sense of nature that is implanted in the very essence of the human condition. This condition can be expressed as man's desire not to be circumscribed by the necessities of the given nature and inevitability of the universe's facticity, and to see the universe according to man's will and in its own image. To know thyself means

21. Martin Rees, Astronomer Royal, assessing the reliability of cosmological theory with respect to the early universe, wrote in one of his papers that "I would now place 99% confidence in the extrapolation back to *one second* [from the ultimate beginning of the universe, AN]." M. Rees, "Our Complex Cosmos and Its Future," in *The Future of Theoretical Physics and Cosmology: Celebrating Stephen Hawking's 60th Birthday*, ed. G. W. Gibbons, E. P. S. Shellard, and S. J. Rankin (Cambridge: Cambridge University Press, 2003), 24. Whatever is beyond this one second toward the Big Bang is not subject to a strict verification even by theory itself and as Rees expressed himself, he leaves a 1 percent chance of being deluded by theory of what was before one second. In this sense the ultimate "originary" origin of the universe, as its preexistent past, cannot be brought to presence in presence even by means of theory, thus leaving us with its unavoidable absence.

to know the universe. To know thyself means to understand the indestructable presence of the immanent self-consciousness that is always intending its own origin but, failing to fulfill its intentions, appeals to the indefinite context of totality to which it desires to refer itself. Consciousness functions in the conditions of its intentional immanence to the universe; thus the searched foundation of consciousness as such, that is, the ground of its facticity, implies the ground for the universe to which the immanent intentionality is directed. Then, if in the natural attitude consciousness thinks of its own origin in temporal terms (as birth), the origin of the universe also acquires some features of temporality and this leads to varieties of the Big Bang models. In a complete incapacity to establish the origin of the incarnate transcendental subjectivity in rubrics of ground-grounded relation, the understanding falls back into the natural attitude and physicalizes its own origin by displacing it to the beginning of the universe whose idea (psychologically, because of the fear of contingency of birth which implies death, one wishes to establish a reference to a sort of stability that attaches some sense of existence forever). One argues then that the tendency to search for the origin of the universe is deeply embedded in the human condition, being a kind of an *innate* idea, donated to every human being at the event of birth.[22] Theologically, this is an idea of the Divine image in man and an archetype of its lost likeness when man, like God, knew all because he was "all in all."[23]

22. J.-L. Chrétien, discussing the sense of the unforgettable (immemorial) in human life, and referencing Malebranche, invokes an idea of being as something similar to that innate idea which we put under the rubrics of the "originary" origin of the universe: "The mind breathes only through being which is more original to us than ourselves. We are of being more than of ourselves. This unforgettable and incessant presence of being to mind is not an object for the mind, but the mind's very opening, its only light, and its condition of possibility." J.-L. Chrétien, *The Unforgettable and the Unhoped For* (New York: Fordham University Press, 2002), 86.
23. Cf. 1 Cor. 15:28.

We have already analyzed in detail the dynamics according to which this *innate* idea realizes itself in cosmology. Let us discuss here briefly the nature of that hidden aporia in attempting to think the Big Bang, which G. Marcel called a "meaningless pretence."[24] When cosmology predicates the idiosyncratic past of the universe, a characteristic displacement of the self of cosmologists takes place when they consciously descend into a condition of the universe in which no incarnate conscious life was possible. The more a cosmologist's mind, in its eidetic reduction of the empirical, advances toward the Big Bang, the more it intends something that is fundamentally *nonhuman*. By circumscribing conceptually the allegedly physical content of the early universe, the self, its consciousness, works in a mode whose intentions will never be filled and fulfilled because they are contradictory to the conditions for this consciousness to function at all. From a philosophical point of view one finds here not only an urge of the human psyche to find the impersonal "foundation" of the world at the expense of losing the sense of uniqueness and identity of every particular human person, but also the loss of the sense of humanness. Characteristically those philosophizing cosmologists who believe that through studying the alleged origin of the physical universe they touch upon sacred truth related to the mystery of one's birth, in fact dissolve themselves in the abyss of nonhuman physics, which, although a very interesting eidetic exercise, turns out to be devoid of any spiritual and soteriological meaning. A corresponding cosmology begins to predicate the universe in esoteric (mathematical) and nonexistential (not having immediate references to the realities of the life-world) language that is accessible only to those who follow this cosmological gnosis (in this it recalls all the mystical philosophical systems of

24. G. Marcel, *Being and Having* (London: Collins, 1965), 24.

antiquity). Such a cosmological gnosis ignores obvious philosophical doubts about the ontological universality of its claims and objectivity based in the natural attitude. That is why from the critical stance of a phenomenological stream of thought it seems paradoxical to claim the object-like status of the Big Bang models (or different preexisting universes/multiverse) that exclude incarnate consciousness.[25] Gnostic cosmologists sometimes believe that the asserted ultimate reality of the past universe, where all forms of matter were present in a kind of undifferentiated soup, has more relevance to the truth of existence than the variety of empirical experience, and that it is the meditation of this primaveral "reality" that fills one's life with content and meaning. In spite of the implicit teleological intentionality hidden behind such cosmological constructions there is one element in them that remains disturbing and existentially irrelevant, namely, the so-called "lure of cosmos"[26] driving the cosmological mind in the search for the cradle of life in the "cosmic heaven," a lure that is deeply interwoven with the discomfort and anxiety arising from the

25. One could suggest along the lines of the anthropic inference in cosmology that the nonhuman past of the universe was a necessary condition for the later appearance of life, so that there is no contradiction between what the Big Bang cosmology affirms as the nonhuman physical state and what emerged from this states afterwards. The naïveté of this argument is based in the belief in the continuity of cosmological as well as biological evolution, which led to emergence of the consciousness that articulates this same evolution as well as its origin. The difficulty lies in the part of this argument that is supposed to deal with the sufficient conditions of the emergence of consciousness. These conditions are not part of physics but rather belong to the realm of human will and destiny. It is in this sense, that when F. Dyson argues, along the lines of the anthropic argument, against J. Monod's apology for the accidental coming of intelligent humanity in the universe, by saying that "I do not feel like an alien in this universe. The more I examine the universe and study the details of its architecture, the more evidence I find that the universe in some sense must have known that we are coming," he effectively invokes a *teleological* argument by reference to the existence of another, parallel sense of the universe's future as the unfolding of a transcendental history through which the physical history is articulated. But this "knowledge" by the universe that we were coming cannot be consistently placed in the framework of scientific explanation. It is rather an axiological and soteriological argument that refers to the *teleology* of human reason. See F. Dyson, *Disturbing the Universe* (New York: Basic Books, 1979), 250.

26. This term was used by N. Berdyaev, who discussed the theme of "cosmic temptation" in *Slavery and Freedom* (London: Centenary, 1944), 93–102.

transience and mutability of forms of matter and life of this earthly world, thus contradicting the human longing for all-encompassing knowledge and immortality. This anxiety inheres in the event of birth so that it comprises the innate *idea* presupposing a hidden *teleology* of the human spirit in explicating this event. All speculations on the origin of the universe are thus unavoidable as pertaining to the very essence of humanity in articulating its own purposiveness in the universe.

It is phenomenology that makes it possible to change the attitude to the theories of the beginning of the universe by explicating the hidden *teleological meaning of some cosmological ideas* by referring them to the event of birth.[27] Teleology appears as an endless commitment to a theoretical task (understood philosophically), so that each particular scientific result or theory is considered as temporary and provisional and must be put aside while looking for successive developments. The existential meaning of cosmology, its own ground and its sheer possibility, can only be understood if cosmology is referred to its roots in human subjectivity and to the cultural dimensions of the life-world, and thus to the hidden teleology of reason. Correspondingly, the persistent exercise by cosmology of that intentionality which is directed toward the removal of the contingent facticity of everything and the search for the foundation of the universe (its originary origin, as if "all was in all")[28] reveals the meaning of the notion of the Big Bang as the *telos of cosmological explanation* (as well as a *telos* of the human reason enquiring about the foundation of its facticity in the life-world in general).

27. The ideas about teleology of scientific research were developed by E. Husserl in *The Crisis of the European Sciences and Transcendental Phenomenology*. See also Rizzacasa, "The Epistemology of the Sciences of Nature in Relation to the Teleology of Research in the Thought of the Later Husserl," in *The Teleologies in Husserlian Phenomenology*, ed. A.-T. Tymieniecka, *Analecta Husserliana* 16 (Dordrecht: Reidel, 1979), 73–84.
28. Cf. 1 Cor. 15:28.

In the natural attitude the meaning of the Big Bang was to describe the temporal origin of the universe as if it took place in the physical past: this was the ideal and *telos* of cosmology, that is, to find an expression of the original state of the universe that would allow one to describe in terms of species the variety of cosmic objects observed in the sky. The Big Bang, being the *telos* of cosmological research that advances and unfolds the sense of the universe forward in time, is allegedly placed within the natural attitude of the human mind in the physical past. Here we see the competing tendencies of human subjectivity, which cannot function properly if this subjectivity is not "purified" by being placed within the phenomenological attitude. For that one who is in the natural frame of mind there is a paradox: How can we talk about the Big Bang as a *telos* of explanation, that is, something that is supposed to be in the future, if this Big Bang is, by definition, supposed to be in the past?

As we argued above, the natural attitude to the Big Bang, that is, putting the origin of the universe and time under the rubrics of datum, must be abandoned on the simple grounds that its construct is achieved by means of a series of eidetic reductions, so that its allegedly physical characteristics, even if one admits the high degree of coherence in its theoretical description, make no sense in the context of what it is supposed to describe, namely, the present-day universe and life-world. Then, by being a fundamentally unfinished and unfinishable construct, all existing and future theories of the Big Bang have equal weight and importance if they are treated from within a strictly scientific discourse: the idea of the Big Bang can only be a "normative form situated at infinity."[29] Thus each historical

29. To understand this means to avoid mythology in questions of the beginning of the universe. As it was expressd long ago by Hannes Alfvén: "It must be absolutely clear that if a scientist makes a guess about the state of the universe some billion years ago, the chance that this guess is realistic is negligible. If he takes this guess as the starting point for a theory, this is unlikely to be a scientific theory but very likely will be a myth." H. Alfvén, "Cosmology: Myth or Science?,"

realization of cosmology in its theories still has the idea of the Big Bang (as the hidden totality of all) within its historically contingent scope. The philosophical importance of cosmology becomes evident if it is appropriated phenomenologically, when the reduction of all historical facticity of its theories is performed and the essence of cosmology as a search for the foundation of contingent facticity of all is realized. In this case the presence of the Big Bang theories in cosmology reflects the deep and inherent propensity of humanity to enquire into its own origin and facticity, and then into the facticity of all that is given to humanity in the event of birth. This is what Husserl called *entelechy* of the reason, the reason that attempts to disclose the meaning of the universe in the perspective of its *telos*, when it will be united to the entire universe, that is, when for this reason "all will be in all."

But physical cosmologists do not anticipate their own aspiration for the Big Bang theory as a "teleological principle" of the working of their mind. The natural attitude of a cosmologist treats the Big Bang in terms of an objectivity pertaining to a remote hypothetical past; its consciousness does not realize that, in fact, in allocating the Big Bang to the past, it acts from the present and its actions are initiated by the *telos* (of explanation) with which humanity is endowed (as having a future) at the event of birth. The philosophical sense of the oblivion of the human presence behind all objectivizing theories is that cosmology, as well as any science, is incapable of contemplating the purposes and ends of nature—which are accessible to humanity through its innate assignment to explicate the universe through explicating the sense and purpose of its own existence.

One can clarify the meaning of the Big Bang as a *telos* of scientific explanation by reminding oneself that the observable universe is

in *Cosmology, History and Theology*, ed. W. Yourgrau and A. D. Breck (New York: Plenum, 1977), 13.

always turned to us by its past: because of the finitude of the speed of light, signals travel from distant space to reach us, so that we detect the information from the sky not as it is here and now, but as it was at the time of its emission in the past. In the language of theory of relativity, the universe is observed along the past light-cone, where the human observer, being at its apex, while progressing into the future still faces the universe only in its past and the realm of the unconcealed past expands together with the progression of the observer into the future. In spite of the fact that our capacity to observe the universe is restricted by the era when matter decoupled from radiation and the universe effectively became transparent for propagation of light, in its exploration of the early stages of the universe theoretical cosmology goes beyond this limit in the past and attempts to model the processes in the universe up to the point of its temporal origination 13.7-8 billion years ago. The advance of cosmological research, as the process directed to the future (the future that is also phenomenologically concealed) explores in more and more detail the allegedly existent past. Thus our knowledge of the past expands in the future. It is in this sense that one can claim that the future of the cosmological research results in the expanding of our knowledge of the past. The "past of the universe" becomes the "future of cosmological explanation," so that the Big Bang as the ultimate goal of explanation becomes the *telos* of this explanation.

In different words, the human observer progresses in its knowledge of the universe by moving into the so-called future by its backward glance, that is, by always turning its eyes toward the past. The Big Bang then appears as an ongoing *ideal* of cosmological constitution at which the human gaze is always turned. The content of its knowledge is represented by the ever-expanding base of the "cone of knowledge" so that future knowledge will acquire more and more content of the past (see Figure 4). Thus the *telos* of cosmological explanation, its ongoing goal and anticipation, is the past of the universe, its alleged origin, the Big Bang.

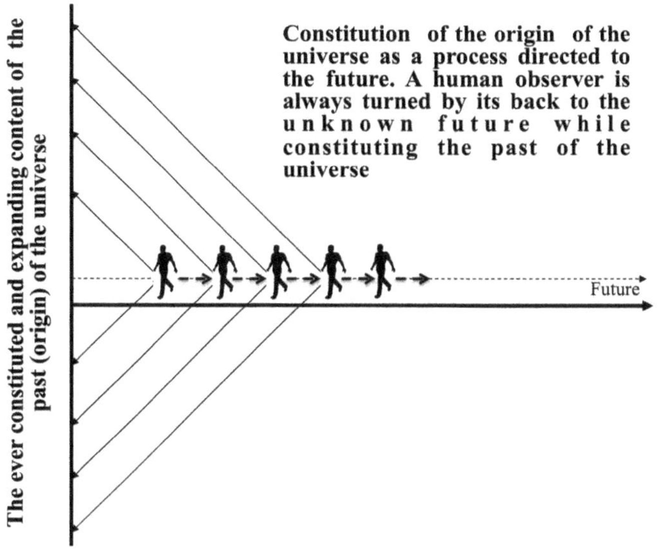

Figure 4

Conclusion: Phenomenological Parallelism between Birth and Origin of the Universe *versus* Genetic Similarity

In conclusion, we would like to rearticulate major points discussed in this chapter by making a certain contrast with another line of thought, introduced in the philosophical treatment of cosmology under the title of "genetic similarity" between the development of the universe and development of a particular human being.[30] This

30. See A. N. Pavlenko, "A Place of 'Chaos' in the New World 'Order,'" in *Philosophical Problems of Cosmology: The Universe from "Nothing" or the Universe from "Non-Being"?* (Moscow: URSS/Librocom, 2012), 66-87 (in Russian); "Universalism and Cosmic Harmony: A Principle of Genetic Similarity," *Skepsis* XV, no. 1 (2004): 389-401; "European Cosmology: Between 'Birth' and 'Creation,'" in *Scientific and Theological Thinking of Ultimate Questions: Cosmology, Creation,*

idea appears as a natural inclination to understand the universe along a certain teleology related to the fact of existence of human being and thus to the fact that the universe is observable and knowable. Correspondingly, a hidden teleology implied in the formation of an individual human organism is arbitrarily transferred to a sort of similar "teleology" in formation of the universe. The idea is simple: there were two major stages in the development of the universe. The first one is the famous "inflation," or the exponential expansion of the universe near its very beginning, which took place in the region of 10^{-36}–10^{-34} seconds after the "origin" at the cosmological singularity corresponding to the absolute zero of time.[31] The remarkable feature of this stage of expansion is that the universe, depending on different scenarios, inflates in size by at least a factor 10^{27}, thus expanding the universe approximately from a size of 10^{-27} cm to 1 cm. This remarkable figure shows that matter in such a universe is subjected to an incredible transformation in terms of space: being initially in a grain of space it experienced a change in size that is "astronomically" huge in comparison with what it had started. Then within the next 300,000 years the universe expanded by approximately the same factor to the size of 10^{25} cm when matter decoupled from radiation and became transparent, that is, that visible universe we face as sky. In other words, within 3×10^{6} years (which constitutes approximately a grain of time of 1 out of 2×10^{5}, corresponding to about 13.7 billion years of the predicted age of the universe), the universe expanded by a factor of 10^{52} whereas for effectively the same last 14 billion years it has only expanded by a factor of 10^{3} to its estimated present size of 10^{28} cm. What stuns the imagination is a strange nonuniformity

Eschatology, ed. A. A. Grib (Moscow: St. Andrew's Biblical Theological Institute, 2008), 128–30 (in Russian).

31. The absolute numerical values of the beginning and of the inflationary phase of expansion vary in different books on cosmology. See, for example, A. Liddle, *An Introduction to Modern Cosmology* (Chichester, UK: Wiley, 1999), 102–3.

in the rate of change of the universe's scale (at least 27 orders of ten) during an infinitesimal transition from 10^{-36} to 10^{-34} seconds on the one hand, and the long 14 billion years for nearly the same order of scale growth up to the present state of the universe.

This incommensurability of rates of change in the development of the universe which, from the point of view of physics, is a contingent fact, is brought, quite arbitrarily, into correlation with the fact of biological development of a human being, namely that it develops in its prenatal stage from the size of a chromosome (or the order of 10^{-7} cm) to the size of the newborn baby of, let's say, 50 cm, and then during the whole span of life its size grows only 3-4 times. In other words, there was an "inflation" in the prenatal development of a human being, that is, a growth in size of 10 million times within 9 months, and then a very slow development to the adult state, let's say within 20 years, when the size of a human being reached, let's say, 2 m. This contingent analogy was proclaimed as a "principle of genetic similarity between anthropo-genesis and cosmo-genesis."[32] The author of this "principle" recognizes that it probably does not advance either cosmology or anthropology since it has a heuristic character. Indeed this principle functions only on the level of consubstantiality between human being and the universe by making some particular suggestions on how this consubstantiality develops through the prism of either the universe's evolution or an individual human development. The analogy is developed in the natural attitude, assuming in advance the presence of developed intentionality grasping this analogy. This implies that "the principle of genetic similarity" does not establish any phenomenological insight in the event of birth and the endowment of a human being with personhood, that is, the future and freedom of its creative acts, which transcend nature, including the universe itself. Indeed the

32. See Pavlenko, "A Place of 'Chaos' in the New World 'Order,'" 79.

hypothesis of genetic similarity does not venture to make any speculation on the origin of *personhood*, that is, on the act of birth as coming into existence of otherness and freedom from the bonds of nature, which tends to subordinate personhood to its uniform conditions.

In distinction with this idea of "genetic similarity," our analysis of the phenomenology of birth was mostly concerned with the formulation of its phenomenality as not related directly to the natural conditions of birth. It does not mean that birth is a singular and unrelated event: on the contrary, birth itself is a response to "a preexisting invitation-to-relationship."[33] In this sense birth is an event in the mode of nature, where its potentiality for freedom and future points to a pre-natural factor, a hidden personal center that transcends all datum and meaning, an existential factor of referential reciprocity or relationality. A relation that is implied here is a relation to that transcendent other with respect to which the very event of birth affirms itself as an existential response of generating a new other. On the one hand the event of birth of a person is free from any antecedent definition in the realm of natural conditions (the necessary conditions for sentient life do not provide the sufficient conditions for creation of persons); on the other hand the birth of a person is a response to invitation, which as such is not naturally conditioned but is related to that nonoriginary origin which man attempts to constitute through its movement to the future granted at birth. And it is here that the unique and incommunicable mode of personal existence manifests itself as a unique way of constituting the sense of birth. The uniqueness of a newborn's personhood lies exactly in the fact that it is not a duplication of parents and has an independence from the constitutive principle of life in general. And here the

33. C. Yannaras, *Postmodern Metaphysics* (Brookline, MA: Holy Cross Orthodox Press, 2004), 171-82.

implicit teleology in explicating one's birth, which cascades up toward a teleological way of explicating the origin of the universe, comes in contrast with a naturalistic teleology implied in the hypothesis of genetic similarity. But the latter, having gotten an arbitrary character, is subject to Kant's critique of the purposiveness of nature in his *Critique of Judgment*.

And finally, in spite of these obvious differences in the phenomenological parallelism between an event of birth and the Big Bang, an interesting observation can be made related to a hypothesis of genetic similarity. The prenatal stage in the development of human beings, the stage that immediately precedes physical birth, is very rich in terms of communion of an embryo, that is, a future full-bodied human being, with physical reality on spatial scales that are not accessible to our empirical contact in the state of adulthood. In other words, if one suggests that there is a sort of memory of prenatal states of consciousness (even undeveloped) at the microscopic scales of communion with the physical universe, this memory can act as the overwhelming and bedazzling experience of coming into being from something "other" than that from which the developed adult realizes itself. This other can be the universe, and communion with that other is the communion with the universe. In this sense the phenomenology of birth, as the constant inerasable presence of the nonoriginary origin of personhood, can point toward the phenomenality of the universe and its nonoriginary origin.

6

Cosmology and Teleology

Purposiveness in the Study of the Universe

The time in which I was not yet human, the time before my birth, lost for me through my birth, will remain irremediably lost, and will never be regained as such. Yet everything that I recognize and recapture . . . is rendered possible only by this loss. I am always already in the truth, which is itself always and forever. But for human beings, this "always already" is that of forgetting: the immemorial consecrates it to the future, it comes from the future itself, though without ceasing to be immemorial and without us being able to either return to the origin of our being or coincide with it

–Jean-Louis Chrétien, The Unforgettable and the Unhoped For, p. 19

Introduction

In this chapter we continue to investigate the delimiters in cosmological research that originate in the structure of the human knower—in particular, how the purposiveness of human actions cascades toward the purposiveness of cosmological research.

Purposiveness of research is not purposiveness related to the alleged object of this research, that is, the universe. In this sense we are not dealing with a traditional teleology, which would imply the assertion in the purposiveness of the universe's evolution. We rather deal, as we could say together with Kant, with a "formal" purposiveness of cosmology which, because of the specificity of its subject matter, has to conduct its research under the assumption that there is a goal of research, a motivational purpose, related to the explicability of the universe. However, we argue here that this explicability originates in the human condition, that is, in the human intentional search for the sense of its own existence in the universe. Thus the purpose of explanation in cosmology is related to the explication of the human condition. Correspondingly, the purposiveness of cosmological research acts as a certain delimiter in the explicability of the universe related to the human condition.

In discussing matters related to teleology, we need to start by noticing that contemporary natural sciences consider any sort of teleological explanation as inacceptable and inappropriate. The objections are perennial and can be reformulated as follows. The first one is that teleology is treated as if it represents the causal operation in the present of future events. If there is a process that is subject to this teleology it must be purposive, that is, progressing toward a goal, so that every event in in this process must be explained by reference to this goal, which determines the course of the whole process. Contemporary physics maintains that it is impossible to understand how a future material event (the goal) can causally influence an event that precedes it. Teleology seems to be incompatible with the efficient physical laws that govern the universe of inorganic bodies. Certainly there remains an open question about the facticity of the laws of efficient causality themselves, namely whether these laws, being related to the boundary conditions of the universe (which are beyond space and time), are in place according to the transcendent *goals* of the universe as a whole.

In contradistinction to the physical causality of the processes of nonliving matter, the case of human action is understood in terms of consciousness and intention. In the realm of human affairs causality is complemented by intentionality. In scientific research, for example, scientists are aware of their purposes and aim at them consciously, so that their actions are caused not by some future events related to nature, but by their present awareness and the intention to act which they consciously form from within an intellectual intuition. In other words, in all aspects of being where consciousness is involved, teleology is permissible as related to the intrinsic teleology of any human activity.

However, being concerned with cosmological research, there is one particular feature of such research activity (as "goal-directed") that will make the case of cosmology special in comparison with other purposive human actions, which usually assume that the determining cause of such an action comes at the end. In any conscious purposive action it is assumed that the prevision of this "end-state" is characteristic of the process and is the inspiration to action. It is usually the case that the end-state of a purposive process is the goal at which the action is directed, so that the action is supposed to terminate when the goal is attained and will mark the end of the process in time. Cosmology positions itself aside from this, so to speak generic case, because the goal of its research is related to what is called the "universe as a whole," the concept of which cannot be attained in any finite time. Correspondingly, the purposive activity in explicating the universe is an open-ended process, an *open-ended teleology*.

Indeed, the open-endedness of purposive activity is obvious since cosmological research goes in phases of success and failure, so that the intermediate "end" of such a process is not necessarily that at which the activity aimed. The fulfillment or "satisfactoriness" of cosmological theory can have only a temporary character, so that the acting *telos* of research (as directed to the universe as a whole) does not cease to function as long as consciousness itself is in existence. In cosmology any research action has a conscious aim, often accompanied by the imagination of the disclosed state of the universe which the researcher is trying to bring about, but this imagined wholeness of the universe is not the state of affairs in which the activity is supposed actually to terminate even in the case when the activity is successful within the stages of research. In fact, for cosmologists, it is impossible to imagine what it would be like to attain the end of research at which they are aiming. One may not even know clearly what is that something which a cosmologist tries to bring about, because this something is being brought into existence in the course of research in spite of the fact that this same research is driven by the anticipation of that which it wants to articulate.

Whatever awareness a cosmologist may have of the goal he seeks in his purposive research action, this goal is not just an event or set of events of the ultimate disclosure of the universe, conceived as in the future. This goal is something that, by his own effort and activity, a cosmologist is engaged in bringing about. The achievement of the *telos* of research and thus the formation of that which this research attempts to disclose and articulate is not conceived as preexisting and thus disclosed in the future without a cosmologist's active participation. In a sense, cosmological research is dealing with its goal in the making, and the goal is what this activity is all the time generating. Cosmological research thus can be described as a process in which the outcome is all along immanent and potential, or through which the "final state" (the universe as a whole) is being generated. The purposiveness of such an action can be seen as a deliberate use of the knowledge of laws of physics in order to bring about desired "effects," that is, in our case, the picture of the universe as a whole.

In order to appreciate the peculiar character of teleological activity in cosmology, if human purposiveness in general is taken as a model, one must consider the relation between the subject of research (the agent) and the intended objective pole, that is, the universe as a whole (the end), the properties of the latter, in virtue of which this universe as a whole appeals to the subject as worthy of pursuit, that is, indeed, as that *telos* which drives cosmological research. This issue is closely related to the basic function of the human thought. The idea of the universe through the sense of belonging to it organizes our desires and directs our energies to the pursuit of theoretical models and their alleged physical references in such a manner as to attain what, through its activity, one anticipates as the total satisfaction of the enquiring person because the objects are consciously and persistently pursued only as contributing to the alleviation of the existential anxiety of the person. Elucidating the sense of the universe means to elucidate the sense of one's own existence, which is always encapsulated in any form of quest for the universe.

It is because cosmological theories naturally compete and conflict even in a single researcher (one can recall Einstein's blunder over the use of the cosmological constant), and because no single researcher can satisfy the criteria for "objectivity" of their research intentions without the assistance of the scientific community, that the satisfaction of the self of a cosmologist involves a complex process of organization of knowledge, which one usually calls education and training in which some kinds of scientific strategies are given up as incompatible with those accepted and agreed-upon views of the scientific community. Finally, one can speak of the cosmological enquiry as a system of activities, or a way of life, which is found to be the most satisfactory, on the one hand for purely scientific and communal reasons and, on the other hand, for the maintenance of personal existential commitments. Being placed in the system which involves the communal aspect of science and which provides for a given researcher the criterion by which one judges the coherence of research strategy and its objectives, the existential anxiety and objective in disclosure of the sense of the universe is subjected to correction and choice. Thus conscious purposive activity in cosmology is not impelled by any simple force or drive toward an imagined object, nor is it drawn toward some single event (ultimate disclosure) that is supposed to occur in the future, but is the continuous endeavor to maintain, in a self-conscious personality, a dynamic system of research activities and satisfactions that are continuous with a corresponding system of the never-ending human enquiry into the sense of the human's own existence. One must point out that the "object" (the universe as a whole) that research activities pursue is satisfactory only to the extent that the research maintains a system that does not frustrate existence. Correspondingly, the pursuit of knowledge of the universe as a whole is chosen for the purpose of maintaining the sense of life, and this qualifies as conscious purposive action. It is, therefore, the property of the universe to be a special sort of the whole of life that renders the objects of conscious purposive action, that is, cosmological research, satisfactory. Then the aims of such action are not "ends" in the temporal sense of finality, *because the real and ultimate aim is the persistent maintenance of the system of enquiry into the sense of existence*. Correspondingly, cosmological enquiry is not simply a linear process that aims to terminate in a state of satisfied rest, but a continuous activity, all of which is intrinsic to the satisfactory state of consciousness of those who enquire into the sense of existence. Hence the cosmological quest is at one and the same time both means and end. And thus is our main thesis: there is a *teleology* of explication in cosmology that drives research but has an existential origin, related to the purposiveness of any human action—in particular, to the desire to explicate the sense of existence. But this teleology, *de facto*, determines the strategy of cosmological research and thus acts as a delimiter relating the unconcealment of the universe to the parameters of human existence.

To assign to the thus-outlined view of teleology in cosmological research a more rigorous philosophical status, we turn, as a matter of a historical philosophical reference, to some classical writings of Kant on the paradigm of *teleology* as contraposed to that of causal physical explanation. The aim of this reference is to invoke Kant's argument for the demonstration that the prevailing mathematical physics' trend in cosmological research implicitly contains its own teleology, so that the teleology of research not only does not contradict the mechanistic-like approach to the study of the universe, but on the contrary it is this teleology which ordains physico-mathematical study of the universe. The methodology of research, which implicitly drives cosmologists to the disclosure of the universe, contains teleology as its inevitable component related simply to the purposiveness of any human action.

It is known that Kant in his investigation of teleology in his *Critique of Judgment* made a shift in emphasis away from metaphysical aspects of the study of nature in the direction of the methodology of this study. It was no longer a question of the way the world is, but of what it is possible for us to know and to understand. The phenomenal world is not given all at once, but unfolds as we investigate it. One particular aspect of scientific methodology that was promoted by Kant and we want to explicate here, in the context of cosmology, is the *purposiveness* of cosmological research, which implies (in spite of all attempts in modern scientific enquiry of the universe to get away from teleology as related to the final causes of the universe) that teleology is intrinsically preserved, not in respect to the material references of cosmological theory, but in respect to the very way the cosmological enquiry works, that is, the way teleology pertains to the essence of human enquiry into the nature of the universe, but also into the nature of human beings themselves. To make the explication aimed at, one needs to relate cosmological ideas not to Kant's views on the scientific systems as they were presented in his first *Critique* and other treatises on the natural sciences, but, and this is novel and unexpected, to his *Critique of Judgment*.[1]

Before Kant published his *Critique of Judgment* in 1790, his writings on science, including the *Critique of Pure Reason* and *Metaphysical Foundations of Natural Science*, were concerned with the logic and structure of scientific systems, as well as their justification. Kant believed, according to the knowledge of his time, that the ideal model of scientific knowledge was Newtonian physics, so that, in general, the only legitimate explanations of natural phenomena can be provided by mathematical physics. It is in the first *Critique* that Kant formulated the general methodological presuppositions of the possible scientific system of knowledge. However, it is only in the *Critique of Judgment* that he approached science differently, observing it not as an accomplished system, but as a research program and strategy, as a mode of human activity. The motive of Kant's enquiry

1. There are few papers that deal with the purposiveness of research in the context of Kant's third *Critique*. See, for example, R. E. Butts, *Kant and the Double Goverement Methodology. Supersensibility and Method in Kant's Philosophy of Science*. (Dordrech: Reidel, 1984), ch. IX; R. E. Butts, "Teleology and Scientific Method in Kant's Critique of Judgment," *NOÛS* 24, no. 1 (1990): 1-16; C. Fricke, "Explaining the Inexplicable: The Hypothesis of the Faculty of Reflective Judgment in Kant's Third Critique," *NOÛS* 24, no. 1 (1992): 45-62; A. N. Krichivets, "Apriori, Faculty of Judgment and Aesthetics," *Bulletin of the Moscow University*, Ser. 7, *Philosophy*, no. 3 (1996): 41-50 (in Russian); E. Mamchur and Y. Sachkov, eds., *Causality and Teleonomism in Modern Paradigm of the Natural Sciences* (Moscow: Nauka, 2002) (in Russian); D. N. Razeev, *Teleology of Immanuel Kant* (St. Petersburg: Nauka, 2010) (in Russian).

was the question of the *limits* of scientific investigation that could follow from the setting of scientific process in the context of human history. Prior to this, Kant had already given some responses to this question in his epistemology of science, asserting that scientific investigation is limited to study of the phenomenal world, whose objects are to be located in space and time and subjected to the categories of the understanding. Since our interest lies in cosmology and we are concerned with its alleged "object," the universe as a whole, one can easily grasp that cosmology deviates from that pattern of explanation which was prescribed by Kant on the basis of Newtonian physics. Kant's epistemological prescription, related to space and time, as well as to the exhaustion of the object through categories, cannot work in cosmology and Kant himself placed the enquiry on the wholeness of the world in the context of his cosmological antinomies of reason which explicate the limits of discursive thinking (as well as to cognitive faculties in general). Nevertheless, one wonders whether there are any features, related to our structure as human knowers such that, in spite of constraining the ways and scope of our investigation of the universe, they allow the justification of research strategies of present cosmological theories that pretend to understand the universe as a whole. To answer this question it seems reasonable to refer to some ideas of Kant's *Critique of Judgment* dealing with teleology in understanding nature; for we believe that it is the underlying teleological propensity of human subjectivity which allows it to assert the universe as a whole not as a metaphor or as a limiting pole in human understanding, but as that intrinsic existential *telos* which encourages humanity toward self-understanding in the background of its ambivalent position in the universe. However, before we turn to a more detailed employment of Kant's ideas, one may, as an example, give a hint of how *teleology of cosmological explanation* revealed itself in a methodological move that by its historical and methodological essence was *anti-teleological*.

Cosmological Principle and Explicability of the Universe

The fact that the universe looks isotropic from our earthly position does not entail that the universe should look isotropic from every possible location in the universe. However, if one assumes that it is only from our position that the universe looks isotropic (but not from others), this assumption carries within itself a suspicion in a kind of *teleology*, namely that our position in the universe is somehow biased, or intentionally adjusted in order for us to observe the universe as isotropic. In order to remove this teleological flavor, cosmology postulates the cosmological principle, that is, that the observed isotropy of the universe must be extended to all possible locations, which entails that the universe is uniform in space. In this case the enquiry into our absolute position as observers in space loses any sense because all locations in space become equivalent. As sometimes said, humanity's position in the universe becomes indifferent, or mediocre. Here one notes that the cosmological principle must be understood not simply as a cosmographic *ad hoc* statement, but in terms of that which this principle effectively rejects, that is, the *teleological* assumptions about the biased, selected position of humanity in being, including its special cosmographic place.[2]

As we have already discussed in chapter 3, it is only under the assumption of the uniformity of space and matter that any scientific methodology of studying the universe as a whole is possible. The uniformity of the universe is needed in order to predicate its

2. See E. McMullin, "Indifference Principle and Anthropic Principle in Cosmology," *Studies in History and Philosophy of Science* 24, no. 3 (1993): 373.

properties in terms of the same physical laws in locations that are inaccessible to our reach. The global picture of the universe would not be possible if in every corner of it physical laws were to be different and distinct objects or phenomena infinitely varied. The inexplicability of the universe would follow from incommensurability of the universe with our finite rationality. Indeed, astronomers claim that the number of distinct objects in the sky is finite[3] and some guess that their finitude is linked to the given "finite" cognitive faculties, whose "finitude" cascades down to the integrity of the human picture of the universe.[4] The cosmological principle, by postulating the uniformity of the universe, reduces the number of parameters of its description (related to the large-scale structure) to a minimum and this corresponds to the human ability to explicate the universe as a whole. The contingent facticity of various objects in the universe is replaced by the undifferentiated substance called the "cosmological perfect fluid" (related to clusters of galaxies). The "cosmological principle," implied in the specific form of equations for matter distribution in the universe, provides a cosmologist with a necessary physico-mathematic tool, describing the universe in terms of the efficient causality of the laws of general relativity. However, one must remember that this efficient causality is ordained through the anti-teleological cosmological principle, the principle that, as such, is related to "another teleology," namely to the purposiveness of cosmological research. To amplify the latter point, one must understand that the introduction of the cosmological principle, being de facto a different way of formulating the contingent facticity of what is observed, does not explain the contingent facticity itself, that is, it does not explain *why* the universe

3. M. Harwit, *Cosmic Discovery* (Brighton, UK: Harvester, 1981).
4. W. I. McLaughlin, "Kantian Epistemology as an Alternative to Heroic Astronomy," *Vistas in Astronomy* 28 (1985): 611-39.

is uniform. One needs uniformity of the universe as a principle of its explication, but this demand for the uniformity comes from human researchers, who realize the purposiveness of their actions in explication of the universe through the postulate of the cosmological principle. Thus, being anti-teleological in its initial intentions, this principle remains teleological, but on a different, so to speak, epistemological level.

One must notice that the "cosmological principle" changes our perception of the contingency of our spatial position in the universe. If the universe is uniform in terms of matter and space this means that the fragment of the universe that is visible to us, while being limited in its particular image from our vantage point, makes a fair representation of that which is possible at all. However, one must remember that the actual observability and the possibility of theoretical explication of the universe as a whole requires more than the uniformity of the universe formulated at the level of clusters of galaxies. One expects that the conditions for existence of life in one particular galaxy and particular planetary system are much more sophisticated and are controlled by cosmology only on the level of necessary conditions. The explication of the actual appearance of life and conscious beings on Earth, as well as their perseverance through the hardship of physical existence, requires an appeal to a different type of explanation, not only in the realm of physics. Indeed, since life on our planet is now dependent on and controlled by socio-economic as well as ethical factors, which are secondary with respect to the natural conditions, the continuation of life requires a certain spiritual commitment (as humanity's *telos*). Thus the explication of the universe intrinsically contains elements of the acting *telos* of explication of life itself: the purpose of understanding the universe lies within the purpose of understanding life, and vice versa.

As we have argued, the cosmological principle was introduced in modern cosmology as an "anti-teleological" move that fights teleology on the level of material poles, *nexus finalis*, in the universe. The universe as such is stripped off of any purposiveness, any intention behind its structure, any design founded in extramundane reasons. However, the treatment of the cosmological principle as a transcendental principle, as a principle of explicability, inserts teleology back as the formal purposiveness of research. Indeed, the purpose of cosmology is to explicate the sense of the universe. This purpose originates in the innate urges of human beings to understand their place in the universe and, in general, to understand the sense of human existence. In this sense the purpose of research is directly implanted into the purposiveness of all human actions. Thus the purpose of cosmology becomes a particular manifestation of the existential *telos* of humanity. But, to explicate the universe one needs the cosmological principle, which effectively replaces the alleged purposiveness of the material universe by the formal purposiveness of cosmological explanation.

Correspondingly the centrality of human beings in the universe, being deprived of its absolute cosmographic sense, is reinstated to the centrality of disclosure and manifestation of the universe through rationality. Human beings are still in the center of the universe because the universe becomes palpable and self-conscious through human beings who live in a particular period of cosmic evolution when the large-scale structure of the universe is accessible to observation and explication. Claiming the centrality in the universe in this sense does not mean that we imply that the universe was created with the purpose of our coming into being or that it evolved bearing us in mind. What we say is that the universe supports us and this is a fact inherent in physical laws, not in any particular "conscious will" of the universe itself. On purely physical grounds it seems improbable that such physical agencies in the universe as dark matter intentionally organized its forces over billions of years in order to pull atoms together into galaxies containing stars and ultimately us. The balance of cosmic forces did not cradle our galaxy and the solar system for the benefit of our arrival through biological evolution. Probably, if physics is right, the universe had to be exactly that which it is now according to physical laws. Certainly the facticity of this physics is an even greater mystery related to the boundary conditions, which are not controlled by physics itself. In this sense, if the questions of the purposiveness of creation of the universe through special initial (or boundary) conditions are considered as unacceptable by physicists, one has to admit that there is still another efficacious *telos* of the universe, namely its explicability by human beings for whom the purposiveness is implanted in any practical actions. If then the aim of cosmology is to attempt to understand the contingent facticity of the universe, and through this the contingent facticity of life itself, this aim, referred in cosmological research to the initial (boundary) conditions of the universe, becomes teleological *per se*. Correspondingly, the past of the universe as that "grandfather" or "grandmother" of humanity receives a different meaning: it is not only knowledge of this past that matters, but most of all the spiritual leap to this past as respect for that which constitutes an inherent teleology of human life.

Purposiveness in Scientific Research and Cosmology

In general, any strategy of scientific research aspires to have a guarantee that the patterns of its enquiry will provide satisfactory answers to its questions. Applied to the wholeness of the universe, this aspiration is frustrated by recognizing that the universe appears as an aggregate of contingent laws that apply across the universe, but the origin of these laws (which sometimes is attributed to the boundary conditions of the universe) is beyond comprehension. Even if we grant, along the line of Kant's thinking, that there are synthetic universal and necessary principles that are employed in our enquiry of nature, it does not follow that any particular discovered natural law is also synthetic and a priori, and therefore universal and necessary: it can be the case that these laws are active only in that part of the universe where we live. For Kant all natural laws, although some of them are in a restricted sense "necessary," are synthetic and a posteriori, which means that although they are indeed the laws *in* the universe, it is logically and conceptually possible that the universe as a whole might have been governed by quite different laws. The framework of scientific cosmology implies that the universe must be coherent with the categories of the understanding, but it does not tell us in advance of how and along which ways this could be discovered. To know in advance that the universe must be exactly that which is actually discovered would involve a human subject in the impossible kind of intuition of the universe as a whole, which recalls a mystical communion. Expressed in modern terms, based on the understanding that scientific synthesis rooted in mathematical physics presupposes computational synthesis, this assumption would be tantamount to the possibility of an infinite computational synthesis, which is impossible. In Heideggerian language, the foreknowledge of the whole universe would imply the completion of the unconcealment of the universe and removal of all delimiters in its knowledge related to humanity as a measure of this unconcealment.[5] In other words, humanity must have been considered in this case as commensurable to the universe as a whole, which is philosophically a weak point. Correspondingly in Kantian terms, the intuition of the universe as a whole would amount

5. M. Heidegger, *Nietzsche*, Volume IV: *Nihilism* (San Francisco: Harper, 1991), 91–95.

to divine foreknowledge; but critical philosophy rigorously opposed to any claim for the possibility of mimicking this god's-eye view of the universe.

The brute empirical fact is that we continually encounter *items* in experience that cannot be fully understood, and for which no theoretical concepts are at hand. In cosmology this situation is particularly manifest in its assertion that 96 percent of the material content of the universe is described by the so-called dark matter and dark energy; the observations, to fit into theory, demand one to postulate their existence. However, what these dark components of the universe are nobody knows at this stage, for there are no experimentally evidenced carriers of these two types of matter. In this sense, both these components of the universe represent *particulars* whose theoretical generation and structure cannot be fully understood with reference to mechanical principles based on patterns of efficient causality simply because this very causality, exercised at the theoretical level, cannot be applied to all things in the realm of experience. However, the contemporary model of the universe operates with these notions of dark matter and energy as if they were determinate from their concepts on the level of material causes. Unfortunately this is not the case, and in spite of any realistic commitments with respect to these notions they still remain those particulars for the appropriating of which one requires *judgment*, in a Kantian sense, that is, a faculty intermediate between the understanding and reason. The same is true for the concept of the universe as a whole (in spite of the fact that to think of the universe as a particular seems strange). The universe as a whole cannot be a subject of theoretical knowledge restricted to the schematism of the forms of sensibility and the understanding. The notion of the universe as a whole represents a "particular" that escapes any determination on the grounds of mechanistic-like causality, because by its definition the universe is beyond any *relationality* based on worldly causality. In its sheer givenness to us, the universe is an escaping, irregular "particular," which demands from us an exercise of a faculty of *judgment* in order to attempt to understand it. In this sense the exercising of judgment with respect to the universe ultimately vindicates the very enterprise of physical cosmology: to initiate the process of the study of the universe (based in the faculties of reason) we need to have prior *judgment* of it.[6]

The fact that the universe as a whole falls under the rubric of an irregular "particular" follows from the very definition that the universe as a whole cannot be in the chain of causality that we usually ascribe to its visible part. Here is the essence of Kant's critique of the concept of the world. In this respect the universe is freely generated by escaping all transcendental delimiters, and what it turns out to be could have been otherwise: the universe is presented to us in its contingent facticity.

It is worth making some terminological clarifications in order to elucidate the meaning of that which is meant by the phrase "judgment of the universe." Kant defines the faculty of judgment as "the faculty of thinking the particular as contained under the universal."[7] The faculty of judgment with respect to a given particular that is to be thought as included in a general rule, principle, or law can be *determining* or *reflective*. Any judgment that includes an individual or event in the rule, or under a categorical scheme, is determining. This is true of any judgment that subsumes a particular under a given law or concept as a principle.[8] If the universal rule, principle, or law be given, the judgment that includes the particular

6. Cf. I. Kant, *First Introduction to the Critique of Judgment* II:204–5, trans. J. Haden (Indianapolis: Bobbs-Merrill, 1965), 10–11. Hereafter we will use the abbreviation: FICJ.
7. I. Kant, *Critique of Judgment*, Introduction IV, trans. J. H. Bernard (New York: Hafner, 1951), 15. Hereafter we will use the abbreviation: CJ.
8. CJ § 69, 232.

under it is *determining*.[9] In contradistinction to that latter, the *reflective* judgment is to subsume the particular under a law (the universal) that has to be found.[10]

Reflective judgments seek to subsume particulars under laws that are not yet known, and thus can be thought of as principles that are formulated from themselves. It is because this judgment is not objective (as one would say, not related through certain rules to empirical reality), it acts as a subjective principle for the purposive employment of our cognitive faculties, that is, for reflecting upon objects.[11] The reflective faculty of judgment attempts to find a concept for the manifold of empirical data by means of which this is not represented in a discursive unity of an *objective concept*, but by means of how it appears in the form of a *systematic unity* of a whole organized according to the representation of a *purpose*. Such systematic unity is characteristic for the notion of the universe as a whole containing objects and laws that are supposed to fit in a logical system. The same can be made more precise by saying that the reflective faculty of judgment subsumes the representation of a particular (empirical aspects of the universe) under the *representation* of the universal, that is, the universe as a whole, in spite of the fact that this universal is not given. Correspondingly, it is because no universal is given to the reflective faculty of judgment related to the totality of the universe (or the Big Bang as an encapsulated initial condition of such a totality), that the task of this faculty is to reflect on a given representation of the universe (in its pieces and moments) and to produce a possible concept,[12] that is, "to compare and combine given

9. Kant talks about the determinant faculty of judgment in the *Critique of Pure Reason* (A130/B169ff.). The faculty of judgment is the faculty of subsuming under rules; that is, of distinguishing whether something does or does not stand under a given rule (A132/B171). Thinking of a given particular as contained under a given universal means to subsume the representation of the particular under the representation of the universal.

10. CJ § 69, 232; Introduction IV, 15.

11. CJ § 69, 232.

representation either with other representations" (that is, to provide a *coherent* picture that unites different empirical representations of the visible cosmos), "or with one's cognitive powers, with respect to a concept which is thereby made possible"[13] (the cognitive faculty of the unity of consciousness, which is capable of integrating the variety of empirical data under a conceptual symbol of the totality of all in all). The purposive employment of human cognitive functions is manifest either through achieving systematic unity through the coherence of explanation, or through establishing the relationship of all aspects of the universe with the unity of subjectivity.

Determining judgments are thus objective and are ultimately based upon a priori principles. Reflective judgments apply *maxims* which for Kant are always subjective and are only to be employed regulatively.[14] According to Kant's definition: "I entitle all subjective principles which are derived, not from the constitution of an object but from *the interest of reason in respect of a certain possible perfection of the knowledge of the object*, maxims of reason."[15] (These maxims, in modern terms and in the context of our interest, articulate the *research strategies* deployed in order to understand the universe as expressed in its empirical laws.[16]) Kant's insistence that the maxims are subjective can be nowadays qualified as "intersubjective," or "collective" as related to the community of cosmologists. It is the community of cosmologists that determines what maxim, or methodology, to use in order to fit the variety of data in a reasonable theory, or alternatively adopts the criteria for reality of that which stands behind abstract mathematical theories, in particular, in those cases when the principle

12. FICJ V:211, 16.
13. Ibid.
14. I. Kant, *Critique of Pure Reason*, 2nd ed., trans. N. K. Smith (London: Macmillan, 1933), A666-68/B694-96. Hereafter we will use the abbreviation: CPR.
15. CPR A666/B694: 547 (emphasis added).
16. CJ § 69, 232-23.

of correspondence with empirical reality does not work and one appeals, for example, to coherence arguments. This happens, as an example, in the cosmology of the early universe when cosmologists choose as a maxim (that is, a research strategy) the model of the inflationary universe on the basis of a possible resolution of some predicaments of classical cosmology.[17] Being a theoretical model of the remote past of the universe, inflationary cosmology cannot be tested empirically, so that its adoption as a strategy of theoretical research has a precarious status deeply based in a belief-motivated commitment supported by the collective opinion.[18] Thus inflationary cosmology functions in the conditions of a reflective judgment imbued with the *purposiveness* of explanation of the variety of astronomical facts as well as resolution of some meta-empirical puzzles.[19] In addition one can mention another purposive dimension of inflationary cosmologies, namely the desire to address the issue of generic initial conditions of the universe, whose specialness (according to classical cosmology) points toward a fundamental contingency of the observable universe. It is by means of this "inflationary maxim" that cosmology reflectively seeks a knowledge of the kinds of theoretically expressed laws that can be used for

17. The bibliography on technical aspects of the inflationary cosmology is vast, and we give references only to two books: R. Penrose, *The Road to Reality* (London: Vintage, 2005); S. Weinberg, *Cosmology* (Oxford: Oxford University Press, 2008).

18. Both Penrose and Weinberg stress a point that the whole inflationary scenario is based on the assumption of the existence of a fundamental field, inflation (φ). Penrose while commenting on inflationary cosmology points out that the introduction of a new field φ into the "menagerie of known (and conjectured) physical particle/fields" was dictated solely by the desire to have an exponential expansion, so that no other physical motivation of relating this field to other known physical laws was established (Penrose, ibid., 751). Weinberg on his side while commenting on the hypothetical predictions of eternal inflation about the existence of many disjoint universes asserts that the validity of this idea "will probably have to come from progress in fundamental physics, which may verify the existence of a suitable inflation field, rather than from astronomical observation" (Weinberg, ibid., 217).

19. The puzzles of flatness, horizon, and monopole (see, e.g., Penrose, ibid., 742-57; Weinberg, ibid., 202-7) are not empirical since there is no direct empirical predicament that could invoke these puzzles. They appear on the level of reflection while generalizing different aspects of physics, in particular, cosmology with the physics of elementary particles.

the purpose of conceptual management of those *particulars* (three predicaments mentioned above, which are not part of empirical research) that as such escape the rubrics of efficient causality available to preinflationary cosmology. Correspondingly, reflective judgment on the unity and integrity of the visible universe, based on an intersubjective maxim (methodology), is *neither true nor false, not even probable or improbable*. Such a judgment is rather a *rational estimate* of the way the universe operates, and expresses a normative research strategy to render the universe explicable. The purposiveness of the strategy implied in this judgment is related to the demand for explicability.

If one follows Kant, one must think of two major research strategies or maxims of reflective judgment that can be applied in cosmology:

1) The maxim of mechanism: "All production of material things and their forms must be judged to be possible according to merely mechanical laws."[20] In other words, all aspects of the universe, including its alleged wholeness, must be described by using physical laws established by earthly physics, so that the comprehension of the universe as a whole can be established by means of ascending through the empirical set of causations in the visible universe;
2) The maxim of teleology: "Some products of material nature [universe, AN] cannot be judged to be possible according to merely mechanical laws (To judge them requires quite a different law of causality, namely, that of final causes)."[21] It is problematic to think of the universe as a whole as being subjected to physical laws, for these laws act across the universe.

20. CJ § 70, 234.
21. Ibid.

> In this sense if one enquires into the contingent facticity of the universe as a whole and its laws, one has to appeal to another type of "causality," that is, causality as *origination* (not in a temporal sense) from some foundation, which itself is not part of that which is subjected to this alleged causality.

Correspondingly, in accordance with Kant, if we apply these maxims of judgment in respect of knowledge of the universe in cosmology, it becomes quite clear that the maxims cannot be formulated and treated as *constitutive claims* about the universe. The inflationary model remains only a model! The conflict between that which is expressed by the propositions "the production of the universe is possible on mere mechanical laws" and "the production of the universe is not possible on mere mechanical laws" is irresolvable because there is no a priori way of determining the possibility of the production of the universe only through the empirical laws of nature: the universe cannot be rerun in the laboratory. Instead, the maxims express two different research strategies, that is, the ways of studying the universe, which, as we argue, are not only non-incompatible, but, in fact, are both unavoidable in order to sustain the wholeness of research.

Indeed, on the one hand, in view of Kant, unless the principle of mechanistic explanation is applied ". . . there can be no proper knowledge of nature at all."[22] In accordance with this, in cosmology, one must push the investigation of the universe along the lines of mechanistic explanation (that is, the explanation based on causal physical laws) as far as possible. In our empirical investigation of the universe as causally connected, we should proceed in terms of the universe's purely causal physical laws as far as we can, for in these laws lie the true grounds for a physical explanation of the universe,

22. CJ § 70, 234.

which constitutes that scientific knowledge of the universe which we exercise through reason. According to Kant the efficient causality whose necessity was demonstrated in the Second Analogy of the *Critique of Pure Reason* authorizes every event.

However, the universe as a whole is not an object of possible experience and it cannot be labeled by the term "event" in the sense of an ordinary happening in space and time; one cannot know anything about the ultimate causal features related to the universe as a whole (the antinomies in the *Critique of Pure Reason* accentuate this point). We find that the universe as a whole (in contradistinction with the finite products of nature) represents what Kant would call "genera," in which the efficient causes acting upon the universe and giving it coherence and integrity are contemplated by us (through study of the universe's constituents) as based on the concept of a *purpose*, if we want to experience the universe in terms of a principle appropriate to its inner possibility. Suppose that we tried to judge the universe's form and its possibility merely in terms of mechanical laws, where it is not the idea of the effect which is regarded as the basis that makes the possible cause of this effect, but the other way round, that is, the cause is regarded as the basis that makes the effect possible. If we tried this, we could not acquire (regarding the specific form of the universe) any empirical concept that would enable us to establish the transition from cause to the effect. For the effect we see in the universe as a whole is caused by its parts not insofar as each part on its own contains a separate basis, but only insofar as all of them together contain a joint basis (related to the whole of the universe) making these parts of the universe possible. But it seems to be quite contrary to the nature of physical-mechanical causality that the whole should be the cause that makes possible the causality of the parts; rather, here the parts must be given first in order for us to grasp from them the possibility of a whole. Further to this, one can assert that the presentation of a whole as preceding the possibility of its parts functions as a mere *idea*; and when this idea is regarded as the basis of the causality it is called a *purpose*. Then one can conclude that in the case of the universe as a whole we cannot investigate its character and its cause as given in experience (that is, explained by reason) without presenting it, its form and its "causality," as determined according to a principle of *purpose*. But the employment of the maxim of teleology in cosmology does not eliminate or replace the maxim of mechanistic explanation, it just demonstrates that in some cases the maxim of mechanism cannot be straightforwardly applied. This happens in the case of the universe as a whole, whose idea acts as a *telos* of cosmological explanation initiating *de facto* all particular models that are based on the efficient physical causality. The universe itself can only be understood as meaningful in its integrity if it is seen as being "produced" for the purpose of its explicability by human beings.

"The Universe as a Whole" as *Telos* of Cosmological Explanation

The implementation of the maxim of teleology does not imply that it constitutes an objective pole of that which could correspond to the notion of the universe as a whole. Rather the maxim of teleology outlines the strategy of research as the explicability of the universe. For example, to understand the universe we need to find such a unified description of it in a minimal set of physical laws related to its initial or boundary conditions, so that this ideal description will act as the *telos* of cosmological research. The idea of the Big Bang in this case appears to be exactly the *telos* of explanation, the *telos* which is paradoxically referred to the physical past of the universe, and not its future.

In anticipation of a possible skepticism or objection to this conclusion, which could arise from the camp of cosmologists, one could make a general comment that teleology operates as an a priori presupposition in any scientific enquiry: this teleology is related to the goal of scientific explanation, its objective to disclose nature along the ways humanity is capable of doing so, in order to understand humanity's own "end," a "purpose" of its presence in the universe. In this sense, if teleology is implicitly present in the foundation of any scientific enquiry, it naturally subordinates mechanistic explanation ("ordains" this explanation) as the way to achieving the goal of this explanation, the goal which, as such, transcends all possible particular mechanistic explanations. Hence it is teleology, as the principle of purposiveness of research, that vindicates the use of the mechanistic trend in cosmology's attempts to deal with the universe as a whole in spite of clear understanding that this mechanistic trend will never be able to accomplish the goal for which it is used. It is because we must necessarily think of the universe as having a purpose explicable by us that we are justified in applying the principle of causal efficiency to its visible parts. In a characteristic way teleology of explanation in cosmology functions as a transcendental principle, as a demand for the universe's explicability that cascades down toward some particular theoretical requirements, such as, for example, the cosmological principle, which receives its specific incarnation in the mathematical formalism enabling the universe's explication.

Now, it is clear that the concept of purposiveness of the universe implied by the teleology of cosmological explanation serves us merely *for reflecting* on that which we would like to treat as an object, not for *determining* the object through the concept of a purpose, and that the teleological judgment about the possibility of contingent production of all possible objects in the universe is a merely reflective and not a determinative judgment. What we say is that our thought, in producing a concept of the universe as a whole, includes the thought of the "presentation of a purpose" (that is, the universe as a whole as a material pole), because this thought serves us as a principle by which we can guide our investigation of the universe as a physical entity, and also because thinking the presentation of a purpose here might help us devise theoretical and experimental means to further the effect of the anticipation of the universe. However, expressing ourselves this way we do not attribute to the universe a cause that acts in terms of the presentation of purposes, that is, in terms of some material pole. If we did so, we would be making a determinative teleological judgment which would be a *transcendent* judgment, since it would suggest a causality that lies beyond the bounds of the physical universe as such.

Using a *teleological principle of explanation* of inner possibility of the universe as a whole, we leave undetermined whether the purposiveness of the universe implies or does not imply any transcendent *intentionality*. If a judgment asserted either of these alternatives it would no longer be merely reflective but would be determinative; and the concept of a purpose would no longer be a mere *concept of the power of judgment*, but would be connected with a *concept of* reason, that is, the concept of a cause that we posit beyond the universe and which acts intentionally, whether we are in this case judging affirmatively or negatively. For example, if we imply this purposiveness as related to the intrinsic property of the universe that we do not experience, so that the purpose becomes, in Kantian parlance, a thing in itself, then this purpose is not a constitutive concept of understanding or of reason, but it can function as a regulative concept for the reflective judgment.

If the universe, as an object of our teleological judgment, is to be thought of as employing *reason*, so that reason produces the concept of a purpose, this would be something more than we could expect from judgment. All that judgment can do, as a separate cognitive faculty, is to consider the relation in which two faculties, that is, imagination and the understanding, are related in a representation prior to the emergence of any concept, thereby perceiving the *subjective* purposiveness of the universe relative to its apprehension by the cognitive faculties. Hence, while judgment can indeed have a priori principles for the form of intuition, yet it cannot have an a priori principle for the concepts concerning the production of things, and so the *concept of a real purpose* of the universe, or its end in a physical sense, lies completely beyond the realm of the faculty of judgment. If in dealing with the teleological purposiveness of the universe, this purposiveness is presented through concepts, judgment will have to put the understanding (applicable to the realm of experience, that is, the visible universe) in a relation to reason, which, in fact, delimits the understanding, thus making impossible any speculations on the universe as *natural* purpose. If the universe is treated as a concept, that is, as a rational idea, it is subject to the Kant's first *Critique*, and thus can function only as a regulative concept for the reflective judgment, to guide our investigation about the universe as a whole by a distant analogy with our own causality according to purposes and in our meditations upon the universe's ultimate ground.

It is true that the concept of purposes and of purposiveness can be a concept of reason in the case where we attribute to reason the basis that makes an object possible according to the implied purpose. In *products of art* we can become conscious of reason's causality as giving rise to objects, which are therefore called purposive or purposes. But the analogy with art is very limited, because the concept of the purposiveness of the universe relates reason, as cause, to things (the universe) where no experience informs us that reason is the basis that makes them possible. To attribute purposiveness and purposes to the universe is to use a concept, whose correlate cannot be found in experience. The concept of purpose is employed by judgment only for reflecting with its help upon objects, so that, in the experience of disjoined appearances and objects we are directed to the possibility of uniting them in a system. This is the reason why we talk here about the *teleology of explanation* of the universe, and not teleology of the universe as its natural end.

Indeed, the idea of a *telos* of scientific explanation comes through the reference to our own purposive behavior in art, poetry, and in practical matters: it is there that we can reasonably claim that effects of the activity are also its causes, that is, those purpose-based projects that brought the effects into being. For example, if the purpose is to produce a painting of a landscape, correspondingly the effect is a particular reflection of the surrounding nature in a piece of art. This piece of art is an effect caused by the purpose to produce the image of the landscape, but the effect here becomes, through the *telos* set up at the beginning, the *telos* of producing an image of the landscape. In cosmology the situation is similar in that a cosmological theory can be considered as an effect caused by the purpose to produce the "landscape" of the universe, so that the effect becomes through the *telos* of explanation set up at the beginning. The difference is that the desired "landscape" of the universe cannot be accomplished, so that the effect of the cause is an ongoing intertwining with this cause. Correspondingly the *representation* of the universe as a whole, being a cause for the unfolding theoretical explication of the universe, according to Kant's definition, forms a purpose: ". . . the product of a cause whose determining ground is merely the representation of its effect is called a purpose."[23] But this purpose is itself being constituted through the fact that it is not something preconceived and is always in the making through the research that is launched by this very

23. CJ § 77, 256.

purpose. Since it is understood that this purpose cannot be achieved as material knowledge, the question is as to how one can conceptualize "purposes" of cosmological explanation as fundamentally distinct from purposes related to technical, artistic, or practical contexts.

The representation of "the universe as a whole" is fit for judgment as a purpose of explanation because it satisfies the following condition: the representation of its existence and the form of its parts must be possible only in relation to the whole. In modern cosmology this is manifested through a perception that the large-scale structure of the universe effectively controls all physically isolated objects through the sheer fact that necessary conditions are determined by the global cosmological parameters (an example of such a conceptual causation is the famous anthropic principle in cosmology linking the necessary conditions of biological life on this planet with the large-scale parameters of the universe). This is not empirical material knowledge, but the representation of the wholeness of the universe as related to its parts. As a purpose and hence as a potential end, the "thing" must be understood with reference to an idea determining a priori all that is to be contained in it. With respect to the whole of the universe it is impossible a priori to determine all its contingent components. However, the theoretical hypothesis that all its contingent components originate in the idiosyncratic singular state called "Big Bang" (in conjunction with various versions of the so-called "Theories of Everything") attempts to represent the universe a priori as the potentiality of everything. Such a definition is usually made in the form of mathematics, which is devoid of any *intuitive* content (for this definition is hardly to be related to the world of experience); it is constructed on the *trans-logical* premise of the universal validity of intuition-free mathematics. This is also true with respect to the theories of multiverse, which extend the idea of a unique singular state to the underlying ensemble of universes, similar to the perennial idea of the plurality of worlds. The above-mentioned trans-logical premise corresponds to what Kant would describe as an attempt to establish accordance between the contingent things in the universe with our judgment in terms of necessity. To achieve this we must "think of *another understanding*, by reference to which and apart from any purpose ascribed to it, we may represent as *necessary* that accordance of natural laws with our judgment, which for our understanding is only thinkable through the medium of purposes."[24] The "another understanding" is that transcendent "subject" which contains as its intentional correlate the idea of the multiverse. Cosmology wants to assert the multiverse as that necessary foundation of this contingent world, where the contingency is related to the fact that this world is just one among the infinite many others.

Interestingly, the idea of the Big Bang, as an initial state of the universe, that is, the nonoriginary origin of the universe, effectively contributes to an argument of why the universe as a whole can be attributed the title of the purpose of explanation. If one uses the Kantian phrasing, the universe has parts that "combine in the unity of a whole such that they are reciprocally cause and effect of each other's form."[25] Cosmology treats the Big Bang as the cause of the visible universe, where this visible universe becomes an "effect" of the Big Bang. However, in cosmological knowledge as a process of purposive human action, it is the display in the visible universe that initiates the theoretical ascent to the notion of the Big Bang as the unifying origin of the variety of the visible. In this sense the Big Bang can be treated as an explanatory effect of integration of the visible universe in the whole system. It is then not surprising that the Big Bang, as the purpose of cosmological explanation, becomes its

24. CJ § 77, 255 (emphasis added).
25. CJ § 65, 220.

efficacious *telos*. In the case where we seek to estimate the universe's causal dependence in such teleological forms, we are *constructing* the nexus of efficient causes productive of the universe's form—that is, the Big Bang—as a concept that is not the constitutive concept of the understanding and reason, but a regulative concept for the reflective judgment in order to guide our investigation of the universe.[26] A similar situation concerns any knowledge: every empirical cognitive judgment assumes the objectivity of determining concepts and laws that fit in a logical system. Such cognitive judgments consider their objects hypothetically as purposive, that is, as the purpose of explanation; their objects are products of intentional actions (their intentional correlates), without assuming that they are artifacts (that is, material creations).

The Explicability of the Universe as Its *Telos*

As we have discussed above, the reflective judgment, unlike the understanding and reason, can conceptualize any "particular" that escapes exhaustion by its discursive signifiers. This can be done either under the principle of mechanistic explanation or under the principle of teleology. Kant observes that this point involves a principle of reflection, which in a contemporary language can be qualified as a principle of explicability: ". . . for all things in nature empirical

26. Cf. CJ § 65, 222. If one thinks naturalistically, that is, treats the Big Bang as a material final cause of the universe, then it clashes with Kant's view according to which only organisms satisfy the above-mentioned conditions for being the ends of nature. However, despite his remark that organisms provide "objective reality" to the conception of an end of nature and allow us to distinguish such ends from merely practical ones, he insists that the idea of an end has no constitutive meaning, but only provides us with a rule for guiding investigation of organisms. Here one can point out that the ideas of the Big Bang as nonoriginary origin of the universe can be paralleled in an organism-like fashion with the idea of birth of any human being as that phenomenologically hidden nonoriginary origin of any individual human life which acts as an "end" in itself, because the whole sense of the human life originating in this event is ultimately directed to the explication of that mystery of birth (see previous chapter).

concepts can be found."[27] In other words, there is a correspondence between the universe and our understanding of it: there must be in the universe the conditions of its explicability. This means that one can reflect upon any object in the universe as organized in a sense that it is fit for knowledge by human beings. It is this purposiveness of the universe for being known, that is, the adoption of a teleology as being hidden in the foundation of any knowledge, that encourages further investigation of the universe along the lines of efficient causality implanted in physical laws. The physical laws are applicable in cosmology only as being ordained by its intrinsic teleology, that is, the teleology of explanation: one must apply empirical physical laws in order to understand the foundation of the visible universe. The strategy of cosmological research presupposes that either one of its parts, that is, that which estimates the universe teleologically, as an ultimate goal of explanation, and that part which estimates it, so to speak, "mechanically" (in order to fulfill the intention of having a systematic unity of the universe), are related to each other.[28]

If teleology ordains a "mechanical part" of explanation in cosmology, there must be epistemological assurance that the laws will be found, which fulfill the task of mechanistic explanation. Kant's attempt of such an assurance involved viewing nature as a whole in analogy with organisms as self-preserving: the affinities between species and members of species are provided through principles of preservation. One can extend this conjecture by introducing a principle of preservation of self-identity of persons in the course of their life from birth to adulthood and death. And it is this self-identity which is the ultimate purpose in understanding the uniqueness of a person. The explicability of the uniqueness of a person is driven by the desire for explicability of its unique origin, that is, birth.

27. FICJ V:211, 16.
28. Cf. CJ § 77, 258.

This explicability presupposes a form of logic that is preserved in the course of life. Cosmology in turn presupposes that the contingent laws of nature have analogous affinities to their nonoriginary origin, and this affinity is provided by the logic of mathematical physics. The universe is to be treated as a system of empirical laws in which the kinship relations are seen as holding between those propositions we produce in order to obtain understanding of the empirical that contains us: by understanding nature we understand ourselves. Kant, by developing the analogy between kinship of properties preserving life and the integrity of properties preserving nature, emphatically states: "Therefore it is a subjectively necessary, transcendental *presupposition* that this dismaying, unlimited diversity of empirical laws and this heterogeneity of natural forms does not belong in nature, that, instead, nature is fitted for experience as an empirical system through the affinity of particular laws under more general ones. This presupposition is, then, the transcendental principle of the faculty of judgment, which is not simply a capacity of subsuming the particular under the universal whose concepts are given, but also the converse, of finding the universal for the particular."[29] Continuing this thought, one can add that indeed the explication of the contingent variety of empirical laws and phenomena in the universe implies a transcendental principle of the faculty of judgment of finding the universal (Big Bang) for the particular (aspects of the visible universe extended in space and time) and this is analogous to finding the universal (the event of birth) for the particular (aspects of person's life extended in time) in the context of human personhood.

The latter quote from Kant, in the preceding paragraph, can be translated in the context of this enquiry as a principle which makes research strategies in cosmology intelligible and not as a metaphysical

29. FICJ IV:209, 14–15.

claim about the universe: in cosmology we are not subsuming all particular astrophysical discoveries under the known universal laws (we are not simply describing galaxies, their clusters, or the microwave background radiation in terms of the laws of physics established on earth as if these entities would be disjoint from each other); and vice versa, the universe as a whole is that universal which is being found for the particulars (all astrophysical objects in the universe are related to each other, for example, through their common evolution from the Big Bang). It is this transcendental principle of the faculty of judgment which alone makes possible the explication of the universe as a whole. Whereas Kant employs an idea of affinity of laws for one another by referring to the analogy with the life-preserving properties of organisms that provide stability both for the species and for the individuals, in cosmology one can correspondingly talk of stability of properties of things and of forces that hold astrophysical entities together, all of them subjectively assured by the principle of affinity (galaxies and clusters of galaxies are held by the affinity to the law of gravitation[30]), which ultimately expresses the unity of the universe.

As a matter of methodological comparison with Kant's example of the inductive generation of the law of universal gravitation[31] one can point toward the construction of a scenario of the so-called inflationary cosmology. As we have mentioned above, historically, the emergence of inflationary cosmology is usually accounted for as a response to three meta-empirical problems (puzzles) in the standard hot Big Bang cosmology. They did not appear as a result of formal disagreements on the level of theory and observations, and by its

30. There are some hypotheses, however, that the stability of galaxies, which is now accounted for by reference to the so-called dark matter, may be sustained by laws of gravitation that deviate from the Newtonian one (to which Kant referred in his writings). If this turns out to be true, then the hypothesis of dark matter could be abolished.
31. CPR A662–63/B690–91.

origin they are not due to physical causation among the elements of theory and observations but rather represent the influence of the intentional motives originating in the integrating tendencies of human subjectivity to account for the contingent facticity of the initial conditions in the universe. Before the advance of the inflationary model in the 1980s it was believed that the problem of the initial conditions of the universe needed a quantum description. The discovery that one can use classical gravity to address the problem of the initial conditions made the whole theory less speculative and in this sense "realistic." Thus, the three problems (or puzzles) that led to the advance of inflationary cosmology are known as the flatness puzzle, the monopole problem, and the horizon problem. Seen at the meta-empirical level these problems originate from different angles of view of the universe and thus have a different weight in the inflationary argument which attempts to resolve them. Without going into detail one can state that under the assumption that all these problems exhibit affinity to the same laws of physics, there was made a hypothetical suggestion that all problems can be resolved if the general law of expansion of the universe were to be exponential in time. Correspondingly, the overall substance of the universe is thought of as an indefinite field (called "inflaton"). Thus, using the Kantian language, the universals, like the inflaton field, which is responsible for another universal—the exponential geometric expansion—were introduced in order to subsume "recalcitrant particulars"[32] (that is, three cosmological puzzles). Exponential expansion of the universe is treated as the "natural kind" that establishes affinities between the kinds of meta-empirical puzzles, thus providing unification of otherwise unrelated sorts of observed phenomena. In analogy with the Kantian appeal to the principle of affinity which discovers ". . . a unity in the generic forms of the orbits,

32. This is the expression from the paper (Butts, ibid).

and thereby a unity in the cause of all the laws of planetary motion, namely, gravitation," the assumption of inductive affinities in the standard cosmological model which conciliates otherwise disparate and unrelated facts (three problems) discovers (predicts) the law of the exponential expansion of the universe. What is important here is that there is a *teleological commitment* implanted in the whole history of construction of inflationary cosmology related to the desire to explicate the unity of otherwise disjoint aspects of the universe. It is this sought unity (as well as the desire to construct generic initial conditions in the universe that remove the problem of their contingency) that becomes a *telos* of explanation in inflationary cosmology, so that the implicit teleology ordains the usage of the "mechanistic-like" trends (that is, causal physical dynamics) for explication of this sought unity, thus producing a physico-mathematical model of the inflationary universe.

Another important aspect of Kant's explication of the function of reflective judgment in science is that it requires the presupposition that the universe's empirical laws are related in the form of a deductive system, that is, the laws discovered on the assumption of inductive affinity also form a deductive hierarchy ranging from most to least general propositions (in contemporary usage this would amount to the assertion that empirical laws have their mathematical doubles organized in structures). The unity of natural kinds (empirical objects like galaxies and their clusters) is preserved by inductive affinity; the unity of the truth of the system of laws is preserved by deductive logic (unfolding from the mathematical hierarchy underlying physical laws). It is not a matter of theoretical knowing that entitles us to think that nature forms a set of inductively and deductively interrelated empirical laws; for knowledge itself is possible because of the underlying correspondence between inductive affinity and deductive hierarchy. Both the "mechanistic-like" explanation and the teleological pattern of reasoning do coexist with no serious contradiction. In both cases, in order for human patterns of judgment to be possible at all, the whole of the universe must be treated as being "designed" in a very special sense (related to what has been said above): the correspondence between inductive affinities of empirical laws and deductive patterns of mathematical explanations is the manifestation that the universe can be treated as being produced in order to become explicable. The principle of the universe's explicability, being a transcendental delimiter in knowledge of the universe, thus reveals itself as that hidden *telos* of cosmological explanation, from which all cosmological theories receive their theoretical content. The *telos* of research is to explain the universe.

As a result of our deliberations we can see that the maxims of teleology and of mechanism initiate different stances with respect to research in cosmology that turn out ultimately to be intertwined. Our intention was of showing that human faculties of judgment cannot operate without teleological commitments, and that, even though mechanistic-like explanations promote the interests of categorical knowing and make possible knowledge of nature "in the true sense," that is, knowledge in mathematical physics, the maxim of mechanism is ordained by the active *telos* of cosmological explanation related to its explicability.

Let us now elucidate the point of constructing the concept of the universe as a purpose of explanation of nature in general. According to Kant those things that are considered to be possible only as purposes of nature provide the "best proof of the contingency of the world-whole."[33] The universe as a whole, taken to be the purpose of our quest, and found by us in the given form of the visible

33. CJ § 75, 246.

universe, could have been otherwise. Modern cosmology understands that this is the problem and attempts to escape the contingency of the universe (and hence the design-like pattern of its explanation) by constructing the most generic *initial* conditions that would follow from the immanent physics.[34] Sometimes the initial conditions are replaced by the "boundary" conditions, or, more precise, by a form of the dependency upon "natural" laws enacted *by* or *in* the so-called multiverse—a new version of "intelligent and extramundane" being that operates as a designer of our universe. This move of thought definitely does not mean that we can have cognitive access to such a being. It is conceptually impossible for us as human knowers to conceive of the possibility of an ordered universe (its so-called fine tuning) without also conceiving of it as designed (that is, ordered or fine-tuned), and, as a result, that where there is design there is a designer. However, Kant insists that "we can form absolutely no concept of the possibility"[35] of such a designer. Human beings either construe their reflective commitment to a designer of the universe as involving facts about human cognitive capacities, thus upon psychological laws (which manifests the contingency of the intelligibility as such), or they construe on the basis of the tautology that thinking of design is logically also thinking of designers (this corresponds to the natural attitude, when designer is thought of as something extraneous to the facticity of the universe and consciousness). When the conceptual item of the ground of the universe's facticity (or an idea of extended reality of that which "orders" the universe) extends human thought beyond the limits of possible experience, theoretical knowledge becomes impossible, and one must look to the special conditions that adjust our thought

34. See A. Albrecht, "Cosmic Inflation and the Arrow of Time," in *Science and Ultimate Reality: From Quantum to Cosmos*, ed. J. D. Barrow, P. C. W. Davies, and C. Harper (Cambridge: Cambridge University Press 2003), 363–401.
35. CJ § 75, 246.

to fit this idea. The understanding is not precisely suitable for this purpose, for it determines (through application of the categories) only those particulars that are given in sensible intuition. Judgment, however, requires that all (not only given) particulars be subject to understanding, that they be kept intelligible and explicable under some law. In respect to the universe, our limited cognitive capacity makes it impossible for us to fulfill the demands of judgment in any theoretical way: knowledge remains apophatic, that is, incapable of exhausting the reality of that which it signifies in terms of signifiers alone. In spite of this we are able to think (within the demands of mathematics) that the universe is organized so that we will be able to understand it. This presumption, or better, a certain belief-based commitment, establishes the intrinsic *teleology of research*: the research aims toward its *telos*, that is, the allegedly objective pole of the concept of the universe as a whole (theoretically explicated, for example, in the idea of the Big Bang as its encapsulated origin). However, in order for this thought to be fully coherent, we must also presuppose (again as a matter of logic), the possibility of an understanding different from our own, an intuitive understanding for which all particulars and all individuals, including the universe as a whole, are given at once. This understanding either proceeds from the theological teaching on the *creatio ex nihilo* (the universe is created by God) or, alternatively, as an infinite compendium of universes with all possible logically admissible structures that can sustain universes, including our own (the concept of the multiverse is implied here). This understanding (an intelligent designer of the universe or multiverse) then stands as the rational ground of our (God-given) expectation that the universe as a whole (the infinite) can be captured by our (limited, that is, finite) form of understanding.[36] According to Kant, the metaphysical infinitude of human beings proceeds from the awareness of our practical rationality, which inherently transcends all that is available to our senses. We are left with a sense of admiration and awe, focused on our own rational agency and its freely created moral law. This is how we, finite beings, "feel" about the metaphysical aspect of our infinitude. The question is: how to link the sense of metaphysical infinitude, having a practical character, with our epistemological limitedness related to our finitude in space and time.[37]

Here, together with Kant, we need to invoke the representation of the *will*. According to one of his formulations of the moral law: "Act as if the maxim of thy action were to become by thy will a *universal law of nature*."[38] In spite of the fact that the scientific action is not exactly related to the moral one, the urge to conduct research with the aim of disclosing the unity of nature in the concept of the universe as a whole, is something that originates in *human will* (it is not a necessity that is subordinated to the biological order of existence). One can say, it originates in man's desire to imitate in themselves the divine image which aspires to avoid any circumscription of humanity to physical nature and to "create" nature according to the human will exercised in a God-like fashion. It is this will, realized in posing the *telos* of cosmological explanation—the unity of all in all, or the Big Bang—which ordains reason for exercising all possible mechanistic patterns of explanations to asymptotically approach its goal.

To amplify the point discussed above, one must stress that the affinity between human understanding and the unexamined aspects of the universe can only be either an inductive or a logical connection

36. CJ § 77, 255.

37. This problem is discussed in A. W. Moore, "Human Finitude, Ineffability, Idealism, Contingency," *NOÛS* 26, no. 4 (1992): 427–46; *The Infinite* (London: Routledge, 2001).

38. I. Kant, "Fundamental Principles of the Metaphysic of Morals" 48, in *Kant's Critique of Practical Reason and Other Works on the Theory of Ethics*, trans. T. K. Abbot, reprinted from the 6th edition (London: Longmans, 1959), 39 (emphasis added).

between propositions (laws). But the expectation of a unified systematic nature cannot be realized by means of either induction or deductive logic (indeed we cannot accomplish this synthesis). Instead, it is only on the presupposition either of a designer of the universe or our extended cognitive God-given abilities (divine image) that we can trust induction, and that we can anticipate the application of logic as the form of organization of empirical laws of nature. There is a hidden purposiveness related to the explication of the very possibility of this knowledge. The postulation of understanding that is capable of a God's-eye view cannot be viewed as any kind of comprehension of a being actually possessing intuitive understanding and the capacity to create an ordered universe. The concept of a designer is in this sense replaceable by the regulative idea of an ordered universe in principle always accessible to human comprehension. The postulation of a designer of the universe thus amounts to nothing more than rational acceptance of an assumption about the systematic order of nature and the affinity between our cognitive capacities and that nature (expressed in the transcendental conditions of knowledge). Finally, it is not difficult to comprehend that Kant's argument in § 78 of *Critique of Judgment*, that the principle of mechanistic explanation must always be subordinated to the principle of teleology, becomes seen, in the context of cosmology, as something that goes without saying: indeed the judgment of the existence of the universe and its unity initiates the cosmological research that attempts to tackle the object corresponding to this concept discursively, along the lines of mathematical-physical explanation.

If either teleological or mechanistic-like explanation were to be applied in disjunction with each other, they could be seen to be "competing" explanations each excluding the other. However, understood as principles related to the overall intelligibility and explicability of the universe, they can be employed in harmony when one complements another. What is meant here is that even the mechanical investigation of the phenomenal world as the only methodology for the production of positive theoretical knowledge *presupposes* the continuing *ability* of human understanding to uncover the secrets of the universe in the form of laws. In other words, the intelligibility of the universe is postulated on the basis of a rationality that in its practical function manifests itself as the transcendental condition of explicability. By arguing that the methodology of cosmology based in mathematical physics is subordinated to the intrinsic teleology of explanation, we, together with Kant, are not arguing that the maxim of teleology is more basic than the maxim of mechanism. That conclusion would be inappropriate because it is impossible to know beforehand whether this particular research strategy is more successful than another: the justification of the rules of investigation is grounded in research practice. Success in generating physical theories in cosmology is the test of those methodological programs that are drawn from the *telos* of explanation, that is, the universe as a whole. In addition to this, the maxim of teleology holds only with respect to the process of investigation but not to its result. To think that this maxim is related to the physical referents *per se* would be equivalent to understanding that the entire body of empirical laws is to be judged to be an organism with visible or invisible ends (in the style of Plato); but this has no justification whatsoever. Indeed if the teleology of cosmological explanation were to be wrongly associated with some remote physical reference in the future, that is, as if cosmology is destined to disclose the goal of the universe's evolution in the future, the whole novelty of our claim for the Big Bang as a *telos* of cosmological explanation would be obscured. Teleology is efficacious in the process of investigation of the universe because this teleology originates from within humanity which, being desperate to disclose the sense of its own existence, "humanizes" the universe, becoming not simply microcosm, but rather "macro-anthropos" (the humanization of the universe is sometimes described as a continuous embodiment of humanity in the universe). Following the *telos* of cosmological explanation in research, humanity does not lose itself in the process of articulation of the universe. This articulated universe does not "swallow" humanity as its infinitesimal part. Humanity remains what it is, being affected by cosmology only in that the wonder and anxiety of its existence in the universe becomes more expressive. The goal in the cosmological explanation proceeds not from the universe as such, but from that humanly inferred aspiration to order the process of knowledge toward the explication of the human condition. In this sense the present teleology of cosmological explanation, in agreement with Kant, does not abandon the crucial epistemological standing of the mechanistic methodology in actual cosmological research, and it is the efficacy of teleology that generates new discoveries and theories of the universe through employment of the formalism of space and time, as well as the categories of the phenomenal world. In this sense cosmology as such does not manifest any teleological laws of the universe; all particular cosmic phenomena are subject to empirical laws as expressions of causal mechanisms. Correspondingly when Kant writes at the end of § 78 in *Critique of Judgment* that in spite of the necessity to explain all products and occurrences in nature by the principle of mechanism, he admits that there are things "which we cannot even state for investigation except under the concept of a purpose of Reason,"[39] and these things "must, in conformity with the essential constitution of our Reason . . . be subordinated by us finally to causality in

39. CJ, 197.

accordance with purposes."[40] We must understand this teleological principle in application to cosmology as being, not the maxim of teleology related to the material pole of the universe (as organism), but as the presuppositional principle guaranteeing that human understanding is fit for making the phenomenal world intelligible in order to explicate the sense of humanity's existence. The principle of teleology of explanation in cosmology, which originates in the contemplation of purposiveness in the very fact of human existence, reveals itself as more general than either the maxim of mechanism, or of teleology understood in an old-fashioned matter-referred style. The purposiveness of human actions as related to their underlying condition, which is imbued with will, becomes that decisive tool which makes discoveries and generates the laws of the universe.

Human purposive action (as a living process directed to the future) intends toward discovery of the laws of nature as unified in a single system that includes the "laws" of existence of human beings. In cosmology this single system of laws finds its completion in the idea of the Big Bang as the originary undifferentiated state of matter potentially containing "all in all," which is implied by the explanation. Seemingly paradoxical, however, this alleged "end" of explanation, in fact, lies in the past, that is, in the temporal beginning of the universe, not its unknown end. The fact that this alleged "end" of explanation, is, in fact, at the limits of human comprehension, its infinite task, as Husserl would say, makes it quite clear that this "end" is not the universal Law of Nature which would be treated as a work of the transcendent creator. This "end" of explanation is the constituted end, within the transcendental limits related to human embodiment as the network of communicating observers.

Kant criticized the idea of the world, and he would be even more critical of attempts to create a construct of the universe as a whole.[41]

40. Ibid.

In this sense his transition from the *Critique of Pure Reason* to *Critique of Judgment* was a certain redirecting of one's attention away from the universe *per se*, to the ways by means of which one can make some versions of the universe. One can conjecture that in insisting that there is no sense of talking in terms of positive certitudes about the universe as a whole, Kant would have encouraged us to explicate the underlying motives and methods of thinking of the world. Contemporary cosmology in this sense represents a complex story about what is observed, quantified, measured, calculated, and extrapolated—a story that contains in itself the predispositions to its own effectively endless hermeneutics. Since the storytellers are scientists, the question about the truth of this story cannot be answered, by definition, for this story as such is the explication of the question itself. Correspondingly the questions of teleology in cosmology become matters not of any trans-experiential convictions, but matters of trust and confidence in the ability of humanity to understand the meaning of its existence in the universe. Finally, teleology implanted in the very human condition as a predisposition to knowing the universe retains humanity in a state of apprehensiveness and uncertainty stemming from human capabilities

41. Kant did not discuss, in his reflections on teleology in the *Critique of Judgment*, the universe as "the end of nature," for the universe was treated by him, in his first *Critique*, as a rational idea, but not as an aesthetical or teleological idea. Correspondingly, if one approaches the universe from the side of communion with it in an ancient Greek philosophical sense, that is, the universe (cosmos) as a category of personal relationship with harmony and beauty, as work of art, then the universe represents an end in itself that demands judgment, rather than reason. In judging the form of the universe aesthetically, we are able to find, without presupposing a concept of the object that corresponds to this universe as a source of beauty and harmony, that certain things that are displayed in the universe are purposive (in the empirical apprehension of them in intuition) merely in relation to the *subjective conditions of the power of judgment*. Thus, when the Greeks judged the universe aesthetically, no concept of this universe as a spatio-temporal object was needed, nor was it produced. Correspondingly, in cosmology, by judging the universe aesthetically, for example perceiving the universe as a subject of communion, we do not make an *objective* judgment and do not declare the universe as *natural end* (intentionally created for us to be bedazzled by it), but declare it to be *purposive* only in relation to the subject, namely, for their faculty of representation of the universe as beauty and harmony, as *cosmos*.

not only in trying to know, but in trying to live (as expressed before in different words, humanity by its mysterious, God-given ability to articulate the whole universe, is still capable of retaining its own transcendence with respect to the universe, to be different from it and thus not "crushed under the number of astronomical facts").

The Concept of the Big Bang as an Example of Formal Purposiveness in Cosmology

The *concept of an object* is one of the causes of the actual existence of this object when, firstly, a rational person has with intent brought about this object. In cosmology the concept of the Big Bang was brought into existence through the intent of cosmologists to explain the origin of the observed contingent facticity of the cosmological display. It certainly does not entail the existence of the Big Bang in the same sense as physical objects. Still, the very existence of the Big Bang as an object of theoretical research is based in its concept, which was brought into science through human intent. Correspondingly, the will of a cosmologist acting through using mathematical physics is motivated by the desire to explicate its own origin. This desire is the foundation-stone for the appearance of the concept of the Big Bang as an object of research. A cosmologist's will is determined by the concept of the Big Bang when this person tries to bring this object into existence, moved by a desire to have the object conceived under that concept as physically real. In accordance with Kant one can assert that the concept of the Big Bang by means of which the will of a cosmologist is causal (with respect to its possible material pole) is the representation of the purpose.[42] The concept of the Big Bang can then be one of the causes of its constituted existence when

42. Cf. CJ § 10, 55.

it is a representation of a purpose (the purpose of uniting the manifold of the astronomical display in a single whole).

One must notice that Kant's use of the term "purposiveness" diverges from its ordinary use: for in the latter case the term tends to be applied only when the means to the end is a *material object* or *causal event* and not when it is a conceptual representation determining the will of a person. For a rational person the concept of an object can be the representation of a purpose only if this person has a reason to desire the actual existence of this object. For Kant this takes place if the person considers the object to be pleasant, useful, or morally good; for us this means that the intended Big Bang is the "object" that unites all appearances and aspects of life in a logical system as its ultimate origin. If we would consider an object of pleasure as existing in the world of appearances, it could only be considered as an actual purpose if it were to be an *artifact*. This is definitely not the case of the Big Bang in cosmology, for the Big Bang, as an intended object, cannot be recognized in virtue of the actual history of its origination, for exactly this history is hidden from us: the history of the Big Bang's origination is unknown—it is not a human artifact. Certainly there is a reason for a cosmologist to desire the actual existence of the Big Bang on the grounds of intended search for the foundations of the facticity of all. But in no way is the production of such an "object," that is, its accomplished theoretical construction, at all possible for human beings because not only do they not have a sufficiently exact conceptual representation of it, but they cannot control the ways and means that are appropriate for its "production" once and for all. One can assert with "negative certainty" that the universe is not an artifact but a "product" of nature (or whatever transcendent powers), whereas its intended unity (for example, in a construct of the Big Bang) can be treated as an intelligible and never-accomplished artifact.

According to Kant, judgments in which objects are judged to be actual or possible artifacts are those in which the representation of an object is combined with the representation of an actual or possible conceptual cause of the existence of this object. Kant is concerned not with such judgments, but above all with judgments about relations of purposiveness in which it is maintained that an object in the world of appearances is necessarily to be considered as a purpose or an artifact, because human beings can explain its possibility only by having recourse to a conceptual cause. Such an object is not an artifact; it is not an object that has been or even could be produced by a person according to rules—however paradoxical this might at first appear: "But an object, or state of mind, or even an action is called purposive, although its possibility does not necessarily presuppose the representation of a purpose, merely because its possibility can be explained and conceived by us only so far as we assume for its ground a causality according to *purposes*, i.e. in accordance with a *will* which has regulated it according to the representation of a certain rule."[43] Here one must take note of what Kant meant by the will: in the same paragraph he defines it as "the faculty of desire, so far as it is determinable to act only through concepts." That the possibility of an object cannot be explained or conceived (along the line of efficient causality) means that the object (Big Bang) cannot be an artifact, that it cannot be a product of the intentional act of a person, whose will was determined by the representation (for there is no representation of the Big Bang) of the object. For in order to intentionally produce an object a physicist has to be able to explain the possibility of the object according to laws of nature, that is, the physicist must know and be in control of the causes that can bring about the object. It is obvious that this is

43. CJ § 10, 55 (emphasis added).

not possible in the case where the Big Bang is treated as a certain matter-pole of cosmological theory. The Big Bang, or the universe as a whole, cannot be artifacts in any possible sense because their implied definition runs against the Kantian conviction that artifacts are the only objects whose possibility one can completely explain, "for we see into a thing completely only so far as we can make it in accordance with our concepts and bring it to completion."[44] The universe as a whole, or the Big Bang, whose possibility humans cannot explain by means of laws of nature, but which is nevertheless treated as nature, cannot be considered as an artifact. One can conjecture that they are intelligible artifacts, which are unaccomplished. They are artifacts in the making, that is, mental creations forever being constituted. Together with Kant one can judge "objects" of this type to be purposive because they appear contingent in the light of the laws of nature, or, to be more precise they are contingent for they set up these laws (the universe as a whole as the integrity of its boundary conditions sets up the laws of nature).[45] In the case of the universe as a whole one allows a generalization, for the contingency of the universe is judged not with respect to the given laws of nature, but with respect to the fact that the laws of nature can be considered themselves as contingent. The connection between the contingency of an object according to laws of nature and the judging of the object as purposive can be clarified in the following manner: a cosmologist to whom an "object," that is, the universe as a whole, appears contingent as transcending the natural laws (more precisely, setting these laws), but who does not want to renounce an explanation of its possibility through these natural laws, can attribute to the universe as a whole a relation to a conceptual cause in a hypothetical explanation (for example, the

44. CJ § 68, 231.
45. Cf. CJ § 75, 246.

Big Bang hypotheses realized in different scenarios like those of Hawking, Penrose, or in the concept of the multiverse where the cause of the visible universe is positioned in the realm of intelligible forms similar to that of Plato[46]). This conceptual cause in the hypothetical explanation is that *telos* of cosmological explanation as the universe's explicability.[47] In this way hypothetical judgments about relations of purposiveness arise as hypothetical explanations of objects whose possibility would remain otherwise inexplicable. (The varied contingent facticity of the universe would not be explicable if one would not relate it hypothetically to its conceptual cause.) The purposiveness that is attributed to an object in such judgments Kant calls a "purposiveness without purpose": "There can be, then, purposiveness without purpose, so far as we do not place the causes of this form in a will [that is, in the faculty of desire which is determined to act only through concepts, AN], but yet can only make the explanation of its possibility intelligible to ourselves by deriving it from a will. Again, we are not always forced to regard what we observe (in respect of its possibility) from the point of view of reason. Thus we can at least observe a purposiveness according to *form*, without basing it on a purpose (as the material of the *nexus finalis*), and notice it in objects, although only by reflection."[48] "Without purpose" is the purposiveness that is attributed to an object in these judgments because the conceptual representation of the object (theory of the Big Bang) is not claimed to be one of its real causes. Kant emphasizes that in these judgments, the cause of the object judged to be a purpose is *not* found in a human will (that is, in the

46. See my analysis of Hawking's and Penrose's hypotheses in A. Nesteruk, *Light from the East* (Minneapolis: Fortress Press, 2003), 152–59, 167–77.

47. The universe, being an "object" of cosmology, could not be the lived fact, even if the constancy of a lived experience confers an experiential generality on it. However, the universe can become a type of "lived fact" proceeding from possible idealities in which the laws of eidetic concatenation warrant its rational representation and may be explication.

48. CJ § 10, 55–56 (emphasis added).

faculty of desire which is determined to act only through concepts), thus not in an intentional act of thinking, where intentionality is understood as directedness of consciousness toward a certain material object (Big Bang, for example). We observe in the object (the universe as a whole), whose possibility cannot be explained by natural laws, a purposiveness according to form, without however basing it on a purpose (such as the material of the *nexus finalis*; in our case the universe as a whole understood as an accomplished material object), but this is done not through intentional rational thinking, it is done by reflection.

"Without purpose" is a purposiveness attributed to an object, although this is not an artifact, so that this is purposiveness "without an actual conceptual cause." One can attribute to those "objects" in cosmology whose possibility cannot be explained by natural laws, a "purposiveness without purpose," a hypothetical purposiveness. Kant conceives the judgments about relations of purposiveness in which a hypothetical purposiveness is attributed to objects in analogy with judgments of artifacts. But it should not be overlooked that in the judgments of artifacts one attributes to an object a relation to a conceptual cause quite different to that which is attributed to an object in hypothetical judgments about relations of purposiveness: in a judgment about an artifact the judged object is connected as an effect with its concept according to the causal law, which is why these judgments are an expression of theoretical knowledge. In a hypothetical judgment about a relation of purposiveness, in contrast, an object whose possibility humans cannot explain by means of natural laws is connected only with a *supposed* conceptual cause (the Big Bang), and this connection does not take place according to the causal law—it does not have the status of objective determination.[49]

49. The models of Hawking and Penrose mentioned above can serve as examples where objective determination has itself a hypothetical character.

These judgments, according to Kant, do not belong, then, to the judgments of theoretical knowledge that are concerned with the possibility of the existence of things under the laws of nature. They are, rather, judgments of the reflective faculty of judgment: "purposiveness according to form," that is, hypothetical purposiveness, which can be noticed "in objects, although only by reflection."[50]

By summarizing, we have investigated the delimiters in the strategies of cosmological research, which originate in the fact that all human actions (including scientific research) can be considered as purposive. In spite of a general tendency in science to dispense with teleology, we argued that because of the specificity of the subject matter in cosmology, its research goes on under the assumption that there is a goal of research, the motivational purpose, related to the explicability of the universe. This latter explicability originates in the human condition, that is, in the human intentional search for the sense of its own existence in the universe. Thus the purpose of explanation in cosmology is related to the explication of the human condition. Correspondingly, the purposiveness of cosmological research as its certain strategy acts as a delimiter in the explicability of the universe related to the human condition. The universe is being explicated in the conditions that the goal of this explication is formulated and followed not on the grounds of concepts of the reason, but on the grounds of the faculty of reflective judgment. Here cosmology exhibits an example where both principles, namely that of teleology and that of the causal efficiency of the natural laws, work together. The *telos* of cosmological explanation then is that which is constantly constituted through the interplay of "mechanistic" and teleological principles. It is the anticipated purpose of explanation

50. CJ § 10, 56.

that ultimately ordains the advance of scientific cosmology. It was demonstrated that there are two basic concepts in modern cosmology related to each other, namely the universe as a whole and its encapsulated origin (the Big Bang), that act as the *telos* of cosmological research and explanation. The case of the Big Bang is particularly interesting in this respect, for it gives an example of reversed teleological temporality, for the alleged material pole of the *telos* of explanation is situated not in the future, but in the past of the universe. This result is not surprising in a phenomenological context, where the Big Bang would be treated as a noematic pole toward which intentionality aspires; but being an "object" of the *kosmos noetos* and thus being subject to laws of eidetic concatenation, the Big Bang is devoid of any attributes of physical temporality. Its functioning as a *telos* of explanation refers to the teleological essence of the cognizing subject, its self-elucidation, that light of reason struggling against the self-forgetfulness that is brought by cosmology in its "mechanistic" application, and in which thought would no longer recognize that it is the center of the world. As was said once by E. Husserl, the supreme *end* of knowledge is to not forget. This is the meaning of the resistance to that which G. Marcel characterized as the crushing of man under the weight of astronomical facts. To resist dissolution of the *ego* by the universe is the ultimate *telos* of a humanity destined to commune with God beyond the boundaries of the created world.

7

The Universe as a Saturated Phenomenon

Christian Concept of Creation in View of Modern Cosmology and Philosophy

The saturated phenomenon refuses to let itself be looked at as an object precisely because it appears with a multitude and indescribable excess that suspends any effort at constitution. To define the saturated phenomenon as a nonobjective or, more exactly, nonobectivizable, object in no way indicates a refuge in the irrational or the arbitrary; this definition refers to one of its distinctive properties: although exemplarily visible, it nevertheless cannot be looked at.

–Jean-Luc Marion, "The Saturated Phenomenon," p. 209

Introduction: Creation of the World and Creation of Humanity

One of the tasks of the dialogue between theology and science is to elucidate in the modern scientific and philosophical context the sense of what is meant by creation of the world out of nothing (*creatio ex nihilo*). As is often argued in current discussions on the theme, the adequate theological appropriation of the scientific approach to the study of the natural universe is possible only if nature and the universe are treated not as an "environment" for physical and biological existence, but as creation. This implies not only a dispassionate study of the universe which is contingent upon God, but also of the relationship between human creatures with God through nature.[1] Contrary to what Karl Barth wrote in the first half of the twentieth century, namely that "there can be no scientific problems, questions, objections or aids in relation to what Holy Scripture and the Christian Church understand by the divine work of creation,"[2] a contemporary interpretation of what is meant by the creation of the world is necessary in order to elucidate the modern liberal consciousness of that which is constantly quoted in biblical formulae related to creation of the world.[3] First of all this is related to the vision of the physical universe, as created, in modern natural sciences, cosmology in particular. Secondly it is related to an appropriation of the notion of creation to elucidate the sense of the human condition and mystery of created personhood.

Since all contemporary discussions on science and Christian theology refer to the Divine, they imply the distinction between this world (which is studied by science) and God, who created this world and who is beyond this world. Thus the issue arises of the relationship between God and the world, and how to commune with God—that is, in a way, to transcend the world. The distinction between God and the world in theology encodes the whole problem

1. See, for example, B. Foltz, "Discerning the Spirit in Creation: Orthodox Christianity and Environmental Science," in *Science and the Eastern Orthodox Church*, ed. D. Buxhoeveden and G. Woloschak (Farnham, UK: Ashgate, 2011), 108.
2. K. Barth, *Church Dogmatics* (Edinburgh: T. & T. Clark, 1936-77), III/1, ix.
3. Cf. W. Pannenberg, *The Historicity of Nature: Essays on Science and Theology* (West Conshohocken, PA: Templeton Foundation, 2008), 25.

of creation of the world as its contingent facticity.[4] The discussion between science and Christianity attempts to establish the truth of propositions about actual being either on the grounds of the world's self-sufficient existence according to laws of nature, or as being "rooted" in (or contingent upon) its other, transworldly origin, which nevertheless allows one to *detect* its traces in the world. Phrased in a scientific manner, the latter would mean that this world with its acting contingent laws must have the foundation of their contingency in some "other transworldly laws," as an outcome of the latter. Seen in this way, the question of the facticity of the world would reduce to its supposed inclusiveness in a wider "reality," and it is not important whether this reality is associated with the Divine,[5] or with some transworldly principle similar to the multiverse.[6] However, all attempts to describe creation encounter a fundamental

4. For introduction to the theological discussion of creation see a classical paper of G. Florovsky, "Creation and Creaturehood," in *Creation and Redemption*, the collected works of Georges Florovsky, vol. 3 (Belmont, MA: Nordland, 1976), 43-78. See also his "The Idea of Creation in Christian Philosophy," *The Eastern Churches Quarterly* 8, no. 3 (1949): 53-77. There are many modern books dealing with creation that have a historical, as well as systematic character. See, for example: G. May, *Creation Ex Nihilo: The Doctrine of "Creation Out of Nothing" in Early Christian Thought* (Edinburgh: T. & T. Clark, 1994); K. Ward, *Religion and Creation* (Oxford: Clarendon, 1996); T. Torrance, *Divine and Contingent Order* (Edinburgh: T. & T. Clark, 1998); W. Pannenberg, *Toward a Theology of Nature: Essays on Science and Faith* (Louisville: Westminster John Knox, 1993); O. Davies, *The Creativity of God: World, Eucharist, Reason* (Cambridge: Cambridge University Press, 2004); W. Pannenberg, *The Historicity of Nature*; E. Theokritoff, *Living in God's Creation: Orthodox Perspectives on Ecology* (Crestwood, NY: St. Vladimir's Seminary Press, 2009); M. Barker, *Creation: A Biblical Vision for the Environment* (London: T. & T. Clark, 2010).
5. Talking about Divine presence in the world we imply Christian panentheism: God is present in the world without losing his transcendent essence (he is present in absence). Correspondingly, the language of God as "wider reality" in which the world is embedded is metaphorical: the world is contingent upon God, who is present in its signs in the world, yet remains beyond the world. See P. Clayton and A. Peacocke, eds., *In Whom We Live and Move and Have Our Being: Reflections on Panentheism in a Scientific Age* (Grand Rapids: Eerdmans, 2004).
6. The idea of the multiverse in modern cosmology corresponds to a perennial philosophical problem of plurality of worlds. For a detailed account of the ongoing discussion see B. J. Carr, ed., *Universe or Multiverse* (Cambridge: Cambridge University Press, 2007). For a recent scientific critique of the idea of the multiverse see G. Ellis, "Does the Multiverse Really Exist?" *Scientific American*, August 2011, 38-43.

difficulty. Such a description can be done in abstraction, when the very act of creation, as well as that which is created, are presented as *objects*. J.-L. Marion gives a fine qualification of what is typical to those phenomena that are called objects; he defines objects as those which are poor in phenomenality: to constitute an object means to lose a phenomenal autonomy and spontaneity that a thing manifests from itself, that is, from its event-like appearance. The condition of objects is exactly deprivation of this event-like manifestation and its reduction to the rubrics of "I think"; correspondingly the conditions of the object can be described in four rubrics: quantity, quality, relation, and modality.[7] But this entails an imminent difficulty because creation and the universe as a whole cannot be thought in these rubrics, so that the natural attitude with respect to the universe is impossible. Despite the fact that we have already used the term "natural attitude" many times before, it is worth rearticulating it here once again. The natural attitude is related to the activity of consciousness and, according to M. Natanson: "Within the natural attitude I act in a world which is real, a world that existed before I was born and which I think will continue to exist after I die. This world is inhabited not only by me, but also by my fellow men, who are human beings with whom I can and do communicate meaningfully. This world has familiar features which have been systematically described through the genetic-causal categories of science. The world of daily life is lived within this natural attitude, and as long as things go along smoothly and reasonably well, there arises no need to call this attitude into question." The reasonability and wellness is related to the situation when the question about the contingent facticity of that which is going on around us, including the facticity of the whole world and the "I" that makes enquiries about it, does

7. See more details in J.-L. Marion, *Certitudes négatives* (Paris: Bernard Grasset, 2010), 253–69.

not arise. Natanson continues: “But even if I do occasionally ask whether some things ‘really real,’ whether the world is ‘really’ as it appears to be, these questions are still posed in such a way that they are my questions about the natural world in which I live. I do not really scrutinize my natural attitude in any rigorous manner: I merely mark off a bit of it for more careful study.”[8] Since, as we argue in this chapter, the natural attitude is not suitable for the description of the relationship between God and the world, or between the whole creation and humanity, the elucidation of these relationships can be done only on the grounds of questioning the very facticity of the natural attitude. Indeed, when talking of creation, the one who is talking implies its own creation and the limits of its comprehension following from being created. To represent creation mentally as “an object” one needs to exit one’s own existence in order to “look” at one’s own coming into being (as well as the coming into being of the world) from “outside,” as if there were some antecedents to it. However, as was argued by existential philosophers, as well as modern phenomenologists, such an approach is philosophically untenable, because it contradicts the facticity of the given life as that originary fact and event from within whose horizon the whole world order is unfolded, and whose nonoriginary origin cannot be linguistically and mentally located.

Thus the problem can be formulated: Since the exit from that which is supposed to be created is impossible, how can one speculate about creation with a reference to the transworldly, that is, the transcendent, remaining immanent to the created world? Since in any philosophy that adopts the a priori given of the cognitive faculties, from within which the phenomena are constituted, the transcendence is deeply problematic, and hence the religious

8. M. Natanson, “Being-in-Realty,” *Philosophy and Phenomenological Research* 20, no. 2 (1959): 232.

experience of communion with God cannot receive its exhaustive explication on the level of reason. In spite of the fact that the rationality of communion with God is different in comparison with that which pertains to scientific research, the very actual facticity and possibility of dialogue between science and theology demonstrates that they have a common ground and their different rationality follows from their belonging to the deeds of humanity. There is a rationality that pertains to faith and has its own reason, and there is a rationality that pertains to the scientific quest and is grounded in hidden beliefs.

The rationality of theology implies that one must know how to speak of God's presence in the world and to enquire into the sense of creation of the world or of any particular human being. This in turn implies *existential transcendence*, that is, a perception of commensurability with the whole creation not in terms of space-time, but in terms of one's fundamental otherness with respect to the rest of creation. Existential transcendence makes nonsensical all attempts to think of creatures as overcoming and abandoning this world: this is impossible not only because of consubstantial corporeity, but also because, theologically, the Earth is spiritually central, being humanity's flesh. Correspondingly, all forms of thought of other worlds remain no more than mental images, eidetic variations on the level of intelligible forms, which are posed as the other with respect to hypostatic subjectivity. These other worlds could mimic the creaturely reality of the spiritual order. In this case the invocation of other worlds is dangerous and soteriologically futile, and Christian theology warned against it because we cannot understand the meaning and purpose of other worlds given that the meaning of our own physical world, because of the Fall, remains obscure.[9]

Then the question of creation of the world becomes reduced in its content: one speaks not about creation of the world in general, but about creation of *this world as its factual givenness to humanity*. Thus what is important is not a "dynamic" of creation as its hermeneutics (be it biblical or scientific), but understanding creation as bringing humanity into this world, placing it face to face with this world, so that humanity could see God as present in the world. Speaking of creation and creaturehood, one speaks of knowledge of the Creator, that is, of theology, which "consists not in the conjectures of man's reason or the results of critical research but in a statement of the life into which man has been introduced by the action of the Holy Spirit."[10] The issue of creation is thus not just the question of creation of human beings in their substantial similarity with this world, but in the Divine image, capable of knowing God.

It is known that St. Athanasius the Great noted that in spite of the fact that the Father provided the works of creation as means by which its Maker might be known,[11] this did not prevent men from wallowing in error;[12] because of this the Word of God descended to men in order to "renew the same teaching."[13] Since the Incarnation was initiated by the Holy Spirit, who is not transparent in rubrics of space and time, we can affirm again that the theology of creation is

9. As was stated by V. Lossky, "the mysteries of the divine economy are thus unfurled on earth, and that is why the Bible wants to bind us to the earth [that is, our universe, AN]. . . . [I]t forbids us to lose ourselves in cosmic immensities (which our fallen nature cannot grasp anyway, except in their aspect of disintegration), . . . it wants to win us from usurpation of fallen angels and bind us to God alone. . . . In our fallenness we cannot even place our world amidst these spiritual immensities." V. Lossky, *Orthodox Theology* (Crestwood, NY: St. Vladimir's Seminary Press, 1997), 64.

10. Archimandrite Soprony (Sakharov), *St. Silouan the Athonite* (Moscow: St. Trinity Sergius Lavra, 2002), 171 (in Russian).

11. Athanasius, *On the Incarnation*, 12. ET: (Crestwood, NY: St. Vladimir's Seminary Press, 1998), 39.

12. Athanasius, *On the Incarnation*, 14. ET: op. cit., 42. In modern terms this means that any natural theology is insufficient in order to know truth about God.

13. Ibid.

a statement of the life into which humanity has been introduced by the action of the Holy Spirit in the image of the eternally conceived incarnation of the Word of God. One observes here a Christo-geocentric reduction of the problem of creation because the Incarnation, as an element of the Divine economy, implies the existence of a universe where humanity is possible, and hence the coming of the Son of God in human flesh would be possible. In this case the retaining of the transcendent in the created world would be equivalent to retaining the dual nature of the Word-Logos of God in the structures of the physical world, that is, a perception of the extended space and time of the universe through the prism of its relation to God as the creator and sustainer of the universe.[14] Being created by the Logos and through the Logos, this world manifests the spatial paradox of Christ, that is, his presence *in* space but not *of* space; his historical presence on earth, which is equivalent to omnipresence devoid of any spatio-temporal extension.

The createdness of the world, being the otherness of God rooted in his love, means a global, spatial, and temporal correlation and correspondence between all places in the created universe simply because this world is a "moment" and "event" of the Divine love. Createdness of the world must in this case not only point toward some unique antecedent moment in the history of the universe when "all was in all" and from which all came to be. It must point toward the actual omnipresence of human insight, created in the image of

14. Since "the Divine image of the world always remains above and beyond creation by nature" (Florovsky, "Creation and Creaturehood," 72), what is implied here is the retaining of the image of the world in the archetype of Christ *through grace*, that is, without compromising ontological boundaries between God and creation. In different words, the Divine image of the world can be linked to its enhypostasized identity. However, since there remains a transcendent gulf between That who is enhypostasizing and that which is enhypostasized, all human conjectures about the identity of the universe, even if they are accompanied by graceful enlightenment, do not reflect the image of the world as it is present in the Divine nature. We are capable of "seeing" the image of the world, its identity only, "as puzzling reflections in a mirror" (1 Cor. 13:12).

Christ himself, that is, its presence in all corners of the universe extended in space and time. This would signify a retaining of the transcendent in the immanent creation, a retaining that overcomes the physical representation of the universe as being divided into multiple non-overlapping causally disconnected segments, only one of which is visible to us. The very intuition of the universe as a whole manifests an archetypal trace of "all in all" in human consciousness when the transcendent revealed itself in the immanent without compromising its otherworldliness. The intuition of the universe as a whole—that is, as creation—reveals that quality of human reason which positions itself in *catholicity* with all levels of being,[15] that quality which originates in the Divine image, that is, in the archetype of Christ himself. This quality of reason as its theological commitment to the Divine image is accentuated through the teaching that human beings attempt to understand the underlying sense of beings and things not according to their "nature" (which is disclosed in particular sciences) but according to the final causes of these beings and things (that is, in a global, cosmological context) in relation to the place and goals of humanity, made in the image of God, in creation. This is the reason why humanity wants to recognize the universe as creation not according to its nature (as happens in cosmology), that is, according to its compelling empirical givenness, but as results of humanity's free will.

The very ambition of cosmology to be a science of the created universe as a whole is determined by the human capacity to transcend empirical and astronomical facts and invoke the idea of the wholeness of the world and the universe. Human free will makes itself manifest here through the desire to be commensurable with the universe

15. This co-relation with the "community" of all things in the world was characterized by the French philosopher and writer Paul Claudel in terms of "catholicity." Without having an idea about such a co-relation, that is, without a universal, catholic idea, it is impossible to understand the sense and meaning of life. See P. Claudel, *Positions et propositions* (Paris: Gallimard, 1934), 9.

despite its physical impossibility. The mental image of the whole created world is *de facto* the manifestation of the willing activity of humanity, and "knowledge" of the universe as a whole is possible only as this willing. Since the universe as a whole cannot become a matter of investigation in the phenomenality of objects, all cosmological attempts to grasp the sense of this whole even if they have a complex mathematical shape still appear to be the urges of free will.

Modern cosmological models, or more precisely, the metaphors of creation, including models of the so-called multiverse, hardly shed any light on creation understood theologically. All these models have a common feature, namely, they appeal to intelligible mathematical forms that allegedly correspond to physical reality detached from us by billions of years. It is clear, however, that in spite of uncertainties in their ontological status, these mathematical constructs constitute a part of the created world, the world of ideas or Platonic forms, which are far cry from being transworldly in a theological sense. Seen in this perspective, all cosmological models assert only one thing: in its attempt to approach the boundary of the physical world, human reason inevitably appeals to the intelligible images of this boundary, by building a characteristic hermeneutics of the transition from the intelligible to the physical, that which, by a sheer philosophical mistake, is treated as a hermeneutic of transcendence and the creation in an absolute theological sense.[16] However, in spite of all futility to exercise such a transcendence in the limits of scientific rationality, cosmological speculations on the theme of creation turn out to be very useful and contribute toward a perennial mythology of creation. They manifest the infinite advance to the mystery of creation, using

16. A critical analysis of such a hermeneutics related to cosmological models of Hawking and Penrose was undertaken in A. Nesteruk, *Light from the East: Theology, Science and the Eastern Orthodox Tradition* (Minneapolis: Fortress Press, 2003), chapters 5 and 6.

notations of physics and mathematics, understanding beforehand that the sense of that which is signified by them will never be exhausted. Being interpreted phenomenologically, cosmological models deliver us knowledge of how the human subjectivity acts and structures itself when it approaches the limiting questions of the sense and foundation of its own existence.[17]

The issue of creation becomes a question about the contingent facticity of that which is given to humanity in life. This facticity is stated in a scientific quest, but is never disclosed completely. One has here a certain analogy with theology: the Fathers of the Church taught that theology can claim that God *is*, but it will never ever be able to respond to the question, "What is God?" Scientific cosmology asserts that the universe as a whole *is*. It endeavors to go even further and attempt to respond to the question, "What is the universe as a whole?" Any possible answer to this question will be apophatic by default. The "name" of the universe, its signifier and its "identity" will always escape any attempt of its accomplished definitions, leaving the enquirer with the same unanswered and unanswerable question, "Why is the universe, and why is it as it is?" Sometimes scientific cosmology expresses its hope that by responding to the question "What?," it will, in the future, give a response to the question "Why?" However, this hope seems to be teleological, and may be even eschatological. Teleological, because it drives the scientific quest to the future (whereas the goal of this quest is the supposedly existent past of the universe from which everything "came to be"[18]). Eschatological, because it is linked to the hope of humanity to acquire knowledge of the unity of the universe as a mirror of knowledge of "all in all" of humanity itself, that is, of its lost prelapsarian unity.[19]

17. Cf. the previous chapter, where we argued that the cosmological discourse of creation turns out to be very close to the phenomenological treatment of birth of a human person.
18. Cf. chapters 5 and 6 of this book.

Creation in the Natural Attitude, or How Not to Speak of Creation

When Christian theology through the writings of the Fathers and numerous commentaries asserts that the world is created, that it came out of nothing, that there was no world "before" it came into being, it implies a psychological frame of thought that treats creation of the world in terms of temporality pertaining to this world: to speak of creation one needs to have an intuition of the distinction between "before" and "after" (an intuition of temporal sequence) that this temporality implies. The sense of the words "the world came out of nothing" can only be understood from within the human sense of existence as existence in time. Correspondingly, if creation of the world is represented in thought as a "transition" from that something "when there was no world" to the actual existence of the world, this representation has a precarious and hypothetical character in terms of possible antecedent references, simply because the very process of thought belongs to the already-created world and it is from this created modality of intellection that one attempts to grasp the sense of the created as the facticity of the world and hence the facticity of thinking of the world.[20]

In spite of this last nearly obvious observation, starting from the Fathers of the Church and finishing with contemporary discussions on the applicability of modern cosmological theories to the riddle of

19. The idea of the unity of humanity as humanity's hope to acquire knowledge of this unity in an eschatological limit is discussed in G. Goutner, "The Unity of Humanity in an Eschatological Perspective," in *Theology of Creation*, ed. A. Bodrov and M. Tolstoluzhenko (Moscow: St. Andrew's Biblical and Theological Institute, 2013), 230–36 (in Russian).

20. Hermann Weyl expressed, in a different context, a similar thought, that ". . . the mere postulation of the external world does not really explain what it was supposed to explain, namely, the fact that I, as a perceiving and acting being, find myself placed in such a world." H. Weyl, "The Open World," in *Mind and Nature: Selected Writings on Philosophy, Mathematics and Physics* (Princeton: Princeton University Press, 2009), 50. Weyl linked the question of the world's reality to the question of the origin of its mathematical harmony and the possibility of its grasp by human reason.

creation, theology struggles to express the problem of creation of the world in terms of thought and speech, avoiding any reference to the problem of the facticity of consciousness which attempts to explicate this creation. Speaking philosophically, theology as well as various forms of its dialogue with scientific cosmology often functions in the *natural* attitude, when creation of the world is approached as if one could speculate about it objectively, in terms of logic pertaining to the phenomenality of objects and in terms of the causality of empirical things. For example, St. Basil the Great, speaking of creation of the world, says that God created heaven and earth, that created beings begin with time and end in time. Time originates together with the world, so that the origination of time is its "first" moment, its "beginning." Then he says that one can start from the present and attempt to trace through events in the past that first day which would correspond to the creation of the world out of nothing.[21] In Basil's affirmations it is implied that time is that part of the created reality which pertains to the intellect speculating about creation. On the one hand time is an attribute of the created world; on the other hand, it is, using contemporary language, a transcendental condition for the very possibility of speculating about creation at all. When Basil points toward the possibility of counting time backwards to the past in order to find its beginning, as if this beginning would be given to the human grasp as an "object," as an outward "thing," he implicitly extrapolates the causality of mundane things toward the origin and foundation of these things. However, philosophy before Christianity understood that the causal principle of the world cannot be constructed by means of the sciences and knowledge. Modern philosophy contributes to this by saying that the origin of the world as well as the origin of one's consciousness

21. Basil the Great, *In hexaemeron*, 1.6. ET: NPNF VIII, 55.

is phenomenologically concealed from one's grasp, so that its explication is possible only through the unfolding in the future of that which is already given. Correspondingly, Basil's explication of the origin can only be understood as undertaken from within the phenomenality of the created and thus establishing the retrocedent causality toward the origin, which will never become the explication of creation.

If, as another patristic example, one turns to St. Maximus the Confessor, one finds a similar assertion that the world has a beginning and consequently is not eternal. Maximus, following his predecessors, repeats that the world was created out of nothing because of God's will and goodness, by his Wisdom and Logos, and the createdness of the word implies its noneternity and consequently its beginning in time. However, in spite of the fact that this beginning in time can be understood only from within the already-created world, Maximus points to a difficulty that can arise. In a passage from his *Centuries on Charity* 4.3 he says, "God, who is eternally Creator, creates when He wills by His consubstantial Word and Spirit, because of His infinite goodness." This is a general statement that does not raise any questions because it is a matter of religious conviction. Then Maximus anticipates a possible question on details of this creation: "Nor must you object: Why did He create *at a certain time* since He was always good?" Here the question is formulated from within those categories of sequence and time that pertain to the already-created world. Indeed, if the creation of the world happened several thousand years ago measured by the created time, why is this age of the world such as it is, or, in other words, can we enquire into the nature of this age's contingent facticity as it is contemplated from within creation? Maximus gives a characteristic response—"no": "The unsearchable wisdom of the infinite essence does not fall under human knowledge."[22] It is impossible to transcend the boundaries of the created and to enquire into its facticity on the grounds of the impossibility of knowing the divine volitions and intentions; creation remains a divine mystery connected with divine providence. It is evident that this response has general apophatic overtones related to the unknowability of God.

However, one must analyze further some implications of the question discussed by Maximus. If the question about the "when" of creation is related to the temporal span of the physical universe seen from within this universe, then one can find parallels with contemporary cosmology. Formally, Maximus's question can be translated by using modern cosmological language into a question about the initial conditions of the universe that fix its physical parameters, including its age. But physical cosmology cannot give an account of the initial conditions for the dynamical laws that drive the matter and space of the universe. Correspondingly, cosmology cannot provide a clear explanation why the age of the visible universe is 13.7 billion years. Since we can speculate on the nature of these conditions only from within our universe by extrapolating backward the properties of the observable universe, the "knowledge" of the initial conditions thus achieved does not tell us anything about these conditions, as if there were special transworldly physical laws responsible for these conditions as the outcomes of these laws.[23] Being bounded by the universe, one cannot know the "laws" of the initial conditions of the universe as if they could be attested from beyond the universe (we can only postulate them). In this sense Maximus's

22. This ET: P. Sherwood, *St. Maximus the Confessor. The Ascetic Life. The Four Centuries on Charity* (London: Longmans, 1955), 192.
23. A famous example of such a hypothetical law is R. Penrose's Weyl Curvature Hypothesis postulating a low gravitational entropy in the beginning of the universe. See, for example,Penrose's classical paper "Singularities and Time-Asymmetry," in *General Relativity: An Einstein Centenary Survey*, ed. S. W. Hawking and W. Israel (Cambridge: Cambridge UniversityPress, 1979), 581-638, or a more recent book, *The Road to Reality* (London: Vintage,2005), 726-32, 765-68.

response "no" with respect to the initial conditions of the created universe exactly corresponds to the "no" of scientific cosmology in respect to the initial conditions that fix the contingent facticity of the universe.

However, Maximus's question can be reformulated differently, so that the question about the "when" of creation is posed as if the universe appeared out of something preexistent. One can imagine a preexistent space-time continuum in which our universe appears as some "moment" and "location" of this preexistent continuum. The question "when" of creation will then have another sense as a particular "when" of preexistent time. We are not concerned here with the nature of this preexistence, that is, whether it is related to the multiverse, or something "before" the Big Bang, or to a cycling universe of Penrose's type.[24] What interests us is a possibility to approach creation as an "object," as a transition from something "before" to that which is here and now. This would be typical for the natural attitude to "look" at the creation and ask a question on the specificity of this or that "moment" of its happening in the preexistent scheme of things. Certainly one could refer to Augustine's ways of responding to such a difficulty simply by pointing to the fact that "before" the world was created no entities such as all-embracing space or time could exist.[25] Such an Augustinian response is true in

24. R. Penrose, *Cycles of Time: An Extraordinary New View of the Universe* (London: Bodley Head, 2010).

25. According to St. Augustine this cannot be a "beginning" as if it would be "seen" from the outside of the world. Augustine in *Confessions*, XI addressed the problem of the origin of time, directly affirming that "[t]he way, God, in which you made heaven and earth was not that you made them either in heaven or on earth. . . . Nor did you make the universe within the framework of the universe. There was nowhere for it to be made before it was brought into existence." Augustine, *Confessions*, Book XI:5. This ET: H. Chadwick, trans., *Augustine's Confessions* (Oxford: Oxford University Press, 1991), 225. According to Augustine the universe was not created by God in time, but was created *with* time. Augustine, *City of God*, XI:6. This is the only consistent expression of the Christian affirmation of *creatio ex nihilo*. The *nihilo* could not be something; it could not have any attributes of created things, and therefore must be an absolute philosophical no-thing.

its essence, but it would be useful to confirm it through a negative assessment of modern models of creation with preexistent space-time. Indeed to ask why creation "now" (e.g., six thousand years ago) but not later or before, would imply the possibility of approaching the creation in the objective scheme of things, that is, to position it as an "event," as a particular happening in the series of causations.

As an example of "creation" in preexistent space-time one can point to a model of "creation" of matter in the universe (not space and time) from an initial state with the total energy of matter equal zero. This requirement can be treated as a meta-law, imposed on matter of the future universe in the *preexistent* space and time. Such a model was offered by Tryon,[26] and its major feature was that the universe originated in *preexistent* space and time as a result of a fluctuation of the physical vacuum (a physical state of quantum matter in which the values of all observables of particles are zero). Geometrically the development of such a universe can be presented as a future light-cone, whose apex, symbolizing the beginning of this universe, is positioned completely arbitrarily in preexistent space and time (see Figure 5). It is exactly this arbitrariness of the "place" and "moment" of origination of the visible universe in the background of the preexistent space and time, which constitutes a difficulty similar to that of Maximus: it is impossible to specify and justify why the universe originated at a specific point of space and time (that is, it is impossible to specify the "when" of this origination). In this theory the spontaneous creation of the universe could occur anywhere and at any moment of preexistent space and time. (A variety of different universes could originate at different locations of the preexistent space-time, driving cosmology to face a serious problem of the mutual influence of different universes; see Figure 5.)

26. See E. P. Tryon, "Is the Universe a Vacuum Fluctuation?," in *Modern Cosmology and Philosophy*, ed. J. Leslie (New York: Prometheus, 1998), 222–25.

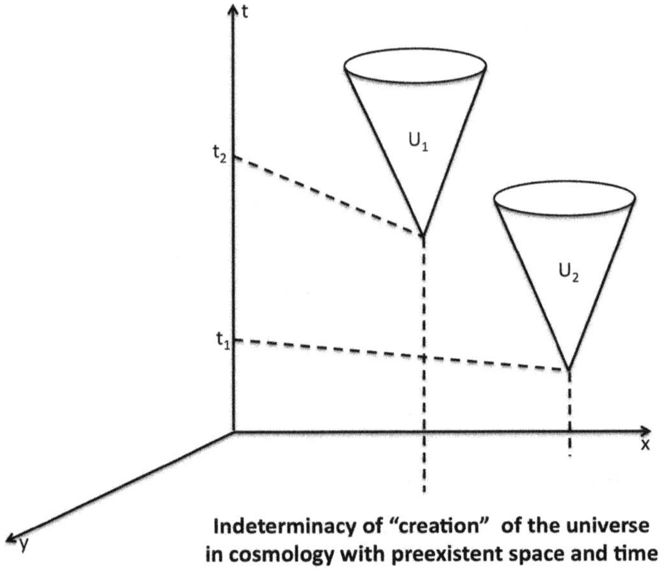

Figure 5

Correspondingly, the question of the "when" of "creation" not only cannot be answered, but, in fact, does not have any sense, for if the preexistent space-time is infinite, an infinite time could have passed since our universe originated. But this makes the question of temporality of the moment of creation devoid of any meaning. There is no need to argue that this kind of model has nothing to do with creation out of nothing in a theological sense, for space, time, the meta-law, and the quantum vacuum are all assumed to be preexistent. It is reasonable to talk about the *temporal origination* of the material universe rather than about its creation out of nothing.[27]

27. It is interesting to note that the first "scientific" ideas on the origination of the universe in preexistent space and time were proposed by Newton, who intended to reconcile the biblical account of creation, where the world had to have a beginning, with his view that time could have neither beginning nor end. Newton asserted that the visible universe was brought into existence by God in the past which is separated from us by finite time, but this took place within the absolute and infinite space and time. E. McMullin points out that the position of Newton

The beginning of the world and its created temporality can be grasped from within the world, so that this beginning is the constituted beginning from within the world. No constitution or objectivization of this beginning is possible from beyond the world, because this "beyond the world" is not an "object" but rather the condition of the very possibility for the world to be manifested *to* and articulated *by* human beings. In this sense the quest for the beginning of the created universe reveals itself as a boundary of human consciousness attempting to grasp the facticity of the world.

What is interesting is that the refutation of models of "creation" of the universe in preexistent space-time leads us to further clarification of what is meant by creation in theology. Indeed, the logical difficulty of models with preexistent space and time is connected with the inability to locate the moment of time when the universe originated, from outside, by transcending beyond the universe itself, into its imaginable preexistent "before." One can argue about the beginning of time within the visible universe by extrapolating its expansion backward in time. But this will never allow one to claim scientifically that there either was or was not a preexistent time "before" our universe came into existence. The situation was described by Kant in terms of his first cosmological antinomy as a logical tension between the thesis, that the world has a beginning in time and is also limited as regards space, and the antithesis, that the world has no beginning and no limits in space; it is infinite as regards both time and space.[28]

was a departure from the medieval Aristotelians, who were not inclined to separate creation of matter and time. "Is Philosophy Relevant to Cosmology?," in Leslie, ibid., 44. The creation of matter in Newton's model is detached from the creation of time. One sees here a fundamental difference not only with the contemporary views based on General Relativity, where space and time are relational upon matter (so that the split in origination of matter and time becomes theoretically inconsistent) but even with Maximus the Confessor, for whom space and time where inseparable elements of the creaturely nature of the world. See H. Urs von Balthasar, *Cosmic Liturgy: The Universe according to Maximus the Confessor* (San Francisco: Ignatius, 2003), 139.

The antinomy that arises in any cosmology with preexistent space and time can be considered from a different perspective, without any reference to space and time, and this brings us even closer to the thinking of Maximus the Confessor. For example, the thesis can be treated as the affirmation that the visible universe is unique and finite as regards space and time (with its particular age), whereas the antithesis is that the visible universe is one particular representative out of the *ensemble* of universes with different boundary conditions (corresponding in the previous logic to different moments of their origination in preexistent time), which are responsible for different ages of these universes as they are seen from within them. The plurality of different boundary conditions corresponds to the logical multitude of a Platonic-like kind, so that the antinomial nature of any predication on the uniqueness (this universe) or multitude (other universes) of these conditions becomes evident because the ontological status of that which is predicated in thesis and antithesis is different: while with respect to the visible universe we can make an empirical inference, an assumption that there is an ensemble of universes implies a reference to the realm of the intelligible. In this case the whole meaning of the antinomy reveals itself as predication about two ontologically distinct realities, that is, the empirical visible universe and a Platonic-like ensemble of universes. If we extrapolate this reasoning back to the problem, discussed by Maximus the Confessor, the question posed by him in the *Centuries on Charity* 4.3 must be transformed in such a way that the temporal aspect of the specificity of the creation of the world is replaced by the aspect of "choice" of this particular world out of many potential possible worlds, namely, "Why did God choose to create this world

28. I. Kant, *Critique of Pure Reason*, A426-27/B454-55. ET: N. K. Smith (London: Macmillan, 1933), 396.

(with a given age) but not the other (whose internal ages could be different)?" (See Figure 6.)[29]

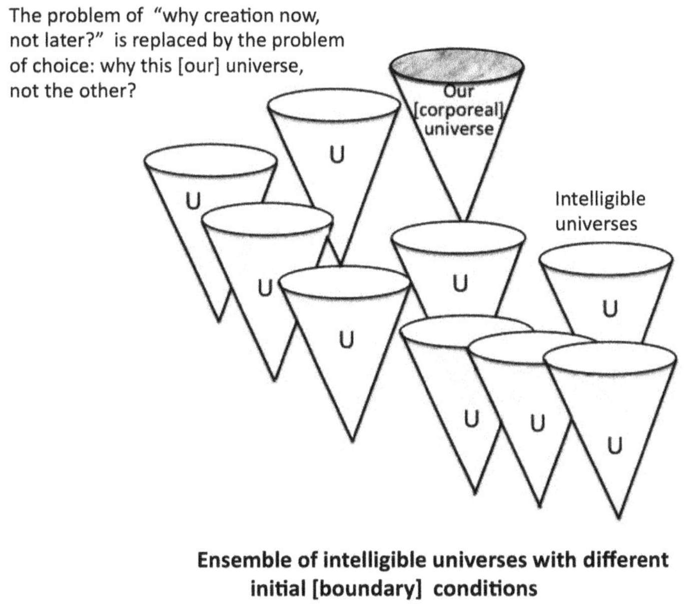

Figure 6

One can conjecture that Maximus's response to the very possibility of asking such a question would probably be negative: one must rely on the wisdom and will of God as an apophatic reference with respect to a given choice of the universe. In modern parlance, one cannot enquire into the facticity of the created universe as one cannot

29. In modern cosmology such an interpretation corresponds to different models of the multiverse. It also has particular connotations with Penrose's old suggestion that the special initial conditions of our universe responsible for the arrow of time in it are set up from outside through choosing them out of many other possibilities, which could lead to different universes (see his *The Road to Reality*, 726–32).

establish a causal principle of the universe that would be required in the natural attitude.

The approach to creation within the natural attitude can be paralleled with the substitution of a metaphysical meaning of the universe with concepts that function according to how the physical facts of the universe are defined, that is, with physics. If the created universe as a whole is understood as a metaphysical and theological concept it cannot be interpreted as a natural fact, but only in the way the signifiers of the created universe define this concept. In theology, words such as "God," "eternity," "creation," which signify cause and purpose, metaphysical sustenance or foundation of existence, refer to mental definitions that do not have direct representations through experience of the senses. However, and this is how the substitution of metaphysics by physics happens, these concepts, in particular that of creation, use the same mental reference to the conditions, delimiters, and contents that form the basis of sensory experience. If God is understood as the "absolutely necessary omnipotent being," something greater than sensible and intelligible reality, devoid of contingencies of the empirical, something that cannot be observed and measured, it is still mentally represented in the mode of ordinary objects. Similarly, eternity is understood as a mental extension of the empirical representation of time, as that which attempts intellectually to circumscribe the overall temporality—that which, by definition, escapes such a circumscription. Finally, creation of the world, in spite of the verbal claim that it takes place out of nothing, is mentally grasped as "transition" or "change" between "that" indefinite and ineffable "nothing" and "everything." This "transition" is presented as if human consciousness could "look" at it from outside of both "nothing" and "everything." Modern cosmology attempts to depict the origin of the visible universe as the actualization of one particular member of the multiverse, the transition from one "object" (multiverse) to another "object" (our universe). As we have pointed out above, this is not the modeling of *creatio ex nihilo*; however, the mental frame of thinking of *creatio ex nihilo* remains unchanged. What is overlooked in this type of reasoning is the simple existential truth that one cannot position oneself outside of creation, which manifests and reveals itself through the fact of life. Since the articulation of creation of the world by human subjects is the process within creation, this process must be included in the explication of creation, so that the explication of creation includes the explication of the transcendental conditions of the possibility of such an explication and thus the explication of the empirically evident embodied consciousness as the manifestation of this creation. In this case the problem of creation out of nothing acquires an existential importance, for it does not say too much about the physical aspects of the universe and its origin (that is, details of creation) but it contributes to the explication of the human condition. One means here not just the biological condition, which is subordinated to the necessities of the natural creation, but another human vocation through which the articulation of creation of the world becomes meaningful, namely the attainment of *immortality*; for it is only through the vision of immortality that all concepts of creation of the world receive their sense as establishing communion with a personal God who created all with a promise of salvation. But this communion is not a matter of the necessity implanted in creation, it is not something subjected to biological instincts and the conditions of embodiment. It originates in the free will of humanity made in the image of That with Whom humanity wants to communicate. Correspondingly, to explicate creation of the world means to explicate the Divine image in man, or, to be more precise, to explicate the impetus in the path of restoration of the Image in the created universe after the Fall. To remove some fallacies in the representation of creation within the phenomenality of objects and to understand the problematics of creation as contributing to the restoration of the Image means to see creation as that saturating givenness of existence which constantly forms all states of human life and its consciousness.

The Created and the Sense of Infinity

When theology and cosmology speak of the created world, that is, the universe as a whole, by these terms is denoted something that is incommensurable with man as a natural being. This incommensurability is conceptually represented through the actual infinity of distance (the volumetric content of the visible universe is 10^{57} times greater than the planet Earth and will grow indefinitely in the future), the actual infinity of time (the visible universe is 13.7 billion years old and unbounded in the future[30]) and the actual infinity of its contingently various objects. All such representations relate to one and the same *possible* extension of that which is sensed and measured, toward infinity through space, time, and qualitative variation. The meaning of the created as infinite is sensed as unlimited and unbounded expanse. The word "infinite" is used as a qualitative metaphor that has its origin in the purely quantitative categorical construction of infinity through unlimited addition. What is specific in such a representation of the infinity of the created is that it is exercised in the natural attitude, that is, under the assumption that the incommensurability with the universe follows only from the infinite values of the contingent parameters of the universe within the fixed type of experience related, for example,

30. Cosmology claims that the universe is expanding with acceleration, so that its "volume" will increase indefinitely.

to scientific practice, but not from the infinite variety of human experience.

Here we come to a fundamental point of a theological conviction about creation: if the created world is indeed thought as brought into existence out of nothing, it was *nothing* that limited its infinite capacity to proclaim itself through the variety of relations between existent things. It is in this sense that human beings possess the immediate experiential knowledge of the infinite variety of modes in which created things can operate. They possess knowledge of the infinite possibilities through the relationship of every human person with any specific expression of the existent. This is achieved through another experiential fact, namely the radical otherness of the human hypostatic existence in comparison with the rest of creation.[31] It is this otherness as irreducibility to any particular mode of existence that makes it possible to experience creation through the infinitely many ways of subjective manifestation of this otherness. This experiencing of the qualitative infinity of creation drastically differs from the sense of its infinity through spatial and temporal extension of the finitely given in the natural attitude. The distinctive feature of sensing the whole creation through the event of personal existence is the convergence of all separate relations to created things in one single hypostatic consciousness of commensurability with creation by the fact of belonging to it and being different from it. All is commensurable with a particular personal existence because this

31. On the one hand, being inseparable from reality in virtue of its embodied intentional consciousness, human persons can exist only in the context of their immediate nondistance from reality. On the other hand, being a hypostatic formation, that is, being fundamentally different from other material things, human persons are "infinitely" distant from those other things. The ability to distance themselves from outer things (even, in abstraction, from one's own body) makes human persons equally positioned with respect to all objects in the universe, so that they can be articulated by human subjectivity as different and uniformly distant from it. Paradoxically the infinite ontological distance from all things in the universe makes human subjectivity to be equally commensurable (and thus close) with respect to all objects in the universe, including the universe as a whole.

existence, being an *event* encapsulating the whole "humankind-event,"[32] makes the whole creation to be an "event simultaneous" with this existence. Such an understanding of experiential knowledge of the infinity of creation presupposes that it transcends the phenomenology of the finite through experiencing the *limitless modes of expression* of this finitude. One can say that this transcending originates in the ontological indeterminacy of all attempts to express rationally that which exists. In other words, reason remains helpless in grasping the contingency of created things: being created out of nothing, they manifest this contingency. Thus the relationship between humanity and the world contains in itself distinct and unrepeatable expressions of unlimited reciprocity, in particular, through the sense of a paradoxical dialectic of commensurability and incommensurability with the universe.

To illustrate what we have said above, one can imagine a spatially finite universe (created in a traditional sense of contingent dependence upon God), whose temporal span is open-ended. The quantitative *finitude* of the universe cannot prevent the disclosure of its qualitative *infinity* through the unlimited expressions of relation to it. In other words, the inexhaustibility of communion with the universe that follows from its radical contingency upon the transcendent uncreated, makes it possible to transcend its geometrical finitude. Another example is the alleged origin of the physical universe in the event of the Big Bang, which has idiosyncratic properties characterized by physical infinities (infinite temperature, infinite curvature of space, etc). What is encapsulated in this construct, is not a metric or an "additive" sense of infinity, but rather a sense of a fundamental separation and distinctiveness from the "originary" event of the universe, the archetypal anticipation of the

32. See my *Light from the East*, chapter 7.

loss of the sense of "all in all" that humanity, treated theologically, ever experiences since the Fall. The sense of infinity of creation is revealed here as the abysmal detachment from the initial unity with the whole creation experienced before the Fall in the image and likeness of God. The sense of the infinite characteristics of the Big Bang thus attests not so much to the scientific construction of infinity through an indefinite asymptotic procedure applied to the finite physical states, but as the infinite distinctiveness and inexhaustibility of the contingent originary state of the universe, which potentially contains the infinite variety of experience of the universe by conscious creatures. The notion of the Big Bang escapes the finite and determined definitions, exhibiting its open-ended, that is, apophatic character. Human reason cannot constitute it simply because it exceeds the possibilities of such an accomplishment in finite time. Through an anticipatory encapsulation of the empirical givenness of creation in the nonoriginary origin, the structures of human subjectivity are constituted exactly to the extent that this subjectivity cannot comprehend this event "of origin." Theologically, one can state that human attempts to comprehend the contingent facticity of the universe as that saturating limit which stops consciousness from grasping its own origin *de facto*, explicates the human condition after the Fall.

The approach to the created universe that we developed above evidently remains incompatible with the modern scientific demand for objectivity, that is, the representation of reality within the phenomenality of objects. In other words, the natural attitude that separates the alleged object from the recipient of its interpretation is impossible in the case of creation. Since an "observer" *of*, or a participant *in*, or a communicant *with* creation cannot be abstracted from it, being a part of this creation, any "glance" at creation is conditioned by the presence of this observer. This condition entails that the very relationship between an observer and what is observed cannot be itself subjected to any quantitative characterization or measurement. The infinity of creation, and thus its inexhaustibility in rubrics of the rational thought, is manifested, on the one hand, as the metaphysical indeterminacy of that which exists and, on the other hand, by the personal (hypostatic) otherness of the human knower with respect to the rest of creation. Being a part of creation and, at the same time, transcending it in the sense of not submitting to it, human beings establish an infinite relationship with creation through the chasm of otherness with it. Correspondingly, every observation made with respect to the created world represents also a mode of experience of self-awareness as a personal mode of existence of the human subject that cannot be subjected to the conditions of nature. It is this mode of existence that makes possible the epistemic coordination of the finite corporeal observer with the universe as a whole in the mode of its actual infinity. Here human freedom of relationship with the whole creation in the mode of infinity manifests itself. The sense of what is real in such a vision of the created refers to the human presence as experience of relationship, which is not entirely determined by the reference to rationality. Thus that which is called "reality" of the created is an existential fact and event of metaphysical relationship between two actual infinities: humanity and the universe, consubstantial, but contraposed through their otherness; contraposed, but cognitively coordinated.[33]

The dynamic of cognition as an existential capacity and experience of relationship between limited conscious humanity in the world

33. As famous Russian scientist and theologian P. Florensky wrote in the same vein: "Nature and man are both infinite. And it is because of being infinite, that they are commensurable and can be parts of each other. . . . Man is in the world, but man is complex to the same extent as the world. The world is in man, but the world is also complex as man." P. Florensky, "Macrocosm and Microcosm," in his *Apology of the Cosmos* (St. Petersburg: Russian Christian Humanitarian Institute, 1994), 186 (in Russian); "Man is the recapitulation of the world, its summary; the world is the disclosure of man, its projection" (ibid., 187).

and the potentially infinite creation is included in the very fact and outcome of scientific observation. The experience of the potentially infinite is rooted not in the relationship of humanity with the world, but, in fact, in its freedom from this relationship. The delimiters in free-thinking of the universe proceed in the long run from the freedom of human beings made in the image of God. All thoughts and articulations of the universe always contain traces of the divine image. Even when cosmology proves the insignificance of humanity in the universe, the divine image remains intact exactly because the human mind always resists all attempts to circumscribe its life in rubrics of the natural, finite, and transient. As articulated above, human beings attempt to understand the underlying sense of beings and things not according to their "nature" but according to the final causes of these beings and things in relation to the place and goals of humanity in creation. Thus humanity wants to recognize beings as results of humanity's *free will*. The image of eternity is retained in any cosmological theory produced through free-willing even if this theory predicts the finitude of all actual forms of existence and life. Free-willing and -thinking of the created universe manifests its existential otherness with respect to the world, that is, the ability to create its own cosmos through art, culture, and history, which contains the physical cosmos as its own fact. That existential otherness which is spoken of here is not understood as an evolutionary consequence of the biologically predetermined capacities of human beings achieved through natural adaptation; it does not have a transcendental fictional origin either. It is an existential response to the need for freedom from nature in spite of its relative contingency upon nature through the limited existence in nature, expressed through place, time, decay, and ultimate death. One does not speak here of absolute freedom from nature, for it would imply that human beings control not only initial substance of their creations, but even the very existential presupposition of their nature. This privilege belongs only to the uncreated, that is, to God, who alone can bring existence out of nothing. However, humanity in the image of God has the privilege of establishing modes of its subjective impression with regard to nature "out of nothing." These subjective impressions as modes of the natural abilities of man can be self-defined in a sense that they are not naturally predetermined. It is here that human freedom becomes an indispensable condition for the experience of the potentially unlimited contingent creation. It is achieved through a mental image and concise symbol of the universe that saturates intuition through an instantaneous synthesis in such a way that this universe enters human life as a work of art, revealing thus not only the content of that to which the signifier of the universe as a whole is intended, but the existential otherness of an artist, that is, of a cosmologist. Since the universe as a whole cannot be presented through the phenomenality of objects, and the scale of the unlimited and potentially infinite creation exceeds any possibility of the natural predetermination of the universe's image, the universe saturates the intuition to such an extent that the faculties of its comprehension become determined by this saturation, exactly to the extent that these faculties cannot comprehend the universe. Thus humanity itself through gazing at the universe is formed by its unconditional givenness and articulates this universe in the conditions of this existential determination. The saturating intuition of the created universe makes the relationship between human beings and the universe unbounded and subjected to communion between the potentially infinite humanity (because of its Divine image) and the infinite creator. The more personal the relation with the universe becomes, the more inexpressible in the limits of cognitive structures and discursive thought it remains. If human beings attempt to express the meaning of personal reference to the metaphysical causal principle of all created things, this meaning always remains incomplete in comparison with the experience of this reference itself. The intuition of the contingency of the created world upon the transcendent God-creator is a natural way of expressing the sense of dependence and originary foundation of all that is. To avoid a suspicion of an eidetic or imaginary unboundedness in such an intuition, one needs reciprocity, that is, the personal response from that which is intuited as the infinite. But the universe, being created and enhypostatically inherent in the Divine, cannot be hypostatic itself.[34] Thus the infinity of the created must be accompanied by the reciprocal response of the Creator who confirms this infinity. This type of reciprocity and communion cannot be subjected to the powers of natural epistemic verification and impression. This confirms an old intuition of some theologians that Christian teaching of creation has never been the narrative of the created world, but rather the narrative of relationship between God and the world, or, more precisely between man and God.[35] It is because the asserted reciprocity can only be achieved and validated experientially that it has an intrinsically precarious character, which must be carefully distinguished from any psychological projections and unjustified fantasies ultimately rooted in the rubrics of the natural.[36] The communal nature of such a validation and the confirmation by relationship transcends any conceptual images, intellectual

34. The meaning of the terms *enhypostsis* and *enhypostatic* is discussed in A. Nesteruk, *Light from the East*, 110-17; A. Nesteruk, "The Universe as Hypostatic Inherence in the Logos of God," in *In Whom We Live and Move and Have Our Being*, op. cit., 169-83.

35. See Florovsky, "Creation and Creaturehood."

36. It is implied here that a glance at nature and the universe as created by God is accessible to humanity only through the Holy Spirit. Correspondingly, for being sure that one indeed deals with the spiritual vision of the universe through the prism of Christian faith it is necessary to make a distinction between Christian spirituality rooted in ecclesial practices implying the invocation of the Spirit, and all other non-Christian and arbitrary appeals to generic spirituality. See in this regard V. Shmaliy, "The Spirit or/and Spirits in Creation?," in *The Spirit in Creation and New Creation: Science and Theology in Western and Orthodox Realms*, ed. M. Welker (Grand Rapids: Eerdmans, 2012), 79-94.

constructions, and ethical systems attempting to express the reference to the whole creation and its creator. The reciprocity implies inerasable presence of the Divine image in any representation of the created universe. This inerasability as a biased position in cognition and relationship to the universe was earlier called theological commitment, which excludes any liberal and secular approach to the issue of creation as being mythopoetical and epistemologically arbitrary. In this respect one must make it clear that in spite of the fact that according to the Christian view the cosmic history and sense of creation is unfolded from within the history of humanity, that is, its history of salvation,[37] the delimiters of historicity do not phenomenalize creation as an object in similarity with the object-like representation subjected to the norms of scientific rationality; this type of phenomenality with respect to creation will be limited and existentially distorted, depriving humanity of freedom following from the Divine image. The delimiters of historicity disclose creation as relationship between humanity and God, and reveal themselves as related to the sense and purpose of human history.[38]

Since the issue of creation of the world is a metaphysical issue it is contradictory to pose questions about metaphysics in the terms and manner of physics; the question of creation demands definitive answers (in contradistinction to the sciences, which provide some positive but uncertain insights on the nature of the created, uncertain

37. O. Clément, *Le Christ, Terre des vivants. Essais théologiques*. Spiritualité Orientale, n. 17(Bégrolles-en-Mauges: Abbaye de Bellefontaine, 1976), 77-94.

38. Here one can point toward Fr. S. Bulgakov, who in his *Bride of the Lamb* (Grand Rapids: Eerdmans, 2002), 221-22, advocated that the adequate description of the relation between the world and God cannot be established on the grounds of mechanical causality, that is, that God is the cause of the world. This relation is that of the Creator and the created. The act of creation of the world is rather an ecstatic transcendence of God through this creation. Creation manifests itself through relationship between God and humanity. This relationship does not subordinate to the constituting constraints of its manifesting historicity. This relationship, as well as the relationship between God and the world, is subordinated only to the promise of God for salvation and eternal life, that is, of renewed creation.

in the sense that they are amendable and corrigible by the very course of the scientific advance). These answers cannot be exhausted through scientific enquiry and thus the acquisition of a metaphysical datum presupposes existential participation in it. What is the sense of such a participation and, to a certain extent, its inevitability? The ultimate issue is humanity's dilemma between a prospect of its ultimate annihilation by the forces of nature, and its liberation from the conditions of nature and attainment of real existence beyond the mode of nature, that is, beyond biological death. Here the metaphysical dimension of the issue of creation of the world becomes imbued with a Christian theological content, namely with the gospel's proclamation of the possibility of attaining immortality. And since it is the church and its experience that represents humanity's deepest need to attain immortality, the ecclesial dimension of the question of creation of the world comes to the fore. Here the human vision of creation as it is aligned toward the eternal creation, that is, creation renewed through the restoration of its unity with the Divine, corresponds to the vision of how mortal life is aligned toward eternal life. The humanity of Christ seeks for the completion of the first act of the creation through the fulfillment of the promise for salvation that is inherent in the very act of creation out of the Love of God. Correspondingly at every individual level, human beings seek for the completion of the act of conception and birth toward the fulfillment of the inherent initial promise for the attainment of salvation and eternal life.

In addressing the issue of freedom from nature and attainment of immortality one cannot appeal to science. One needs a different language, the language of existential otherness and personal mode of existence, which is devoid of any limiting preconditions from nature. This happens in experience of relationship either with other human beings or nature, or in love and art, when communion with other persons unfolds as the wholeness of reality coming not from its differentiated and individual existence in the nowhere of cosmic space and time, but from the stripping off of all extended (Gr.: *diastasis*) dimensions of existence, so that the whole world acquires the sense of being "contemporaneous" or "simultaneous" with the event of communion. The sought language is the language of "horizontal" transcendence, not outwardly beyond space and time, but inwardly inside the theologically understood *heart*, as that center of disclosure and manifestation of the world which forms and constitutes the person. Personal existence can be described, in contradistinction with the individual existence, as experience of the *private absolute*. This experience can be called mystical and amounts to the possibility of sensing the multifarious universality of all that exists, of sensing all that is fragmented and separated in space and time (be it human history or the whole cosmological span) as unique and exclusive experience of the universal in spite of the finitude and limited capacity of the physical and biological sustainment of the individual person. Experience of being privately absolute and unique follows from the inherent sense of commensurability with the whole created universe (distinct from the incommensurability of space-time), not through genetic consubstantiality, but through the God-given ability to see creation as an instantaneous synthesis effected by the Divine image of the Logos, the creator and sustainer of the universe. If science remains silent with respect to the human ability to articulate the whole creation, theology formulates this ability as the actual possibility of experiencing the private absolute as the unique and unrepeatable fact of existence, that is, hypostatic existence, which is ultimately existence in "solitude"[39] understood as the otherness not only to impersonal nature, but also to other human beings.

39. This philosophical motive is developed by E. Levinas. See, for example, his *Time and the Other* (Pittsburgh: Duquesne University Press, 1987), 39–57.

Paradoxically, it is this otherness that opens the ways of communion with others. This solitude is the ability to transcend all partial and contingent relations and attitudes, impressions and feelings, and to retain the core of existence as that private and absolute "cogito ergo sum" or "amo ergo sum" which cannot be communicated in a manner of speech. The private absolute is ultimately that mystery of immanence and incomprehensibility of life which, being caught in thinking and causing the cessation of subjectively extended space and time, leaves a human being speechless and in the state of the prayerful invocation similar to a cry of a newborn child looking for the mother, as that carrier of relationship which consoles for the solitude of being born into this world. This is the reality of being born in such conditions that the motivation and phenomenalization of this birth is profoundly hidden, with no clear antecedents. Every human being has to struggle with the mystery of its own creation as its inevitable commensurability with creation. To uncover the sense of one's existence through the unfolding of life, as a gradual process of filling this sense with a new content of temporally succeeding events, is the same as to uncover the sense of all creation through seeing it as one's own private absolute.

If the sense of the universe as a whole can be experienced as a personal absolute, there is a natural desire of every human being to retain this sense "absolutely" regardless of the conditions of embodiment, and to subject this experience to the liberation from the necessities of nature. The longing for immortality is not an idea of abandoning creation and subjecting it to the annihilating forces of nonbeing. It is a desire to exist, in the sense "to live," in that state of creation which would maximally imitate that invisible foundation which provided this created world with the invitation to exist. If Christian faith in its ecclesial setting is about the attainment of immortality, then the issue of creation becomes an inevitable part of its theology, because the attainment of immortality must be implicitly preordained in the very event of one's conception and birth as the initial creational promise for salvation and eternal life. The transcendence toward immortality as an indefinite retaining of experience of the private absolute does not invalidate the mode of creaturehood: the experience of the private absolute is itself the disclosure and manifestation of being-in-the-world as being created.

The Universe as a Saturated Phenomenon: The Explication of the Sense of the Private Absolute

Experience of "private absolute" introduces a different type of apophaticism in knowledge of the universe, based not in the "universe's darkness" (lack of light), but on the excess of its donation. Here the perception of the universe as extended in space and time is replaced by the perception of the universe in aesthetical and ethical categories. Ancient Greek philosophers called such a universe *kosmos*, that is, beauty and order. However, *kosmos* of the Greeks, unlike that which is understood by the cosmos in modern cosmology, denoted the way by which the natural reality is. It denoted not that which was related to the question "What?" of created nature but rather to the question "How?" *Kosmos* thus is the "ordered" revelation of the existent, that is, the notion related to beauty. But beauty is a matter of personal judgment and observing distinctions that can be justified only within relationship, that is, communion. It is because of this that Plato summarized all pre-Socratic views in his teaching of *kosmos* as a living unity, "animate and intelligent being,"[40] a living totality of animate creatures and inanimate things, gods and people.[41]

40. "Thus then, in accordance with the likely account, we must declare that this world came to be, by the god's providence, in the very truth a living creature endowed with soul and reason": Plato, *Timaeus*, 30 b6-8. ET: J. Warrington, Plato, *Timaeus* (London: Dent, 1965), 19.

The overcoming of disorder and riot as such reveals itself as life so that *kosmos* unfolds as a living whole, the "visible living being."[42] Since life implies soul, the "body" of the *kosmos* is harmonized in the "spirit of friendship" with that who brought it into existence. But then the beauty of the world, that is, the world as *kosmos*, reveals itself as a mode of the living, animated organism whose soul is also intelligent: the order of the world, its measure and commensurability, reveal the beauty of the world but also manifest the intellect.[43] Correspondingly, the beauty of the universe reveals itself not only through the world being animated, but also through its intellect. One can say that the beauty of the universe is not that which is manifested, but the universe as manifestation. It is the "how" of the universe but not its "what."

The approach to the world as a whole based on an attempt to treat the world in categories of beauty, to look at it as a perfect creation, animated and intelligent being, constrains human knowledge of the universe to the limits of its empirical link with the whole of the reality of nature. It does not allow any formal and logical explanation of the world that would depersonify an immediate living communion with it in intellectual abstractions. The dimensions of beauty presuppose that one can recognize and evaluate the way by which cosmic reality

41. "Wise men say . . . that the heavens and the earth, gods and men, are bound together by fellowship and friendship, and order and temperance and justice, and for this reason they call the sum of things the 'ordered' universe (kosmos), . . . not the world of disorder or riot." Plato, *Gorgias* 507e-508a. ET: *Plato Socratic Dialogues*, ed. W. D. Woodhead (Edinburgh: Nelson, 1963), 279.

42. "For this world, having received its full complement of living creatures, mortal and immortal, has become a visible living creature embracing all that are visible, and an image of the intelligible, a perceptible god, supreme in greatness and excellence, in beauty and perfection . . .": Plato, *Timaeus*, 92c 5-9. ET: Warrington, op. cit., 138.

43. "Taking thought, therefore, he found that among things that are by nature visible none that is without intelligence will ever be better that one that is rational, when each is taken as a whole, and further that intelligence cannot be present in anything other than soul. Because of this reasoning, when he constructed the universe he fashioned reason within soul and soul within body, to the end that the work he was accomplishing might be of its nature as excellent and perfect as possible." Plato, *Timaeus*, 30b 1-6. ET: ibid., 19.

exists. One can experience the beauty of the world only through the immediate and intimate *relationship* as personal communion, not as an abstract measure invented by the discursive reason.[44]

Communion with the created universe and comprehension of its beauty is a personal process and achievement. In personal relationship we come to know the universe not as an existent whose phenomenality is limited to the numerical and quantitatively measurable domain, that is, not as nature or essence, but as the unlimited of indefinite differentiations manifested to a person. This mode of personal uniqueness of things is their beauty as the reality of the universe appearing as *kosmos*. In the world as a whole the ontological difference between any existent and being in general, manifesting itself as beauty, comes forth through the difference between the intellectual circumscribing of things and their way of being as their distinction detected by a person. Their beauty claims itself as their unconcealment revealed in personal relationship. The truth (unconcealment) of existent things witnesses of itself, as creative presence and action upon persons. This personal action as an ordering and arranging of the universe making it the beautiful *kosmos* cannot be exhaustively determined by the human reason through logic and quantitative definitions. It rather meets with reason dialogically, in the event of personal relationship-communion. In fact, it is only this relationship that makes possible the process of knowledge of the universe to the extent that we recognize in it *kosmos* as it was defined above. By so doing we first encounter being (in a general philosophical sense) as it is turned to us by its, using Heideggerian terminology, unconcealed "side": "'World' is the *clearing* [aperture, opening, AN] of being into which human being stands out on the

44. One can invoke an allegoric parallel: if one sees stars and their constellations in the sky in a dispassionate curiosity, one does not "hear" the music of the heaven that manifests the universe in its beauty. One sees the script, but does not hear the melody.

basis of his thrown essence."[45] Secondly, we discover the way of the personal givenness of things to us and, as a result, the inner world of persons. If a person contemplates the universe not simply as a conglomerate of different forms of matter arranged in a contingent manner, but as an "object" of art, then in a direct analogy with any human-made piece of art, one realizes that the universe can be seen not as a precisely calculated clock-like mechanism, but as that ecstatic energy in which the creative person can be found and by which the knowing person is constituted. The beauty of the universe thus reveals that present-in-absence "face" of the universe which is enhypostasized by the knowing person in the image of its hypostatic creator, the Logos.

This personal appreciation of the cosmos through communion leads to such an "understanding" of the phenomenon of the universe in which the intuition (based in communion) of the universe gives immeasurably more than intention of the universe ever would have intended or foreseen. The universe perceived in this way naturally falls under the rubric of a phenomenon that is saturated with intuition; to clarify this one may refer to Kant's distinction between the "rational idea" and "aesthetic idea." The "rational idea," for example the cosmological idea, can never become a sensible comprehension because it contains a concept (of the trans-sensible universe) for which no adequate intuition can ever be given. In this sense the rational idea of the universe can be considered as phenomenon that is not only poor, but deprived of intuition. In fact,

45. M. Heidegger, "Letter on Humanism," in *Pathmarks*, ed. W. McNeill (Cambridge: Cambridge University Press, 1998), 266 (emphasis added). The terminology of clearing of being can receive a strictly physical interpretation by referring to the present discussions in cosmology that the present state of the universe, as we observe it, is unique because it is only now that we can observe such features of the universe as its evolution and its enormous large-scale structure. In this sense we indeed have a sort of clearing of being, its unique unconcealment to us. See in this respect, for example, N. E. Abrams and J. Primack, *The New Universe and the Human Future* (New Haven: Yale University Press, 2011), 79–84; L. M. Krauss, *A Universe from Nothing* (New York: Free Press, 2012), 105–19.

the cosmological idea (as a rational idea) is defined as a representation of an "object"—the universe—according to a principle that this representation can never become the cognition of the universe. Correspondingly, to the "rational idea" (of the universe), can be contraposed the "aesthetic idea" (of the universe) as the *cosmos as communion*, that is, a representation according to intuition which itself can never become an intellectual (discursive) cognition, but for an opposite reason: "because it is an intuition . . . for which an adequate concept can never be found."[46] In this case the matter is not that there is the nonadequation of the intuition, that is, its lack, which leaves a concept blind (there is no intuition adequate to the concept of the universe as a whole, or the world); conversely one has here the failure or insufficiency of the concept to clarify the intuition. The excess of intuition related to a particular sphere of experience over its conceptual representation prohibits that any linguistic representation ever reaches it completely and render it intelligible.[47] In other words, the excess of intuition of the universe in communion will never allow one to see the universe as an object. This incapacity to produce an object does not result from a shortage of donation of the universe (as happens in the rational idea of the universe), but from the excess of intuition, that is, from the excess of donation (for it is the intuition that gives). The "aesthetic idea" gives more than any concept can expose, that is, arrange and order the intuitive content according to rules of the understanding. The impossibility of this conceptual arrangement follows from the fact that the intuitive overabundance itself is not accessible to experience within the rules pertaining to discursive modalities of cognition. The intuition is not exposed within the limits of the concept, but saturates it and renders it overexposed, that is, keeps it invisible, blind not

46. I. Kant, *Critique of Judgment*, § 57, note 1. ET: J. H. Bernard (New York: Hafner, 1951), 187.
47. Cf. ibid., § 49.

by the lack of "light" (the universe as a whole is invisible because it remains dark for the "light" of the categories of the understanding) but by the excess of light (there is too much in our intuition of the universe through communion that cannot be discerned). The problem is to find a phenomenological description of the "aesthetic idea" of the universe, that is, communion with the universe, rendering thus the unforeseeable nature of the donation of the universe, the impossibility of seeing it as an object, and its freedom from intentionality of a subject. In spite of all anomalies related to the discursive apprehension of the "aesthetic idea," the universe as a created whole is not disqualified from a phenomenological description. Thus one can use categories of the understanding (in a Kantian sense) in order to characterize the "inexposable representation" of this idea as a saturated phenomenon, although in an apophatic, negative sense.

First of all, the universe as a whole cannot be aimed at in the sense of a successive synthesis of *quantity* applied to ordinary objects. It is because of belonging to the universe and its constant spontaneous givenness to us that the intuition that gives it is not limited; its excess can neither be divided nor put together because of a homogeneous magnitude of its parts. The excess of the donating intuition could not be measured on the basis of its parts since this intuition surpasses the sum of these parts. The "phenomenon of the universe," which is always exceeded by the intuition that saturates, should be called *incommensurable* in the sense of not measurable. This lack of measure does not operate here through the enormity of an unlimited quantity that stands behind the universe. It is rather marked by the impossibility of applying a successive synthesis to it, as if one could foresee a complex whole on the basis of its parts. Since the universe as a whole, as a saturated phenomenon, exceeds any summation of its parts which are in many ways inaccessible to the subjectivity undertaking such a summation, the idea of a successive synthesis has to be replaced by what was called before an instantaneous synthesis (a synthesis of communion),[48] the representation of which "precedes" and goes beyond possible (unobservable and imagined) components, rather than resulting from them according to the extended in time pre-vision. Here a clarification is possible: physically, one can perceive in the universe only a particular side of it, which is "turned" to us by the surface of the past null cone.[49] What we see through the light-cone of the past imposes itself on us with a certain force, which overwhelms us to the point that we are fascinated by it. This imitates a simple human impression of the universe when one looks at the sky and sees patterns of beautiful stars. At this stage the successive synthesis, attempted later in physical cosmology, is suspended exactly at the moment when the first impression from the universe occurred. This happens because another synthesis has been achieved, a synthesis that is instantaneous and irreducible to the sum of all possible parts of the universe. This type of communion with the universe, which is accompanied by amazement and the sense of awe, arises without any common measure with the phenomena that precede, announce, or explain it.

The universe as a saturated phenomenon cannot be accounted for according to quality as an intensive magnitude. The aiming at or foreseeing the universe in perception is heterogeneous in degree for everyone and is marked by a break or discontinuity such that the universe manifests itself in perception as an absolutely singular novelty. The universe seems to be already there, available for our arrival, to live in it and gaze at it. In this sense the universe imposes itself on us as preceding us. It appears to our view in childhood as

48. See chapter 3.
49. Geometrically, the light from remote objects that reaches us on earth forms a light-cone of the past, thus selecting a particular part of the universe that is accessible to observations.

well in an adult state as an unexpected, unpredictable fact, originating allegedly in what we perceive as the uncontrollable past. Supplied by theoretical apparatus, we indeed face the entire cosmos in its past, but this past does not reach us apart from deductions and intuitions. The more we study the universe astronomically, the more splendor we unfold; but this splendor is unexpected and unpredictable, unknown before we glimpsed it in the sky. In this sense the "beauty" of the universe as it manifests itself through picturesque galaxies and nebula formations brings one into a state of awe when one experiences the universe as incomparable and incommensurable with any particular *event* in one's life (in spite of its "simultaneity" with one's life through the instantaneous synthesis). The universe comes to one, engulfs one, and imposes itself without one's control and anticipation: thus it exhibits itself in the phenomenality of events.

What is meant here is that the phenomenon of the universe reaches an intensive magnitude without measure, so that starting from a certain degree the intensity of the intuition exceeds all anticipations of perception. Existentially, while experiencing the immediacy of communion with the universe one cannot predict or measure the intensity of this impression, for it is inseparable from the fact of life, and thus life itself cannot be subjected to any measure: it either is or is not. The intuition of the universe blinds the capacity of its anticipation through perception. It is in this sense that the intuition, which gives the phenomenon of the universe, is unbearable for the gaze: the perception of the universe as a whole is blocked and its comprehension manifests itself as dim and dark, unformed and essentially disturbing. The universe falls under the rubric of the saturated phenomenon that J.-L. Marion characterizes by such a term as "bedazzlement" (Fr: *l'éblouissement*).[50] The universe in its

50. J.-L. Marion, "Le Phénomène Saturé," in *Phénoménologie et Théologie*, ed. J.-F. Courtine (Paris: Criterion, 1992), 109-11. ET: "The Saturated Phenomenon," in D. Janicaud et al.,

pieces and moments can be seen, but the universe as a whole not only cannot be seen, it cannot be borne.[51] The bedazzlement by infinity brings one to the recognition of one's own finitude. The sense of incommensurability with the universe originates through the bedazzlement by its potential infinity. The finitude of humanity is experienced not so much through the shortage of the given before its gaze of the universe (the universe supplies human subjectivity with a potentially infinite set of sense-impressions as different events in space and time), but, above all, through the overwhelming belonging to this universe as a fact of existence, whose magnitude of donation cannot be measured (because it cannot be controlled). Here human consciousness experiences itself in the suffering passivity (that is, ignorance of its own origin, and suspended between subject and object, meaning and being) with no means to evaluate its own givenness in terms of any measure.

The Created Universe in the Phenomenality of Events

The universe as creation cannot be subjected to relational analysis because it is unique and one cannot rerun the universe or stage it as an experimental event. The universe as a whole is identical only to itself, so that its unfolding facticity is characterized not only by irreproducibility but by irreversibility understood not in a technical, thermodynamic sense, but as logical irreversibility, as a coming into the facticity of existence and the impossibility of exiting this existence. The universe as its sheer givenness makes itself a

Phenomenology and "The Theological Turn": The French Debate (New York: Fordham University Press, 2001), 200–202.

51. See also on this my papers: A. Nesteruk, "The Quest for Transcendence in Theology and Science," in *The Blackwell Companion to Science and Christianity* (Chichester: Wiley-Blackwell, 2012), pp. 560–62; "The Universe as a Saturated Phenomenon: The Christian Concept of Creation in View of Modern Philosophical and Scientific Developments," *Theology and Science* 12, no. 3 (2014), 236–59.

phenomenon not arising from our initiative and not responding to our expectations (since it cannot be reproduced); it gives itself to us from its own self to such an extent that it affects us, changes us, and almost constitutes us, and stages us out of its own giving itself to us: hence it has the phenomenality of an *event*.[52]

From the point of view of a physical cosmologist there is an imminent difficulty: How can one treat the universe as an event if, at first glance, it is "a" stable object, that is, the astronomical cosmos out there and what is allegedly beyond it? What is the basis for interpreting the universe (as an object of cosmological research) as "an" or "the" event, if this word has mundane connotations? It is not difficult to realize that the logic of formulating such a question is exercised from within the natural attitude, which thinks of the universe in the phenomenality of objects that by definition have a temporal pattern of stability and thus cannot be events. The universe is out there and it is not "an" or "the" event because it is always over there. However, as we have attempted to argue before, the universe as articulated existence is epistemologically commensurable with the event of one's life and thus has an evential status in this sense. Thus, phenomenologically, one should reverse the question and enquire as to how the essential event-character of the phenomenon of the universe became blurred and disappeared to the extent that it appears no more than an object. The objectivity of the phenomenon of the universe arises from an attempted quantitative synthesis in the style of Kant: to become an object any phenomenon should be expressed in terms of quantity or magnitude. Correspondingly, the totality of the phenomenon is achieved as the sum of its parts through anticipation of a quantitative synthesis.[53] This signifies that the magnitude of

52. More detailed description of the phenomenality of events can be found in Marion, *Certitudes négatives*, 244–308.

53. On the basis of summation of astronomical objects: Kant, *Critique of Pure Reason*, A163; B204.

the phenomenon of the universe is always to be described in finite parameters and depicted in real or abstract mathematical (imaginary) space.[54] In this sense the universe as a whole is intellectually foreseen before it is actually seen. The universe is confined in its quantity, defined through its parts, and brought to a conceptual cessation of any advance of its content by the already-made measurements. This reduction of the universe to its foreseeable quantity turns it into an object as if there were nothing else to be seen in it, nothing other than that which can already be envisaged on the basis of its theoretical construction. This is typical for theoretical cosmologists who no longer need to *see* the universe (that is, commune with it personally) because they foresee the universe in advance. In fact, any speculative cosmology could freely avoid living insights into the universe on the basis of its theoretical foreseeing unless the measurements would contradict this possibility, that is, bring a sort of breakdown to the constructed object. The phenomenon of the universe reduced to an object deprives the universe of its independent and unrestricted appearance, placing thus its event-like character in rubrics of some common laws. When cosmology treats the universe as an object it assumes that everything in it remains seen in advance and nothing unexpected can happen that disqualifies the universe from the status of an object. Then the universe as an object of particular theoretical study remains a phenomenon that has already expired: nothing new can happen to it, since in those rubrics in which it is constituted it appears as that which is devoid of the mode of becoming or

54. Sometimes the universe as a whole is depicted through a geometric shape, as if one could position oneself outside it. See, for example, Hawking's "classical" imagery of the universe with imaginary time in S. Hawking and R. Penrose, *The Nature of Space and Time* (Princeton: Princeton University Press, 1996), 86, 97, 102, or Penrose's recent representation of the universe through cylindrical cycles, in R. Penrose, *Cycles of Time: An Extraordinary New View of the Universe* (London: Bodley Head, 2010), 148.

happening. Metaphorically one can say that the universe as an object appears to be a shadow of the event that is denied in it.[55]

The event-like character of the universe cannot be foreseen since its partial causes, which are invented by cosmology, remain fundamentally insufficient: a typical example is a particular version of the multiverse theory in which the space of all possible initial conditions for universes is postulated.[56] This ensemble of the initial conditions is a necessary condition for *this* universe to be actualized in existence. However, the realized facticity of this universe as the "pinning down" of the initial conditions (in the overall space of possible conditions) that leads to the formation of our universe, is not described by any theory and requires *ad extra* assumptions that do not belong to the sphere of physical causality. The realization of these particular conditions is detected postfactum, when the event of their choice and hence the coming of this universe into existence, happened and was accomplished. But the event of choosing the appropriate initial conditions in this case is not subject to any causation based on the foreseeing of this event. It is not amazing that the postfactum possibility of these initial conditions (that is, of our universe) which are impossible to foresee remains, strictly speaking, an a priori impossibility with regard to the system of previously classified causes: indeed the choice of the initial conditions for our universe is practically impossible since it must be made out of the potentially infinite number of all possible conditions.[57]

55. Cf. J.-L. Marion, "The Event, the Phenomenon and the Revealed," in *Transcendence in Philosophy and Religion*, (Bloomington and Indianapolis, Indiana University Press, 2003), 92.
56. See, for example, Penrose's model in *The Road to Reality*, quoted above.
57. If there are infinitely many different initial conditions, the a priori probability for taking place of those that correspond to our universe is zero: $p = 0$. Correspondingly, the informational uncertainty related to the choice of these conditions is infinity: $I = \infty$, so that one needs to invoke the idea of an omniscient being (Creator in Penrose's terms) who overcomes this informational barrier and makes the choice. However, from the point of view of human beings the choice of such conditions remains a sheer impossibility.

The observation that the universe as a whole cannot be foreseen on the grounds of any causation can be rephrased by saying that the universe imposes itself on perception without one being able to assign to the universe a substance in which it dwells as an accident (or a cause from which the universe results as an effect: there is no causal principle of the world). One could refer to the invisible whole of the universe as that substance "in" which the observable part of the universe (as its accident) dwells. The universe is given to us in its pieces and moments that represent that whole which will never be accessible *per se*. The invocation of the idea of the primordial substance in this context would just mean a conviction that there is an undifferentiated unity of "all in all," and the visible universe represents its particular realization in the ancient sense of demiurgic order out of chaos.[58] This mental split in the representation of the universe as substance and accident does not correspond however to the immediate experience of the universe as an event of life, for it is in this event that it is exactly impossible to make a distinction between "substance" of life and life as an "accident." Correspondingly, it is problematic to look for the cause of the universe (as its effect) if it is perceived as *coaevus universo*, that is, the universe as "simultaneous" with one's life. The question of the facticity of the universe (as an effect of some cause) cannot be even addressed if the universe is seen as the totality of all. All rhetoric about the origination of the universe in the Big Bang has no philosophical significance as the "cause" of the universe, because it does not address the issue of the "cause" of the Big Bang itself. Even the appeal to a "particular Big Bang" taking place in the course of an inflationary generation of many bubble universes does not reach any goal, since the cause of the ensemble of those

58. One could treat the "cosmological fluid" of clusters and galaxies as such a substance, one that gives rise to its contingent accident, that is, our galaxy, solar system, the planet Earth, etc. In a temporal sense such a substance could be associated with the Big Bang, containing in encapsulated form all consequent realizations of matter formations.

bubbles indwelling in the "substance" of the originary inflaton field does not remove the question of the facticity of this field. Indeed, in analogy with ancient Greek philosophies this field can be considered as substance of the same mental kind as, for example, the "water" of Thales from Miletus, from which the actual state of affairs in the world can be produced by a potentially infinite number of ways. Finally, one can suggest that another typical characteristic of the universe as an event amounts to the fact that the number of "possible explanations" of its facticity is indefinite and increases in proportion to cosmological hermeneutics that cosmologists and their interpreters produce.

The treatment of the universe in the phenomenality of events does not deny its temporality. The temporality of the universe as a whole cannot be that of the physical flow in preexisting time; it is a different *temporality of events* that must be elucidated. In order to do this let us start from a simple question: If the phenomenon of the universe giving itself in a mode of an event carries the signs of temporality, does it reaffirm the Kantian position that every phenomenon is a phenomenon if it admits a representation as experience in time-form of sensibility? The response to this is that while in Kant's view temporality serves only to allow the synthesis of phenomena as object with a certain identity, that is, a guarantee of its permanence in presence justified through assigning it a cause or a reason, the event-like character of the universe as a whole cannot rely on this kind of a synthesis and thus permits a corresponding phenomenality of the universe contrary to the objectivity established in physical cosmology. The objectiveness of the created universe as permanence in presence through its evolution becomes a projection or even an illusion of an atemporal event. The universe being described by mathematical laws as an evolving object entails that its notion is emptied of any intuitive content and thus represents a shadow of that event which gives itself in the fact of life. But then there is a question of the internal sense of temporality of life itself. Indeed, temporality belongs to sensibility of subjects articulating the universe and orienting them toward the synthesis of the already-given objects. However, it is this same temporality that is never applied in order to constitute and define the acting agent of this synthesis, that is, the transcendental "I." Then, even if one conjectures that the phenomena temporalized as objects (the evolving universe) preserve a trace of their belonging to the event (an intuition of the universe as a whole), still the transcendental "I" does not phenomenalize itself as an event. This happens because the "I" never phenomenalizes itself at all: it does not appear among other phenomena; that is, it is excluded from that phenomenality which it produces. In this sense the idea that the universe as a whole is an event seems to be counterintuitive: indeed even if the "I" experiences its communion with the universe through the atemporal, that is, nontemporalizable sense of belonging and consubstantiality, any attempt to express this linguistically and discursively puts the event-like sense of communion under the rubrics of eidetic temporality. There is one particular aspect in the phenomenality of the universe that employs such an eidetic temporality, namely the universe's "beginning," its point of origination, the Big Bang. It is here that the universe is explicitly eidetically temporalized as an event. And this happens not accidentally, but because of a deep analogy between the Big Bang and an event of birth of any particular "I," which is the ultimate beginning and end of all possible predications of life as well as the universe. It is here that the "I" phenomenalizes the universe as an event on the same grounds as it phenomenalizes itself as its own event of birth.

When cosmology tells us that whatever we see in the sky points toward the origin of the universe, the Big Bang, it asserts the universe as a phenomenon that shows itself in the mode of the already given to us in its sheer facticity originating in this Big Bang. Such a phenomenon of the universe is properly event-like. The question then is how to understand the origin of the universe (showing itself in its remote consequences as a phenomenon in the presence of humanity) in the conditions where no human subjects have ever seen it by their "own eyes." Cosmologists consider the point origination of the universe as a "phenomenon" because they constantly intend it: these intentions are fulfilled with indirect testimonies relying on intermediate deductions following from observations and belief-laden theories.

The origin of the universe which contains human beings appears in fact as a privileged phenomenon since a significant effort of humanity is devoted to its reconstitution as restoration of the lost memory of it, to giving it sense and even, in a way, responding to its appeal to us as if the universe had its distinctive self-identity. Still, humanity cannot see this undeniable and unavoidable phenomenon directly. The fact that one cannot see the origin of the universe directly and that nevertheless it reveals itself as a phenomenon for which cosmology constantly intends, constitutes an aporia that can be formulated in the following way: *the origin of the universe shows to humanity precisely that its origin cannot be shown*. This aporia urges philosophizing cosmologists to understand how the phenomenon of the origin of the universe that does not show itself directly not only affects humanity as if it did show itself, but, in fact, affects humanity in a more radical way than any other phenomena, since the beginning of the universe forms the necessary conditions for humanity's emergence in the universe. The same cosmologists have to admit that since this indemonstrable origin of the universe reveals itself to them, it "happens" to them in that it endows human beings with a future. The origin of the universe can be called a phenomenon because of its presence in absence, that is, the poverty of demonstrability, so that it comes to pass in human life as an *event*, which was never present in presence, and is always already gone past, whereas it never surpasses the present and, in fact, is always to come. Thus one can say that the origin of the universe phenomenalizes itself but as a pure event unpredictable (there is no sense of temporality before and outside the universe), irreproducible (one cannot rerun the universe), exceeding all cause (there is no physical causation from beyond the universe) and making the impossible possible (the probability of origination of our universe in multiverse scenarios is always infinitely small, that is, the universe is a priori impossible), surpassing all expectations and predictions (the constant advance of knowledge of the universe does not make it possible to assign to the universe some definitive and stable features that could sustain indefinitely the observational tests as well as an epistemological advance). Speaking of the origin of the universe we speak of its donation: it is given to us in the measure that it gives itself and its givenness to us is an apodictic fact-event that is alone responsible for that which we call the constituted phenomenon of the origin of the universe, or the universe as a whole. It is this givenness that initiates an encapsulated temporality of the universe as an event, which being projected onto the object-like temporality manifests the features of a shadow of the universe's wholeness and its nonoriginary origin.

Since the universe as totality cannot be specified in terms of analogies of experience including, first of all, its temporality, the universe acquires the character of an event, that is, a phenomenon that is unforeseeable on the basis of the past, incomprehensible on the basis of the present, and irreproducible on the basis of the future. Cosmology attempts to reconstruct the invisible past of the universe on the basis of its present. In this sense the past of the universe is not "the past in the past" but rather "the past of the present,"[59] so that the unforeseeable future is simply a result of the fact that we do not know the past. However, neither is the universe exhaustively comprehensible on the basis of the present: the contingency of our location in space and time, as well as the contingency of the very factual self-articulation of the universe through the human voice, point toward the incommensurability (not consubstantiality) of the universe as a whole with its particular articulated incarnation, leading to the untestability of many conjectures about the universe and the fundamental uncertainty of cosmology.[60] The universe is an event because it is not reproducible on the basis of the future: there is simply no future with respect to the universe as a whole, which by definition incorporates all past, present, and future;[61] the universe cannot be rerun. Finally one asserts that the universe, by definition, is a unique occurrence; it is that which has a modality of created absoluteness. All cosmological theories that attempt to explore the possibility of non-uniqueness of this universe make a reduction of the phenomenality of this universe to that distorted and damaged phenomenality which is typical of objects constituted by the empirical sciences, the phenomenality which is poor in intuition (foreseeable as reconstructed from the allegedly existing past, exhaustively knowable

59. J. A. Wheeler advocated a view that the temporality of the past is constructed, that is, the "past" is theory. The past has no existence except as it is recorded in the present. By detecting what questions our quantum-registering equipment shall put in the present, we have an undeniable choice in what we have the right to say about the past. See, for example, J. A. Wheeler, "World as a System Self-Synthesized by Quantum Networking," *IBM Journal of Research and Development* 32 (1988): 13. This thought must be placed into an even more general conviction that in the ultimate scheme of things there is no time or temporality at all. Temporality is a human construction: "The word Time came, not from heaven, but from the mouth of man, an early thinker, his name long lost. If problems attend the term, they are of our own making." J. A. Wheeler, "Time Today," in *Physical Origins of Time Asymmetry*, ed. J. J. Halliwell, J. Pérez-Mercader, and W. H. Zurek (Cambridge: Cambridge University Press, 1994), 6.

60. G. F. R. Ellis, "Issues in the Philosophy of Cosmology," in *Handbook of the Philosophy of Science, Philosophy of Physics*, Part B, ed. J. Butterfield and J. Earman (Amsterdam: Elsevier, 2007), 1274.

61. The concept that there is no past and future, but an ever-ongoing present, was developed in A. Comte-Sponville, *L'être-temps. Quelques réflexions sur le temps de la conscience* (Paris: Presses Universitaires de France, 1999).

according to existing theories, reproducible theoretically through the plurality of different scenarios). It is clear that what is left behind in these reconstructions is a historical dimension of the universe as a unique and contingent event that cannot be exhausted by its reduction to human subjectivity, in particular to that mode which deals with the universe's thematization as an object. Here historicity of the universe is understood not as a sequence of its stages of evolution as they are described in cosmology. One implies the historicity of the universe as a whole as its contingent givenness to humanity in its entire span of time and space, that is, as a humanly historical relation and communion. This historicity has a different origin, following not from physical causality but rather originating in intentional consciousness as the intrinsic and mysterious unity of subject and object, being and nonbeing, spirit and matter. It is interesting to note that Christian cosmology is built upon a premise that it is the fate of humanity that determines the fate of the universe, and the whole history of the universe becomes seen as part of the history of salvation.

The Universe as a Saturated Phenomenon: From Analogies of Experience to Plurality of Horizons

Mathematical cosmology works through the analogies of experience understood in a Kantian sense as constitutive principles. However, it is quite otherwise with those principles that attempt to bring the existence of appearances of the universe under a priori rules (that is, formulate the causal principle of the world). For since the contingent facticity of existence of the universe cannot be constructed (we disregard attempts of "constructing" this facticity by appealing to the multiverse models, for the facticity of multiverse itself cannot be scientifically constituted but only postulated), the principles of analogy of experience can only be applied to the relations in the universe as already existent. In this sense they can have only a regulative (not constitutive) character, and in the same sense these principles cannot be mathematical, but only philosophical. Kant comments on this by contraposing the constitutive nature of mathematical predication through a proportion. For example: if there is a proportion as an equality of the ratio of two known quantities to another ratio in which one is unknown, then this unknown is effectively constructed (constituted). In application to the standard model in cosmology it means, for example, that since the ratio of the scale factors at present and at the time of decoupling of matter from radiation (which is equal to a thousand) is proportional to the inverse ratio of temperatures at the same times, by knowing the temperature at present one can infer the temperature at decoupling, that is, it can be constructed. This type of constitution is applied to the already existent and cannot be transferred if the analogy of experience is extended to the issue of the facticity of the universe, for example, its origin. Here the proportion does not work and hence the mathematical constitution of the origin of the universe is simply impossible (the endless series of aberrations approaching this origin points to the fact that the computational synthesis of the origin of the universe is not an accomplished constitution, but an ongoing mode of approximation mocking time itself). In this sense to make an analogy from astronomical experience to the experience of the universe as a whole does not mean establishing a *quantitative* relation between what is given in the limited empirical realm and that which is intended as totality. It is rather a *qualitative* relation: more specifically, one can a priori establish knowledge only of the *relation* to the universe as a whole, but not of what it is. This relation yields a rule for seeking the universe as that member of experience which is in relation to any other experience of ordinary things. Thus that which can be named by an analogy of experience represents only a rule according to which the unity of experience (experience of incarnate hypostatic existence in the universe) may arise from the perception of separate things and astronomical objects. Being a rule, this analogy of experience does not tell us how the empirical or intellectual intuition of the universe comes about: in this it is not a principle that is constitutive of the universe and its appearances, but is a regulative norm for the unity of experience through its particular mode of the intuition of the universe as totality. That which is observed here can be affirmed with respect to the postulates of any empirical knowledge: the facticity of appearance of empirical facts as events cannot be subjected to the criteria of an a priori. Then the analogies of experience that scientific cosmology attempts to stretch across the frontiers of the empirical are distinguished here not from the point of view of their certainty (they are certain in their given facticity), but in the nature of their evidence, that is, as regards the character of the intuitive factors involved. For example, talking about the early stages of the universe which by definition are beyond any empirical verification, cosmology implicitly appeals to the coherence of epistemic justification of its claims related to the communal agreement,[62] which represents a different nature of the evidence related to the analogies of experience. The analogies of experience express and represent here rather the regulative delimiters of the understanding. The important feature is that whatever mode of understanding related to analogies of experience cosmology uses, it places its subject matter in rubrics of time which, according to Kant, guarantees the unity of experience.[63] In this sense the universe of scientific cosmology always stands in relation to any experience in terms of time. It is this feature that guarantees any positive

62. See chapter 4.
63. Kant, *Critique of Pure Reason*, A177/B219–20.

predication of the universe and explication of its structure. However, being a limitation of experience, the analogies of experience that base themselves in time provide us only with a fringe of the universe's appearances, namely only those aspects of the universe that allow their manifestation through relation to time. Time enters experience as that implicit horizon of phenomena without which no appearance is possible: time reveals itself as that which receives this appearance and at the same "time" rejects it to make it appear. The universe that cosmology is speaking about represents thus the breakthrough of its phenomenality, that is, its visibility in the background of the time-horizon which is over there in advance. The universe in its appearing (and thus further theoretical thematization) is thus limited to that portion of the time-horizon (which otherwise remains implicit and invisible) which allows the universe's actualization as temporalization. The question now is whether "the universe as a whole" does fit into such a horizon, or, in other words, whether the universe as a whole can adequately and exhaustively manifest itself within the rubrics of this horizon? A simple semantic reflection directs one to answer in the negative to this question on purely definitional grounds: the universe as a whole, as the totality of "all in all," cannot be subjected to any limitations and conditions of time, for it exceeds all time. Then one must rephrase the same question as to whether the universe as a whole exceeds every horizon. Here we come to the phenomenological understanding of the world as the horizon of all horizons, exceeding all particular horizons. The issue, however, is not to dispense with the horizon in general in cosmology, for in this case the very manifestation of the universe would be prohibited; the question is how to neutralize the delimiting anteriority of the horizon (time) in order to avoid the conflict in claiming the absoluteness of the phenomenon of the universe (as not related to time). The problem is how to avoid the conflict between an approach to the universe as pure event (which is not foreseeable, not knowable or reproducible), that is, as a saturated phenomenon, and the delimiting anteriority of the horizon that makes its particular manifestation possible. On the one hand, the excess of intuition saturates the phenomenon of the universe as communion (as existence and life); in this sense it exceeds the delimitations of ordinary experience, being de facto its own foundation. On the other hand the horizon of time makes possible discursive definitions of experience of the universe as communion, but in doing so this horizon itself must be defined. However, this very definition limits the definition of a horizon, so that the limits of the discursive definition of experience of the universe go together with the limitations in the very structuring of the horizon. The universe represented through the prism of this dialectic of the defining and definable horizon is inevitably manifested in a fringed phenomenality. Finally, one anticipates that if the universe is approached as a pure event, that is, as a saturated phenomenon, it saturates its own horizon, for there is no way out from the totality of the universe and no reference to the other ("sans porte, ni fenêtre, sans autre, ni autrui"[64]). In fact, to ascribe to the universe a single horizon would be a naïve and abstract exercise, because the universe as it appears to humanity in the course of history exhibits itself through a plurality of horizons, none of which can precede or delimit the other. One can even claim that the experience of the universe through different horizons exhibits the persistence of saturation, its multiplication and cross-horizon saturation, which does not relieve the basic factor that the universe saturates the intuition, exceeds the capacity of the understanding and the delimitation through the analogies of experience, rendering its incessant presence at the bottom of any existential act. The plurality of accounts of experience of the universe even in scientific

64. Marion, "Le Phénomène Saturé," 117.

cosmology, such as the universe as an astronomical whole, or the universe as a theoretical construct of the all-unity, the infinite change of theories and endless self-correction of theories through observation, shows that one is dealing here with a saturated phenomenon. If one adds to this the variety of genres expressing communion with the universe starting from philosophical prose and finishing with a mystical poetry (that involves a personal element in communion as an opposite to the impersonal depiction of the universe through scientific naturalism), one then escapes the justification of the fact that as pronouncing the word "universe," one invokes that which saturates intuition and involves one's *ego* in an endless hermeneutics that acts as the constitutive factor for this *ego*. Here cosmology exhibits a very close kinship to theology, whose account of creation of the universe and the divine presence in it involves not only the plurality of the exegesis of the Gospels, including its patristic accounts, but the variety of mystical experience of the universe whose testimony is kept in endless writings of saints, spiritual seers, and liturgical texts, all of which are referring to one and the same experience, but whose different context is irreducible and nonamendable in the course of time. Nowadays, the unrestrained proliferation of popular scientific accounts borrowing ideas from mainstream research in theoretical cosmology exhibits the same, although pseudo-priestly, fervor toward preaching about the universe, in which the horizon of encompassing the subject matter not only does not stop but endlessly corrects and complements itself through an appeal to different analogies and topics from other disciplines. In fact, the very phenomenon of the popular scientific literature as well as science-fiction can be made analogous to biblical exegesis: indeed, whereas the Bible emerges as a main source of interpretation in different intersecting contexts, the scientific papers in narrow and highly specialized journals play a similar role by providing for popular science an indefinite field for various exegeses adjusted for different audiences and serving different social and ideological needs. In the case of cosmology, its content and its texts themselves appear to be a saturated phenomenon simply because they provide a multiple contextual interpretation of that which they speak about and render as their content. Rephrasing it again, the open-endedness of cosmological discourse, as well as its intrinsic incompleteness delimited by the discursive faculties, creates the natural conditions for approaching its pronouncements and descriptions as a narrative, which as such, is subject to a further treatment by "cosmologists," of the second order so to speak—a treatment that deals not with the universe *per se* but through reflection and interpretation in mundane words of that which has been established by scientific cosmology. Thus the universe receives its duplication in literature and becomes a subject of an independent study by "cosmological linguists" and writers who are restaging and rewriting the story, always exceeding the limits of any given context. Finally, one must not forget the impact that the study of the universe in cosmological literature and its mass-media treatment have on human perception of its place in the universe, its happiness and anxiety, and on practical stances on ecology and the value of life.

The signifying conventions of context-language affect those subjects who study cosmology, but subjects are not the agents who transform the language conventions pertaining to this or that context. The possibility of transformation inheres in the very temporal and spatial movement of concepts and the corresponding language, historical sensitivity, and social delimiters that are carried forward but not controlled by the particular context-dependent citations used by subjects. Even though one may intend to transform the meaning conventions of words in a given context, our intentions by themselves cannot establish the desired transformation because one does not control the future course of the signifying chain of questions and answers arising from the human communion with the universe.[65] The transformation of contexts is not subjected to any law in time, as it is not a process implying causality; it reflects events related to the human apprehension of existence here and now and, therefore, is at the mercy of an always-unforeseeable "future."

It is the plurality of horizons or contexts, such that no horizon could delimit or precede the created universe, that justifies the absoluteness of the universe as a saturated phenomenon. The alleged "identity" of the universe "survives" the multiplication of dimensions of its perception. Then the question of identity of the universe as a saturated phenomenon arises in an acute form: If the universe cannot be subjected and held by any horizon, because it gives itself as absolutely free from analogy with any empirical law and free from any network of relations with that which has already been seen (or foreseen), can the universe as creation have an identity at all when the very fullness of multiple contexts pertaining to being manifested to humanity cannot contain or encompass the universe? Can the universe enter into relationship with itself in order to render to it its own identity? The simplest form of identity available to the human grasp is the generic relation between the universe and itself, which can be symbolically presented through a formula: "The universe is the universe." It seems, however, that for the universe as a saturated phenomenon even this is impossible, for the very relationship with itself entails such a differentiation in a single "unity" of the universe that it becomes a definition of this universe from the outside of the initial nonsplit unity.[66] In this sense to speak about the identity of the universe is to enter the tension with its understanding as a

65. This is a natural consequence of the inherent apophaticism in comprehension of the universe: since signifiers do not exhaust that which is signified, the signifying chain of cited words used to express the experience of the universe in a given context cannot be controlled by that which is meant to be signified.

saturated phenomenon. As a saturated phenomenon is neither visible according to quantity, nor bearable according to quality, absolute according to relation and thus unconditioned by the horizon, this phenomenon escapes the conditions of its own ability for being displayed and configured. Thus no positive sign of the universe's identity is possible. However, the withdrawal of the universe's identity and, at the same time, its escaping presence through the act of communion, retains its apophatic manifestation. In spite of humanity's subjection of the universe to the identifying relation with itself, it only remains a symbol, an apophatic manifestation of the impossibility of this relation expressed in rubrics of the identity of human consciousness. The intellectual posing of self-identity of the universe, which effectively splits this identity in itself, creates contexts and horizons pertaining to empirical experience and thus reducing the saturated phenomenon to a glimpse of that which is neither visible according to quantity, nor bearable according to quality, but absolute according to relation.

Creation and Consciousness

Now we come to the climax in understanding creation, or the universe as a whole, as a saturated phenomenon that positions it outside of any general conditions of experience demanded by transcendental philosophy. The question is how the saturated phenomenon of the universe is related to thought in general. According to Kant, the object of experience and the power of knowing must agree in order to be suitable for the source of experience, that is, the transcendental "I." The possibility of a phenomenon depends on its fitness to the conditions of experience

66. This is similar to the Fichtean foundational thought that from the initial unity "A=A" (implied in the identity of one's "I": "I=I") it follows its split in itself posing the "not A."

as originating in the "I": the "I" can know things that this same "I" can access and comprehend in the conditions of this "I's" facticity. Correspondingly, if a phenomenon does not "agree" or "correspond" to the "I's" power of knowing, this phenomenon cannot appear, that is, there will not be any phenomenon at all but just a perceptive aberration.[67] In the case of the universe as a saturated phenomenon, the "I" by definition experiences the disagreement between a "potential phenomenon of the universe" expected to appear in the manner of ordinary objects and its subjective experience through sheer belonging to the universe (communion). Consequently the "I" cannot constitute the universe as an object whose concept would agree with the conditions of experience of the universe as communion. Correspondingly, the failure to objectify the universe does not mean that there is an appearance of nothing and the implied communion is an empty phrase. One has here the intuitive saturation by the universe, which imposes itself by an excess that makes this universe effectively invisible and incomprehensible to the extent that it cannot be tolerated and thus seems to be incommensurable. It is the universe as a saturated phenomenon that resists any regard with respect to itself as an object: it engulfs the subjectivity of the "I" to such an excess that this "I" fails, and any attempt at understanding the universe's constitution is suspended. The universe is *visible* (in its particular pieces and moments) but it nevertheless cannot be *looked at*. It is this feature of visibility as presence and at the same time the impossibility of gazing at it that characterizes the saturated phenomenon.

On the side of a human being, a meeting with a saturated phenomenon of the universe can be characterized as a condition of

67. A typical example of such an aberration is the so-called dark matter and dark energy, which are predicted theoretically, but whose material existence has not yet been confirmed experimentally.

not being adapted to and not being at home in the world. In analogy with J.-F. Lyotard,[68] the meeting with the saturated phenomenon can be described as a return back to the condition of infancy, for as infants, humans are helplessly exposed to a strange and overwhelming environment while lacking the ability to articulate what affects them. By reducing this analogy to the bodily functions, the universe as saturated phenomenon deprives the body of its attunement to the universe. In a trivial sense there is the body's contingent, and literally free-flying, position in the space-time of the universe so that it is displaced and hence not attuned to the universe. In a more sophisticated sense, being in the universe but constituting itself through communion with the universe as a whole, a body is nowhere (according to Kant, as *noumen*, human being is outside of space and time) and thus is not attuned to anything.[69] This condition of not being attuned to the universe signifies a gap between sensibility and the possibility of mental articulation or linguistic expressibility in situations where human beings meet saturated phenomena. If sensibility is a condition of unmasterable and vulnerable openness to excesses of affection, then the passage from sensation to articulation is not guaranteed: what we have here is the suspension of comprehension and its linguistic expressibility, a break between them that indicates that the saturated phenomenon *de facto* breaks that which could be called the conditions of experience corresponding to ordinary phenomena (poor in intuitive content).[70]

Then the question is: What does the "I" "see" in the conditions of the broken link between the overabundance of intuition and the

68. J.-F. Lyotard, *The Inhuman: Reflections on Time* (Cambridge and Oxford: Polity, 1991), 4.
69. Indeed, the body, which is consubstantial only to 4 percent of matter of the universe, can be said to be nowhere.
70. One can draw an analogy with the Kantian sublime: the experience of the sublime indicates not so much the classical dualism of finite sensibility and infinite reason as the irreducible heterogeneity of human faculties.

possibility of discursive expressibility? What is that which appears to the "I" under the rubrics of the saturated phenomenon of the created universe? The answer comes (not unexpectedly from a phenomenological point of view, although extended beyond it) from the recognition that what the "I" definitely realizes is its own incapacity to constitute this phenomenon once and for all, certain and accommodated to its preexistent and finite (as accumulated in time) rubrics of experience. The universe is received through the "I's" sight as pure donation, which cannot be caught as complete in rubrics of thought at any given moment of time. One can say that to wrestle with the saturated phenomenon of the universe is to be in a constant despair of chasing its escaping presence, which constantly reminds the "I" of the obfuscated nature of its handling the universe, a feature implanted in the inevitability of the "I's" created finitude.[71] The presence of the universe in human life and the very definitiveness of the "I" is itself thus subjected to the donation of the universe, and it is the "I" that is not a priori and independent of the universe in its free-willing ambitions of tackling it, but positioned by the nature of the universe's speech into the passive voice, thus becoming a dative of manifestation and the nominative of disclosure only to the extent that the universe itself, being created and enhypostasized by the Divine Logos, makes it possible. The "I" as being unable to constitute the phenomenon of the created universe as a whole experiences itself as being constituted by this phenomenon through inescapable creaturely participation. The hypothetical identity of the universe, sought and intended on the grounds of the free-willing ambitions, remains no more than an unfulfilled intentionality that characteristically returns back to the "I," which is being constituted by the universe as if the "I" is being gazed at by it. However, one

71. The constitution takes place on the grounds of free will through which humanity attempts to overcome its finite circumscription by the conditions of nature.

must not be mesmerized by this conclusion: the "I" is constituted by the universe only in the sense that both the "I" and the universe are enhypostasized by their hypostatic otherness in an intrinsically coinherent way, when one cannot exist without the other. The "I" experiences itself as a subject of the endless dialogue in which it is involved through the pure donation of creation to all humans as creaturely existents in the creaturely universe. In this sense the response of humanity to the invitation to participate in this dialogue has certain modalities of both fear (as non-attunement) and gratitude for the gift of being-in-the-universe. By being in the universe the "I" does not have (it simply cannot have) any dominant point of view over the intuition of the universe as an expression of the very fact of life. The universe as a saturated phenomenon engulfs subjectivity by removing its parts and spatial extension. In a temporal sense, the universe is always already there, so that all events of subjectivity's life unfold from the never-ending donating event of the universe as a constant coming into being, in which the unforeseeable nature of every consequent moment entails the unending historicity and unpredictability of existence.[72] In a spatial sense, the contingency as concrete factuality of an event of appearance of the "I's" life, which is not foreseeable and phenomenologically hidden from the "I's" comprehension, makes its position in the universe *out of tune* (in spite of the fine-tuning related to consubstantiality with the universe implied and explicated by the anthropic inference). This occurs through the fact that the universe engulfs the "I" with the intuitive flood, which ultimately breaks the link between the intuitive and discursive mind and thus deprives the "I" of clear comprehension of the created universe, leaving the human "I" with no place in the

72. This existence was described by P. Brockelman as "the continuous eruption into being of those myriad forms, the active *that-ing* and *is-ing* of everything which emerges into consciousness in the experience of wonder." P. Brockelman, *Cosmology and Creation: The Spiritual Significance of Contemporary Cosmology* (Oxford: Oxford University Press, 1999), 79.

universe. Its "place" is its sheer facticity, and any constitution of the universe's space by the heroic modalities of human free will just reduce the universe to its limited phenomenality, the phenomenality which is not; for the universe is the saturated phenomenon invisible according to quantity, unbearable according quality, unconditioned according to relation, and irreducible to the "I" according to modality.[73] From a phenomenological point of view the universe as a saturated phenomenon represents the conceptual completion of the definition of the phenomenon as that which appears of itself, and starting from itself,[74] since it appears without any limitations by horizons and without being reducible to the rubrics of human subjectivity.

The situation with "knowledge" of the universe as a saturated phenomenon becomes characteristically similar to that of knowledge of God in theology if one remembers that what is called "knowledge" in theology is not related to the discursive faculties of cognition but rather to mystical awareness based on personal participation and communion with the Divine. Correspondingly, in theology one cannot be detached from what is intended as the subject matter of its enquiry; one needs faith and participation in that which is studied. In this case the "knowledge" of God cannot be "objective" (in the sense of scientific rationality) because it depends on a mode of personal experience and involvement. This suggests that theology implies a special understanding of "objectivity," different from the natural attitude where the reason attempts to separate itself from attachments (contexts) in order to be detached from the finite object (which is constituted as freed from attachments). Thus, in theology, no prior assessment of the attachments to its "object" is possible, for the definitiveness of the perceptive intellect is revealed to itself

73. Marion, "The Saturated Phenomenon," op. cit., 211.
74. M. Heidegger, *Being and Time*, § 7 (Oxford: Blackwell, 1998), 54.

only through its relationship with the divine as God-given capacity. This is the moment when the intuition saturates over the intellect in its attempt to grasp its own facticity, and it is this saturation as such that indicates the inseparable union of consciousness and its source in the Divine. As a simple result, any imagined deprivation of this donating intuition of its source (that is, the objectivization of the Divine) would mean the immediate cessation of subjectivity in general. Thus we face an interesting reversal (with respect to scientific knowledge): the "objective" knowledge of God presupposes saturation over all discursive images of the Divine (apophaticism), which paradoxically means the impossibility of detachment from communion with God. If one now turns back to the universe as a saturated phenomenon, one realizes that the commonly accepted objectivity in scientific cosmology based on the presupposition of detachment from all personal, subjective attachments to the universe becomes impossible. By paraphrasing the words of T. Torrance, formulated in a theological context, it is sheer attachment to the universe that detaches us from our preconceptions about it; while detaching ourselves from our preconceptions we become free for the universe, and therefore free for true "knowledge" of it.[75] Then philosophical cosmology is constituted in conversation and communion with the universe, which *communicates itself* to us in acts of donation—and while gazing at us it *requires of us an answering relation* in receiving, acknowledging, understanding, and active personal participation in the relationship it establishes between us.[76] It is this answering relation in receiving, acknowledging, understanding, and active personal participation in the relationship with the universe that constitutes the "I."[77] Certainly within a

75. Cf. T. Torrance, *Theological Science* (Edinburgh: T. & T. Clark, 1996), 36.
76. Cf. ibid., 39.
77. It is worth making a parallel with mystery: the universe is a mystery, for human beings contemplating it are involved in it without any chance of distancing themselves from it.

theological sensibility one must not assign any para-hypostatic features to the universe as actively invoking in us a conscious response. The response is provoked by the fact of our inseparability with the created universe, but as an action and event it is rooted in the ability of the human to articulate its own existence and the existence of the universe, an ability that does not obviously follow the logic of consubstantiality, but originates in the Divine image as a particular mode of enhypostasization of the human formation of the universe by the Logos.

When cosmological imagination positions itself outside the universe either in an attempt to "see" the universe in its entire spatio-temporal span, or, alternatively, when the idea of multiverse is invoked, as if the mind could gaze upon the universe from the otherworldly-multiverse position, consciousness attempts an impossible act. In both these cases, since consciousness cannot quit its communion with this universe, all appeals to other worlds have a rhetorical and simply eidetic character, because their imagery derives from the rubrics of consciousness embodied in this universe.[78] Thus the very logic of transcending this universe has traces of presence of incarnate human subjectivity, so that the transcendence remains in the conditions of immanence. This last thought can be accentuated phenomenologically.

In the natural attitude the universe as a whole, being a synonym of creation, is posited as existing objectively out there, transcendent

"Mysteries are not truths that lie beyond us; they are truths that comprehend us." S. Keen, *Apology for Wonder* (New York: Harper & Row, 1969), 25.

78. What happens in all such abstractions from reality is that consciousness imposes itself as a qualitatively different state of being, including not only the physical world as such, but the conditions of consciousness's hypostatic incarnation. The difficulty here is that consciousness not only disembodies itself, but deprives itself of personal characteristics. In this case the noetic pole of all predications about the universe degenerates into an impersonal and anonymous awareness that wanders at liberty upon that physical reality, which, while being by construction the intentional correlate of this consciousness, cannot limit or condition the very foundation of this intentionality.

to the field of consciousness. Thus the universe as a whole is subject to a phenomenological critique. However, this critique does not simply imply that the question about its reality must be suspended and cosmology is brought to a methodological halt. It is important to realize that transcendence of the universe takes place not through an ascending series of the worldly astronomical phenomena or theoretical causation (which had been critically exposed by Kant), but through observing *teleologies* of explanation that rather characterize the activity of consciousness. Hence the phenomenological reduction exercised with respect to the universe cannot reach its goal, for it disregards the universe as an ongoing accomplishment related to the teleology of the human spirit (implanted in the promise of salvation and eternal life). Here phenomenology does not discern the difference between the universe as a mental construction subject to such an operation of consciousness as reduction, and the universe as communion whose presence in consciousness is exactly that *ontological* link which makes this consciousness possible at all and which can be cut off only in abstraction.[79] One cannot bracket or reduce the *universe as communion* by using this consciousness because by insisting on this, this consciousness deprives itself of the conditions of its embodied existence and hence destroys itself as the intentional consciousness of the universe.

The transcendental reduction of the universe as a whole performed by classical phenomenology in order to neutralize the natural attitude points to a simple fact that the representation of the universe as completely transcendent to consciousness cannot acquire an ontological quality, remaining "transcendent" but only within the immanence of consciousness. Phenomenology rightly suggests the dismissing of all intellectual idols of the universe as pretending to

79. It is the undeniable communion with the universe which, as was asserted by Kant, makes its articulation by reason (which imagines the universe as object) intrinsically antinomial.

exhaust the reality of the universe as communion: any discursive image of the universe remains only an image and is thus incomplete. This returns us back to the principle of *apophaticism:* in the apophatic approach the discursive reason struggles with the idea of creation, or the universe as a whole, because this reason cannot position itself with respect to the universe in terms of "close" or "far." The universe is present in the background of existence through relationship and communion in such a way that allows one to express this presence ecstatically through music, painting, poetry, and so forth.[80] However, this experience cannot be verbalized and expressed in definitions of physics and mathematics. It is exactly this paradoxical "presence in absence" of the universe as a whole that allows the human spirit to make the distinction between what is absent (and hence always suspected in the inadequacy of its expression in concepts and then legitimately bracketed away), and what is present (that is, what is left after the bracketing of conceptual idols) and hence allegorically expressed in mundane language without any risk of being mixed with the ineffable essence of the universe. In fact, one can say that the very bracketing of the conceptual idols of the universe is possible only because the resulting conceptual absence of the universe, which always bedazzles human imagination, is compensated by the reality of its concrete presence, manifested in the very possibility of thinking about the universe.[81] The implicit *presence* of the created universe

80. Cf. with theology, where the expression of the experience of the Divine invokes all forms of visual and performing arts. The art represents an acute form of expression of communion with the universe; see, for example, E. Grosz, *Chaos, Territory, Art* (New York: Columbia University Press, 2008), 23-24. There are numerous examples in which the presence of cosmic immensities caused different poetic expressions either of the sense of awe and unity with the cosmos or the horror of being lost in a hostile universe. For example, one can refer to the lyrics of Ruben Dario for whom to understand humanity, meant to understand the universe and thus to affirm the unity of "all in all" in the universe in a pantheistic fashion. See, for example, C. L. Jrade, "Ruben Dario and the Oneness of the Universe," *Hispania* 63 (1980): 691-98.

81. The presence of the universe is inerasable from the very fact of our consciousness. If one submits what is named "the universe as a whole" to the reduction only in so far as one defines it by transcendence (and insofar as one compares this particular transcendence with that, in fact quite

in all acts of the incarnate human subjectivity cannot be phenomenologically reduced (that is, bracketed as transcendent and "nonreal") because if this could happen, the incarnate consciousness would be bracketed away and hence eliminated. Obviously this would entail the destruction of the factual consciousness itself, and thus lead to a sheer existential contradiction. J. Sharon compared this inseparability of a human being and the universe with its attachment to the universe as its mother, the attachment through "love, and not only its reason" ("*par l'Amour et non par sa seule raison*").[82] Humanity, in spite of its non-attunement and homelessness in the astronomical universe, is predisposed to love the universe through the inherent Divine image in man, for through love of the universe as good creation of a good God the human loves God as its Father. It is in this love as a theological commitment that ultimately lies the foundation of cosmology as the study of the created universe.

different, of the object in the natural attitude) one exercises here an act of a *radical immanence to consciousness*, and in this sense the universe as communion would be confirmed by a reduction.

82. J. E. Charon, *L'homme et l'universe* (Paris: Albin Michel, 1974), 14.

Conclusion

The objective of this book was to reformulate the sense of mediation between theology and cosmology, by explicating a simple truth that both, theology and cosmology, are products of one and the same human subjectivity, so that no intrinsic existential contradiction between two types of experience (that is, of God and the universe) is possible. Correspondingly, it was argued that theology and cosmology are intertwined in the sense that each of them functions in the conditions of the implicit influence of the other. Indeed, for theology to be possible at all one needs theologians, that is, embodied human persons. Hence one needs necessary physical conditions for their existence. These conditions are explicated in cosmology. It follows then that every theological intuition and assertion related to experience of God implicitly contains truth about the world as the condition for this experience. Similarly, for cosmology to be possible, physically finite human beings need to be able to encapsulate the whole world in their personal consciousness. Saying emphatically, the finite humanity must possess the faculty of dealing with infinity. This feature of the human condition cannot be explained through reductions to the physical, and forms the foundational mystery whose elucidation and interpretation can only be provided by theology of the Divine image. Correspondingly, any cosmological assertion is intrinsically theological because it employs the Divine-given propensity of discerning the unity of the world in the perspective of its being created by God.

The tension between theology and cosmology can arise only because of a particular reduction of experience of the world in physical cosmology to its object-like phenomenality when the intrinsic presence of the human insight is lost. Philosophically, this tension is not existential, but rather formal and abstract, because the reduced object-like phenomenality of the universe in physical cosmology is still possible only under the condition of the pre-predicative instantaneous synthesis of the universe as the quality of human consciousness in the Divine image. The possibility of this synthesis can be interpreted theologically by referring to the archetype of Christ, who, while having been present in the earthly location in the universe, sustained the whole universe hypostatically as the Logos. The discursive explication of the instantaneous synthesis of the universe implies the transition from an archetypal sense of *theogenic* uniformity of the universe (following from the equal "distance" of its constituents from the Logos-creator) to the discursively asserted *cosmographic* uniformity of the universe named by the cosmological principle. Thus the cosmological principle *per se* turns out to be a physicalized expression of the Divine-implanted archetypal memory of "all in all" of the Logos himself; that is, it manifests a hidden theological commitment in knowledge of the universe, originating, once again, in the Divine image.

Cosmology needs the cosmological principle in order to speak about the universe as a whole, thus assigning to humanity an indifferent position in it, implicitly contradictory to the fact that this cosmological principle is the result of the theologically understood centrality of humanity in creation capable of asserting the universe's *theogenic* uniformity. Cosmology needs the cosmological principle in order to explicate the universe as a whole in objective terms: thus this principle points toward an inherent purposiveness of all human activity not only to intuit the world, but to understand it. In this sense the cosmological principle rearticulates its intrinsic human origin, this time as a *teleological* principle of explanation of the universe. The anthropology of the Divine image reveals itself, once again, through its teleological mode, as an implicit foundation of the cosmological discourse. However, teleology is implied here not in a traditional sense related to the purposiveness of the universe's physical evolution; rather it is implied as a "formal" purposiveness as explicability of the universe being linked to the human intentional search for the sense of its own existence.

The cosmological principle in its contingent facticity requires from physical reason further explication through the generative steps that have led from the primordial past of the universe to its present. Thus the theoretical representations of the "universe as a whole" and "the Big Bang" act as the *telos* of cosmological explanation and hence as the *telos* of anthropological explanation related to the origin of individual persons at birth. In both cases the sense of comprehension of the given presence unfolds as a process of constitution of its antecedents: to understand the present and disclose its sense, one must enquire into the events preceding it. However, the very process of this constitution as an activity of embodied humanity, subjected to the temporal irreversibility of a macroscopic nature, is directed to the future, thus characterizing the sought *origin* of the contingent state of the universe and the *origin* of any particular human life as their *telos*.

Being imbued with anthropology, cosmology naturally falls under the category of the apophatic method (that is, an inability to exhaust the sense of the universe through its scientific signifiers), thus following subconsciously theology. In its appropriation of the notion of the universe as a whole, as well as that of the origin of the universe, it has to refuse the exhaustibility of truth behind these notions. Thus any commitment to strong physical realism has to be abandoned and the objectivity of the universe should be treated as no more than the objectivity of a construct, that is, of constituted reality. Phenomenologically, the discourse of the universe becomes a constitutive part of the historicity pertaining to human rationality. The result of this is that cosmology exhibits itself not only as a natural science, but also as an enterprise that has features of the human sciences. On the one hand the universe can be represented in the phenomenality of objects (in cosmology as a natural science) as an entity unfolding in space and time. On the other hand it can be experienced through participation *in* or communion *with* it (in cosmology as a human science), such that the separation between it and perceptive consciousness becomes impossible.

The universe as a whole, being a result of the human constitution, that is, a construct whose reference to empirical reality cannot be established by means of the principle of correspondence, has to appeal implicitly to the methodology of coherence of epistemic justification, which relates belief-constructs and corresponding realistic commitments to particular communities of cosmologists. Here cosmology resembles theology through the appeal to the collective (conciliar) establishment of the sense of truth. However, cosmology as such, being devoid of those ways of communication with the source of truth about the universe (which, in theology, sustain the essence of what is called its ecclesiality and communion of the Holy Spirit), can only provide a glimpse of this truth in a precarious and uncertain way.

The phenomenological reversal of the object-like representation of the universe neutralizes the transcendental treatment of such a representation as a *rational* idea. The universe rather enters human subjectivity as a term of personal relationship, as a quality of harmony and beauty, that is, as an *aesthetical* (or even ethical) idea, similar to that of *Kosmos* in ancient Greek philosophy. The existential excess of intuition of the universe over its presentation in categories of the understanding places the universe as a whole in the range of the philosophically qualified *saturated phenomena* in the following sense: it is the created universe, to the extent it cannot be comprehended by the intellect that constitutes human subjectivity as part of this creation, which discloses and articulates creation (humanity acquires truly a status of a microcosm in a very nontrivial theological sense). By resisting the saturating presence of the universe, which physically chains humanity to it and diminishes its God-given destiny of being its hypostasis and its priest, humanity recreates the lost Divine image through the path of a positive appropriation of the "garments of skin." This is explicated in cosmology through the search for the sense of life and the universe by appropriation of the archetypal memory of "all in all" of the Logos himself, by whom and through whom all was made.

Finally, cosmology can be treated as a theological work, as a spiritual and para-eucharistic activity, bringing the universe back to its creator through exploration and articulation. By so doing, human cosmologists endeavor a task of the moral mediation between the universe and God, contributing to the stages of deification through which the universe will be transfigured and seen through the eyes of the Logos-creator himself. The latter implies transcendence, that is, of seeing the universe from its "outside." However, this outside does not have any attributes of spatiality and temporality and is paradoxically linked to the "inside" of human personhood, that is, of the human heart, rather than reason. In this sense the spiritual task and radical theological commitment in our discussion of the sense of cosmology has aimed to advocate the human Divine-given dignity transcending all scientific attempts to "crush man under the weight of astronomical data."

Bibliography

Abbreviations:

PG: ed. J. P. Migne, *Patrologia Graeca*, 161 vols. Paris, 1857-66.

NPNF: *The Nicene and Post-Nicene Fathers*, ed. P. Schaff and H. Wace. Grand Rapids: Eerdmans, 1996.

The Philokalia: St. Nikodimos of the Holy Mountain and St. Makarios of Corinth. The Philokalia: The Complete Text, 4 vols., ed. G. E. H. Palmer, P. Sherrard, and K. Ware. London: Faber, 1979-95.

Classical and patristic writers with the source for the English translation:

Athanasius of Alexandria

On the Incarnation. ET: Crestwood, NY: St. Vladimir's Seminary Press, 1998.

Contra Gentes. ET: NPNF, series 2, vol. 4.

Augustine of Hippo

Confessions. ET: H. Chadwick. Oxford: Oxford University Press, 1991.

City of God. ET: H. Betteson. Augustine. *Concerning the City of God against Pagans*. New York: Penguin, 1980.

On the Trinity. ET: NPNF, vol. 3.

Contra Faustum. ET: NPNF, vol. 4.

Basil the Great

Hexaemeron. ET: NPNF, vol. VIII.

Dionysius the Areopagite

The Divine Names. ET: C. E. Rolt. *The Divine Names and the Mystical Theology*. London: SPCK, 1979.

The Celestial Hierarchies. ET: *The Mystical Theology and the Celestial Hierarchies of Dionysius the Areopagite*. Godalming, UK: The Shrine of Wisdom, 1965.

Gregory of Nyssa

On the Making of Man. ET: NPNF, vol. 5.

Funeral Oration on Placilla. PG 46.

Gregory Nazianzus (the Theologian)

Oration 45. ET: in P. Nellas, *Deification in Christ: Orthodox Perspectives on the Nature of the Human Person*, 203–5. Crestwood, NY: St. Vladimir's Seminary Press, 1997.

Irenaeus of Lyons

Against the Heresies. ET: J. Saward, *The Scandal of the Incarnation: Irenaeus "Against the Heresies,"* ed. H. U. von Balthasar. San Francisco: Ignatius, 1990.

John of Damascus

De fida orthodoxa (Orthodox Faith). ET: F. H. Chase Jr., Saint John of Damascus, *Writings*, 165-406. Washington, DC: Catholic University of America Press, 1958; Reprinted by The Create Space Independent Publishing Platform, 2012.

Maximus the Confessor

Ambigua. ET: P. M. Blowers and R. L. Wilken, *On the Cosmic Mystery of Jesus Christ, Selected Writings from St. Maximus the Confessor*, 211-21. Crestwood, NY: St. Vladimir's Seminary Press, 2003.

Cap. de Charitate. ET: P. Sherwood, *St. Maximus the Confessor: The Ascetic Life. The Four Centuries on Charity*. London: Longmans, 1955.

Cap. Theologicorum. ET: *Two Hundred Texts on Theology and the Incarnate Dispensation of the Son of God*, in *The Philokalia*, vol. 2, 114-63.

Various Texts on Theology, the Divine Economy, and Virtue and Vice, in *The Philokalia*, vol. 2, 164-284.

Mystagogia. ET: G. C. Berthold. *The Church's Mystagogy*, in *Maximus Confessor. Selected Writings*, 181-225. NY: Paulist, 1985.

Plato

Timaeus. ET: J. Warrington, Plato, *Timaeus*. London: Dent, 1965.

Gorgias. ET: ed. W. D. Woodhead, *Plato Socratic Dialogues*. Edinburgh: Nelson, 1963.

Thomas Aquinas

Summa Theologica. ET: New York: Benziger, 1948, reprint in 5 vols. Notre Dame, IN: Ave Maria, 2000.

General Bibliography:

Abrams, N. E., and J. Primack. *The New Universe and the Human Future: How a Shared Cosmology Could Transform the World*. New Haven and London: Yale University Press, 2011.

Albert, H. *Treatise on Critical Reason*. Princeton: Princeton University Press, 1985.

Albrecht, A. "Cosmic Inflation and the Arrow of Time." In *Science and Ultimate Reality: From Quantum to Cosmos*, ed. J. D. Barrow, P. C. W. Davies, and C. Harper, 363-401. Cambridge: Cambridge University Press, 2003.

Alfvén, H. "Cosmology: Myth or Science?," in *Cosmology, History and Theology*, ed. W. Yourgrau and A. D. Breck, 1-14. New York: Plenum, 1977.

Allen, P. *Ernan McMullin and Critical Realism in the Science-Theology Dialogue*. Farnham, UK: Ashgate, 2006.

Aron, R. *Introduction à la philosophie de l'histoire*. Paris: Gallimard, 1938.

Audi, R. *Epistemology*. London: Routledge, 1998.

Bailly, F. "Remarks about the Program for a Formalized Epistemology." In *Quantum Mechanics, Mathematics, Cognition and Action: Proposals for a Formalized Epistemology*, ed. M. M. Mugur-Schächter and A. van der Merwe, 3-8. Dordrecht: Kluwer, 2002.

______. "About the Emergence of Invariances in Physics: From 'Substantial' Conservation to Formal Invariance." In *Quantum Mechanics, Mathematics, Cognition and Action: Proposals for a Formalized Epistemology*, ed. M. M. Mugur-Schächter and A. van der Merwe, 369-88. Dordrecht: Kluwer, 2002.

Bakhtin, M. *Aesthetics of Verbal Creativity*. Moscow: Iskusstvo, 1979 (in Russian).

Balashov, Y. V. "On the Evolution of Natural Laws," *The British Journal for the Philosophy of Science* 43, no. 3 (1992): 343-70.

______. "Uniformitarianism in Cosmology: Background and Philosophical Implications of the Steady-State Theory," *Studies in History and Philosophy of Science* 25B (1994): 933-58.

______. "Two Theories of the Universe: Essay Review," *Studies in History and Philosophy of Modern Physics* 29, no. 1 (1998): 141-49.

______. "A Cognizable Universe: Transcendental Arguments in Physical Cosmology." In *Constituting Objectivity: Transcendental Perspectives on Modern Physics*, ed. M. Bitbol, P. Kerszberg, and J. Petitot, 269-77. Dordrecht: Springer, 2009.

Balthasar, H. Urs von. *Cosmic Liturgy: The Universe according to Maximus the Confessor*. San Francisco: Ignatius, 2003.

Barker, M. *Creation: A Biblical Vision for the Environment*. London: T. & T. Clark, 2010.

Barrow, J. D., and F. Tipler. *The Cosmological Anthropic Principle*. Oxford: Oxford University Press, 1986.

Barrow, J. D., and S. P. Bhavsar. "Filaments: What the Astronomer's Eye Tells the Astronomer's Brain," *Quarterly Journal of Royal Astronomical Society* 28 (1987): 109-28.

Barrow, J. D. "Patterns of Explanation in Cosmology." In *The Anthropic Principle*, ed. F. Bertola and V. Curi, 1-15. Cambridge: Cambridge University Press, 1989.

Barrow, J., S. Morris, S. Freeland, and C. Harper, eds. *Fitness of the Cosmos for Life: Biochemistry and Fine-Tuning.* Cambridge: Cambridge University Press, 2008.

Barth, K. *Church Dogmatics.* Edinburgh: T. & T. Clark, 1936-77.

Becker, O. "Uber der sogenanten 'Anthropologismus' in die Philosophie der Mathematik," *Phil. Anz. Jg.* III (1928): 369-87.

Bello, A. A. *The Divine in Husserl and Other Explorations, Analecta Husserliana* XCVIII (Dordrecht: Springer, 2009).

Berdyaev, N. *Slavery and Freedom.* London: Centenary, 1944.

_______. "Man and Machine," *Voprosy Filosofii* 8 (1991): 147-62 (in Russian).

_______. *Spirit and Reality.* Moscow: AST, 2003 (in Russian).

Bishop, R. L. "What We Bring We Find," *Journal of Royal Astronomical Society of Canada* 80, no. 5 (1986): 224-29.

Bitbol, M. "From the Anthropic Principle to the Subject Principle," *The Anthropic Principle*, ed. F. Bertola and U. Curi, 91-100. Cambridge: Cambridge University Press, 1993.

Bitbol, M., P. Kerszberg, and J. Petitot, eds. *Constituting Objectivity: Transcendental Perspectives on Modern Physics.* Dordrecht: Springer, 2009.

Bohm, D., and D. Peat. *Science, Order, and Creativity.* New York: Bantam, 1987.

Bondi, H., and T. Gold. "The Steady-State Theory of the Expanding Universe," *Monthly Notices of the Royal Astronomical Society* 108 (1948): 252-70.

Bowker, J. *The Sacred Neuron.* London: I. B. Tauris, 2005.

Brisson, L., and F. W. Meyerstein. *Inventing the Universe: Plato's Timaeus, the Big Bang, and the Problem of Scientific Knowledge.* Albany: State University of New York Press, 1995.

Brockelman, P. *Cosmology and Creation: The Spiritual Significance of Contemporary Cosmology.* Oxford: Oxford University Press, 1999.

Bulgakov, S. *Philosophy of Economy.* Moscow: Nauka, 1993 (in Russian).

______. *Bride of the Lamb*. Grand Rapids: Eerdmans, 2002.

E. Butts, *Kant and the Double Goverement Methodology. Supersensibility and Method in Kant's Philosophy of Science*. (Dordrech: Reidel, 1984).

______. "Teleology and Scientific Method in Kant's Critique of Judgment," *NOÛS* 24, no. 1 (1990): 1-16.

Carr, B. J. "On the Origin, Evolution and Purpose of the Physical Universe." In *Modern Cosmology and Philosophy*, ed. J. Leslie, 152-57. New York: Prometheus, 1998.

______. ed. *Universe or Multiverse*. Cambridge: Cambridge University Press, 2007.

Carr, D. *Paradox of Subjectivity*. Oxford: Oxford University Press, 1999.

Charon, J. E. *L'homme et l'universe*. Paris: Albin Michel, 1974.

Chrétien, J.-L. *The Unforgettable and the Unhoped For*. New York: Fordham University Press, 2002.

Claudel, P. *Positions et propositions*. Paris: Gallimard, 1934.

Clayton, P., and A. Peacocke, eds. *In Whom We Live and Move and Have Our Being: Reflections on Panentheism in a Scientific Age*. Grand Rapids: Eerdmans, 2004.

Clément, O. *Le Christ, Terre des vivants. Essais théologiques*. Spiritualité orientale, n. 17. Bégrolles-en-Mauges: Abbaye de Bellefontaine, 1976.

______. *On Human Being: A Spiritual Anthropology*. New York: New City, 2000.

Collins, R. "The Multiverse Hypothesis: A Theistic Perspective." In *Universe or Multiverse*, ed. B. J. Carr, 459-80. Cambridge: Cambridge University Press, 2007.

Comte-Sponville, A. *L'être-temps. Quelques réflexions sur le temps de la conscience*. Paris: Presses Universitaires de France, 1999.

______. *L'esprit de l'athéisme. Introduction à une spiritualité sans Dieu*. Paris: Albin Michel, 2006.

Compton, J. J. "Natural Science and the Experience of Nature." In *Phenomenology in America: Studies in the Philosophy of Experience*, ed. J. M. Edie, 80–95. Chicago: Quadrangle, 1967.

Copleston, F. *A History of Philosophy, Vol. II, Medieval Philosophy: Augustine to Scotus*, London: Burns, Oates & Washbourne, 1950.

Cornwell, J. "Hawking's Quest: A Search without End," *The Tablet* 27 (March 2004): 4–5.

Chryssavgis, J. and B. Foltz, eds. *Toward an Ecology of Transfiguration: Orthodox Christian Perspectives on Environment, Nature, and Creation.* New York: Fordham University Press, 2013.

Dancy, J. *Introduction to Contemporary Epistemology.* Oxford: Basil Blackwell, 1989.

Davies, O. *The Creativity of God: World, Eucharist, Reason.* Cambridge: Cambridge University Press, 2004.

Davies, P. C. W. *The Physics of Time Asymmetry.* London: Surrey University Press, 1974.

Dawkins, R. *The God Delusion.* London: Black Swan, 2007.

De Broglie, L. *Sur les sentiers de la science.* Paris: Albin Michel, 1960.

De Duve, C. *Vital Dust: Life as a Cosmic Imperative.* New York: Basic Books, 1995.

De Laguna, G. *On Existence and the Human World.* New Haven and London: Yale University Press, 1966.

De Waelhens, A. "Science, Phenomenology, Ontology," *Cross Currents* 17 (1957): 167–74.

Dejours, C. *Le facteur humain.* Paris: Presses Universitaires de France, 1995.

Disney, M. J. "The Case Against Cosmology," *General Relativity and Gravitation* 32 (2000): 1125–34.

Dondeyne, A. *Contemporary European Thought and Christian Faith.* Pittsburgh: Duquesne University Press, 1958.

Dyson, F. *Disturbing the Universe.* New York: Basic Books, 1979.

Einstein, A. *Ideas and Opinions.* London: Souvenir, 1973.

Ellis, G. F. R. "Issues in the Philosophy of Cosmology." In *Handbook of the Philosophy of Science, Philosophy of Physics*, Part B, ed. J. Butterfield and J. Earman, 1183-283. Amsterdam: Elsevier, 2007.

______. "Does the Multiverse Really Exist?," *Scientific American*, August 2011, 38-43.

Farber, M. "First Philosophy and the Problem of the World," *Philosophy and Phenomenological Research* 23, no. 3 (1963): 315-34.

Feinberg, G. "Physics and the Thales Problem," *The Journal of Philosophy* LXIII, no. 1 (1966): 5-17.

Filaret Metropolitan of Minsk and Slutsk. *The Way of the Life-Asserting Love.* Kiev: Duh I Litera, 2004 (in Russian).

Florensky, P. "Macrocosm and Microcosm." In his *Apology of the Cosmos.* St. Petersburg: Russian Christian Humanitarian Institute, 1994, 184-201 (in Russian).

Florovsky, G. "Creation and Creaturehood." In *Creation and Redemption*, The collected works of Georges Florovsky, vol. 3, 43-78. Belmont, MA: Nordland, 1976.

______. "The Idea of Creation in Christian Philosophy," *The Eastern Churches Quarterly* 8, no. 3 (1949): 53-77.

______. *The Byzantine Fathers of the Sixth to Eighth Centuries*, in the *Collected Works of Georges Florovsky*, vol. 9. Vaduz, Liechtenstein: Büchvertriebsanstalt, 1987.

Foltz, B. "Discerning the Spirit in Creation: Orthodox Christianity and Environmental Science." In *Science and the Eastern Orthodox Church*, ed. Daniel Buxhoeveden and Gayle Woloschak, 105-18. Farnham, UK: Ashgate, 2011.

Fricke, C. "Explaining the Inexplicable: The Hypothesis of the Faculty of Reflective Judgment in Kant's Third Critique," *NOÛS* 24, no. 1 (1992): 45-62.

Friedman, Y. "Scientific Theory as One of the Fine Arts," *Leonardo* 26, no. 4 (1993): 359-63.

Fromm, E. *Man for Himself: An Enquiry into the Psychology of Ethics.* London: Routledge & Kegan Paul, 1967.

Gale, G., and N. Shanks. "Methodology and the Birth of Modern Cosmological Inquiry," *Studies in History and Philosophy of Modern Physics* 27, no. 3 (1996): 279-96.

Goutner, G. "An Eschatological Aspect of Ethics and Epistemology." In *Scientific and Theological Thinking of Ultimate Questions: Cosmology, Creation, Eschatology*, ed. A. Grib, 210-20. Moscow: St. Andrew's Biblical Theological Institute, 2008 (in Russian).

______. "The Idea of Humanity. Epistemological and Ethical Aspects." In *Methodology of Science and Anthropology.* Moscow: Institute of Philosophy of the Russian Academy of Science, 2012, 170-92 (in Russian).

______. "The Unity of Humanity in an Eschatological Perspective." In *Theology of Creation*, ed. A. Bodrov and M. Tolstoluzhenko, 230-36. Moscow: St. Andrew's Biblical and Theological Institute, 2013 (in Russian).

Gregorios, P. M. *The Human Presence: Ecological Spirituality and the Age of the Spirit.* New York: Amity House, 1987.

Grosz, E. *Chaos, Territory, Art.* New York: Columbia University Press, 2008.

Gurwitsch, A. *Studies in Phenomenology and Psychology.* Evanston, IL: Northwestern University Press, 1966.

______. "Galilean Physics in the Light of Husserl's Phenomenology." In *Galileo: Man of Science*, ed. E. McMullin, 388-401. New York: Basic Books, 1967.

______. *Phenomenology and the Theory of Science.* Evanston, IL: Northwestern University Press, 1974.

______. "Comments on Henry Margenau's 'Phenomenology and Phyiscs.'" In *Phenomenology of Natural Science*, ed. L. Hardy and L. Embree, 35-44. Dordrecht: Kluwer, 1992.

______. *The Field of Consciousness*, in *The Collected Works of Aron Gurwitsch*, vol. 3, 1-412. Dordrecht: Springer, 2010.

Halvorson, H., and H. Kragh. "Theism and Physical Cosmology." In *The Routledge Companion to Theism*, ed. S. Goetz et al., 241-55. London: Routledge, 2012.

Harrison, E. R. *Cosmology: The Science of the Universe.* Cambridge: Cambridge University Press, 1986.

Harwit, M. *Cosmic Discovery.* Brighton: Harvester, 1981.

Harvey, C. W. "Natural Science Is Human Science. Human Science Is Natural Science: Never the Twain Shall Meet." In *Continental and Postmodern Perspectives in the Philosophy of Science*, ed. B. E. Babich, D. B. Bergoffen, and S. V. Glynn, 121-36. Aldershot, UK: Avebury, 1995.

Hawking, S. *A Brief History of Time: From the Big Bang to Black Holes.* London: Bantam, 1988.

Hawking, S., and R. Penrose. *The Nature of Space and Time.* Princeton: Princeton University Press, 1996.

Heelan, P. A. "Nature and Its Transformations," *Theological Studies* 1 (1972): 486-502.

______. "Experiment and Theory: Constitution and Reality," *The Journal of Philosophy* 85, no. 10 (1988): 515-24.

______. "An Anti-Epistemological or Ontological Interpretation of the Quantum Theory and Theories Like It." In *Continental and Postmodern Perspectives in the Philosophy of Science*, ed. B. E. Babich, D. B. Bergoffen, and S. V. Glynn, 55-68. Aldershot, UK: Avebury, 1995.

Heidegger, M. *Being and Time.* Oxford: Blackwell, 1998.

______. "Letter on Humanism." In *Pathmarks*, ed. W. McNeill, 239-76. Cambridge: Cambridge University Press, 1998.

______. *Nietzsche*, Volume IV: *Nihilism*. San Francisco: Harper, 1991.

Heisenberg, W. *The Physicist's Conception of Nature.* London: Hutchinson Scientific & Technical, 1958.

______. *Physics and Philosophy.* London: Penguin, 1989.

Henry, M. *De la phénoménologie. Tome I. Phénoménologie de la vie.* Paris: Presses Universitaires de France, 2003.

Hogg, D. W. "Is Cosmology Just a Plausibility Argument?," arXiv: 0910.3374v1 astro-ph.co, 18 October 2009.

Horujy, S. "How Exactly Is Spirit Present in Creation? The Hesychast Reception of Natural Theology and Its Modern Implications." In *The Spirit in Creation and New Creation: Science and Theology in Western and Orthodox Realms*, ed. M. Welker, 95-108. Grand Rapids and Cambridge: Eerdmans, 2012.

Housset, E. *Personne et sujet selon Husserl.* Paris: Presses Universitaires de France, 1997.

Hoyle, F. "A New Model for the Expanding Universe," *Monthly Notices of the Royal Astronomical Society* 108 (1948): 372-82.

Husserl, E. *Erste Philosophie* 2. The Hague: Nijhoff, 1959.

______. *The Crisis of European Sciences and Transcendental Phenomenology.* Evanston, IL: Northwestern University Press, 1970.

______. *Zur Phänomenologie der Intersubjectivität*, vol. 1-3. The Hague: Nijhoff, 1973.

______. *Experience and Judgment: Investigations in a Genealogy of Logic.* London: Routledge & Kegan Paul, 1973.

______. *Phenomenological Psychology: Lectures, Summer Semester, 1925.* The Hague: Nijhoff, 1977.

______. "Foundational Investigations of the Phenomenological Origin of the Spatiality of Nature." In *Husserl Shorter Works*, ed. P. McCormick and F. A. Elliston, 222-33. Notre Dame: University of Notre Dame Press, 1981.

______. *On the Phenomenology of the Consciousness of Internal Time.* Dordrecht: Kluwer, 1991.

______. *Ideas Pertaining to a Pure Phenomenology and to a Phenomenological Philosophy.* First Book. *General Introduction to a Pure Phenomenology.* Dordrecht: Kluwer, 1998.

Ignatius IV Patriarch of Antioch, "Three Sermons," *Sourozh: A Journal of Orthodox Life and Thought* 38 (1989): 1–14.

Ingarden, R. *Man and Value.* Washington, DC: Catholic University of America Press, 1983.

Isham, C. J. "Creation of the Universe as a Quantum Process." In *Physics, Philosophy and Theology: A Common Quest for Understanding*, ed. R. J. Russell et al., 375-408. Vatican City State: Vatican Observatory, 1988.

Isham, C. J., and J. C. Polkinghorne. "The Debate over the Block Universe." In *Quantum Cosmology and the Laws of Nature*, ed. R. J. Russell, N. Murphy, and C. J. Isham, 139–47. Berkeley: CTNS, 1996.

Jaki, S. L. "Christology and the Birth of Modern Science," *The Asbury Theological Journal* 45, no. 2 (1990): 69–77.

______. *Is There a Universe?* Liverpool: Liverpool University Press, 1993.

Janicaud, D. *Le tournant théologique de la phénoménologie française.* Combas: Éditions de l'Éclat, 1990.

______. "The Theological Turn of French Phenomenology." In *Phenomenology and "The Theological Turn": The French Debate*, 16–103. New York: Fordham University Press, 2000.

Jaspers, K. *Weltgeschichte der Philosophie. Einleitung.* München: R. Piper, 1982.

Jeans, J. *Physics and Philosophy.* Cambridge: Cambridge University Press, 1945.

Jrade, C. L. "Ruben Dario and the Oneness of the Universe," *Hispania* 63 (1980): 691–98.

Kant, I. *Critique of Pure Reason*, 2nd ed. Translated by N. K. Smith. London: Macmillan, 1933.

______. *Critique of Practical Reason.* Translated by T. K. Abbot. London: Longmans, 1959.

______. *Fundamental Principles of the Metaphysic of Morals*, in *Kant's Critique of Practical Reason and Other Works on the Theory of Ethics*. Translated by T. K. Abbot. London: Longmans, 1959.

______. *Critique of Judgment.* Translated by J. H. Bernard. New York, Hafner, 1951.

______. *First Introduction to the Critique of Judgment.* Translated by J. Haden. Indianapolis: Bobbs-Merrill, 1965.

______. *Critique of the Power of Judgment.* Translated by P. Guyer and E. Matthews. Cambridge: Cambridge University Press, 2000.

Kazyutinsky, V. V. "Worlds of Culture and the World of Science: An Epistemological Status of Cosmology." In *A Socio-Cultural Context of Science*, 101–18. Moscow: Institute of Philosophy of the Russian Academy of Science, 1998 (in Russian).

______. "No, Cosmology Is a Physical Science and Not a Human One," *Epistemology and Philosophy of Science* XI, no. 2 (2007): 125-29 (in Russian).

______. "Physical Cosmology and Culture: Introduction." In *Cosmology, Physics, Culture*, ed. V. V. Kazyutinsky, 3-15. Moscow: Institute of Philosophy of the Russian Academy of Science, 2011 (in Russian).

Keen, S. *Apology for Wonder.* New York: Harper & Row, 1969.

Kerszberg, P. "The Phenomenological Analysis of the Earth's Motion," *Philosophy and Phenomenological Research* 48, no. 2 (1987): 177–208.

______. *Critique and Totality.* Albany: State University of New York Press, 1997.

______. "From the Lifeworld to the Exact Sciences and Back." In *Husserl and the Sciences: Selected Perspectives*, ed. R. Feist, 189–212. Ottawa: University of Ottawa Press, 2003.

Kierkegaard, S. *Repetition* and *Philosophical Crumbs.* Translated by M. G. Piety. Oxford: Oxford University Press, 2009.

Kline, M. *Mathematics in Western Culture.* New York: Penguin, 1977.

Köhler, H. "The Relation between Man and the World." In *Astronomy and Civilisation in the New Enlightenment: Passions of the Skies*, ed. A.-T. Tymieniecka, *Analecta Husserliana, CVII*, 37-46. Dordrecht: Springer, 2011.

Kockelmans, J. *Phenomenology and Physical Science.* Pittsburgh: Duquesne University Press, 1966.

______. "The Mathematisation of Nature in Husserl's Last Publication, *Krisis*." In *Phenomenology and the Natural Sciences*, ed. T. J. Kiesel and J. Kockelmans, 45-67. Evanston, IL: Northwestern University Press, 1970.

Koyré, A., *From the Closed World to the Infinite Universe.* New York: Harper & Brothers, 1958.

Kragh, H. *Cosmology and Controversy: The Historical Development of Two Theories of the Universe.* Princeton: Princeton University Press, 1996.

Krauss, L. M. *A Universe from Nothing.* New York: Free Press, 2012.

Krichivets, A. N. "Apriori, Faculty of Judgment and Aesthetics," *Bulletin of the Moscow University*, Ser. 7, *Philosophy*, no. 3 (1996): 41-50 (in Russian).

Kuiper, F. B. J. "Cosmogony and Conception: A Query," in his *Ancient Indian Cosmology.* New Delhi: Vikas Publ. House Pvt Ltd, 1983.

Küng, H. *Does God Exist? An Answer for Today.* London: SCM, 1978.

______. *The Beginning of All Things: Natural Sciences and Religion.* Grand Rapids: Eerdmans, 2008.

Ladrière, J. "Physical Reality: A Phenomenological Approach," *Dialectica* 43, no. 1-2 (1989): 125-39.

______. "Mathematics in a Philosophy of the Sciences." In *Phenomenology and the Natural Sciences*, ed. T. J. Kiesel and J. Kockelmans, 443-65. Evanston, IL: Northwestern University Press, 1970.

______. *Language and Belief.* Dublin: Gill & Macmillan, 1972.

Landgrebe, L. "The World as a Phenomenological Problem," *Philosophy and Phenomenological Research* 1, no. 1 (1940): 38-58.

J.-C. Larchet. *La Divinisation de l'homme selon Saint Maxime le Confesseur.* Paris: Les Éditions du Cerf, 1996.

Leprince-Ringuet, L. *Science et bonheur des hommes.* Paris: Flammarion, 1973.

Leslie, J. *Universes.* London: Routledge, 1989.

______. *The End of the World: The Science and Ethics of Human Extinction.* London: Routledge, 1996.

Levinas, E. *Time and the Other.* Pittsburgh: Duquesne University Press, 1987.

______. *Éthique comme philosophie première.* Paris: Éditions Payot & Rivages, 1998.

Liddle, A. *An Introduction to Modern Cosmology.* Chichester, UK: John Wiley & Sons, 1999.

Longo, G. "The Mathematical Continuum: From Intuition to Logic." In *Naturalising Phenomenology: Issues in Contemporary Phenomenology and Cognitive Science*, ed. J. Petitot, F. J. Varela, B. Pachoud, and J.-M. Roy, 401-25. Stanford: Stanford University Press, 1999.

______. "The Constructed Objectivity of Mathematics and the Cognitive Subject." In *Quantum Mechanics, Mathematics, Cognition and Action: Proposals for a Formalized Epistemology*, ed. M. Mugur-Schachter and A. van der Merwe, 433-62. Dordrecht: Kluwer Academic, 2002.

Lossky, V. N. *The Mystical Theology of the Eastern Church.* London: James Clarke, 1957.

______. *Orthodox Theology: An Introduction.* Crestwood, NY: St. Vladimir's Seminary Press, 1997.

Luijpen, W. A. *Existential Phenomenology.* Pittsburgh: Duquesne University Press, 1960.

Lyotard, J.-F. *Postmodern Fables.* Minneapolis: University of Minnesota Press, 1997.

______. *The Inhuman: Reflections on Time.* Cambridge and Oxford: Polity, 1991.

Lyre, H. "Structural Realism and Abductive-Transcendental Arguments." In *Constituting Objectivity: Transcendental Perspectives on Modern Physics*, ed. M. Bitbol, P. Kerszberg, and J. Petitot, 491-501. Dordrecht: Springer, 2009.

Mall, R. "The God of Phenomenology in Comparative Contrast to That of Philosophy and Theology," *Husserl Studies* 8 (1991): 1-15.

Mamchur, E., and Y. Sachkov, eds. *Causality and Teleonomism in Modern Paradigm of the Natural Sciences.* Moscow: Nauka, 2002 (in Russian).

Marcel, G. *Journal Métaphysique.* Paris: Gallimard, 1935.

______. *Du Refus à l'Invocation.* Paris: Gallimard, 1940.

______. *Metaphysical Journal.* London: Rockliff, 1952.

______. *Being and Having.* London: Collins, 1965.

Margenau, H. "Phenomenology and Physics," *Philosophy and Phenomenological Research* 5, no. 2 (1944): 269-80.

______. "Physical versus Historical Reality," *Philosophy of Science* 19, no. 3 (1952): 193-213.

______. "Physics and Ontology," *Philosophy of Science* 19, no. 4 (1952): 342-45.

______. *The Nature of Physical Reality: A Philosophy of Modern Physics.* Woodbridge, CT: Ox Bow, 1977.

Marion, J.-L. "Le Phénomène Saturé." In *Phénoménologie et Théologie*, ed. J.-F. Courtine, 79-128. Paris: Criterion, 1992.

______. "The Saturated Phenomenon." In *Phenomenology and "The Theological Turn": The French Debate*, 176-216. New York: Fordham University Press, 2000.

______. *Being Given: Toward a Phenomenology of Givenness.* Stanford: Stanford University Press, 2002.

______. *In Excess: Studies of Saturated Phenomena.* New York: Fordham University Press, 2002.

______. "The Event, the Phenomenon and the Revealed." In *Transcendence in Philosophy and Religion*, ed. J. F. Faulconer, 87-105. Bloomington and Indianapolis: Indiana University Press, 2003.

______. "*Mihi magna quaestio factus sum*: The Priviledge of Unknowing," *The Journal of Religion* 85, No 1 (2005), 1-24.

______. "Je suis un sceptique instruit," *Philosophie Magazine* 39 (2010): 58-63.

______. *Certitudes négatives.* Paris: Bernard Grasset, 2010.

______. "La foi et la raison." In *Le croire pour le voir. Réflexions diverses sur la rationalité de la révélation et l'irrationalité de quelques croyants*, 17-29. Paris: Éditions Parole et Silence, 2010.

Mathews, F. *The Ecological Self.* London: Routledge, 1991.

May, G. *Creation Ex Nihilo: The Doctrine of "Creation Out of Nothing" in Early Christian Thought.* Edinburgh: T. & T. Clark, 1994.

McCabe, G. "The Structure and Interpretation of Cosmology," *Studies in History and Philosophy of Modern Physics*, part 1: v. 35 (2004): 549-95; part 2: v. 36 (2005): 67-102.

McLaughlin, W. I. "Kantian Epistemology as an Alternative to Heroic Astronomy," *Vistas in Astronomy* 28 (1985): 611-39.

McMullin, E. "Indifference Principle and Anthropic Principle in Cosmology," *Studies in History and Philosophy of Science* 24, no. 3 (1993): 359-89.

______. "Long Ago and Far Away: Cosmology and Extrapolation." In *Bang: The Evolving Cosmos*, ed. R. Fuller, 105-45. Saint Peter, MN: Gustavus Adolphus College, 1994.

Mensch, J. R. *Postfoundational Phenomenology: Husserlian Reflections on Presence and Embodiment.* University Park: Pennsylvania State University Press, 2001.

Merleau-Ponty, M. *Phenomenology of Perception.* London: Routledge, 1962.

______. *The Visible and the Invisible* Evanston, IL: Northwestern University Press, 1968.

______. *Sense et Non-Sense.* Evanston, IL: Northwestern University Press, 1982.

______. "Eye and Mind.'" In *The Merleau-Ponty Aesthetics Reader: Philosophy and Painting*, ed. G. A. Johnson, 121-49. Evanston, IL: Northwestern University Press, 1993.

Meyerson, E. *Identity and Reality.* London: Allen & Unwin, 1964.

Minkowski, E. "Prose and Poetry Astronomy and Cosmology." In *Phenomenology and the Natural Sciences*, ed. T. J. Kiesel and J. Kockelmans, 239-47. Pittsburgh: Duquesne University Press, 1966.

Misner, C. W., K. S. Thorn, and J. A. Wheeler. *Gravitation.* New York: W. H. Freeman, 1973.

Moltmann, J. *Science and Wisdom.* Minneapolis: Fortress Press, 2003.

Moore, A. W. "Aspects of the Infinite in Kant," *Mind* XCII (1988): 205-23.

______. "Human Finitude, Ineffability, Idealism, Contingency," *NOÛS* 26, no. 4 (1992): 427-46.

______. *The Infinite.* London: Routledge, 2001.

Morin, E. *Science avec conscience.* Paris: Fayard, 1982.

Munitz, M. K. "Kantian Dialectic and Modern Scientific Cosmology," *The Journal of Philosophy* 48, no. 10 (1951): 325-38.

______. *Space, Time and Creation: Philosophical Aspects of Scientific Cosmology.* Glencoe, IL: Free Press, 1957.

______. "Approaches to Existence," *Philosophy East and West* 25, no. 3 (1975): 335-46.

______. *Question of Reality.* Princeton: Princeton University Press, 1990.

Murphy, N., and G. F. R. Ellis. *The Moral Nature of the Universe: Cosmology, Theology, and Ethics.* Minneapolis: Fortress Press, 1996.

Natanson, M. "Being-in-Reality," *Philosophy and Phenomenological Research* 20, no. 2 (1959): 231-37.

Nellas, P. *Deification in Christ: Orthodox Perspectives on the Nature of the Human Person.* Crestwood, NY: St. Vladimir's Seminary Press, 1997.

Nesteruk, A. V. *Light from the East: Theology, Science and the Eastern Orthodox Tradition.* Minneapolis: Fortress Press, 2003.

______. "Human Image: World Image: Patristic Insight into Cosmological Anthropology." In *Studies in Science and Theology*, Yearbook of the European Society for the Study of Science and Theology 2003-2004, 9 (2004): 3-26.

______. "The Universe as Hypostatic Inherence in the Logos of God: Panentheism in the Eastern Orthodox Perspective." In *In Whom We Live and Move and Have Our Being: Reflections on Panentheism in a Scientific Age*, ed. P. Clayton and A. Peacocke, 169-83. Grand Rapids: Eerdmans, 2004.

______. "The Problem of Faith and Scientific Knowledge in Russian Religious Thought of the 19th-20th Centuries." In *Interpreting Nature and Scripture*, ed. J. M. van der Meer and S. Mandelbrote, 371-404. Leiden: Brill Academic, 2008.

______. *The Universe as Communion: Towards a Neo-Patristic Synthesis of Theology and Science.* London: T. & T. Clark, 2008.

______. "From the Unknowability of the Universe to the Teleology of Reason: A Phenomenological Insight into Apophatic Cosmology." In *Knowing the Unknowable: Science and Religion on God and the Universe*, ed. J. Bowker, 63-86. London: I. B. Tauris, 2009.

______. "Transcendence-in-Immanence in Theology and Cosmology: A New Phenomenological Turn in the Debate." In *Studies in Science and Theology*, Yearbook of the European Society for the Study of Science and Theology 2009-2010, 12 (2010): 179-98.

______. "Cosmology at the Crossroads of Natural and Human Sciences: Is Demarcation Possible?" *Journal of Siberian Federal University. Humanities & Social Sciences.* Part 1, vol. 4, no. 4 (2011): 560-76, Part 2, vol. 4, no. 5 (2011): 644-66.

______. "The Origin of the Universe and Event of Birth: Phenomenological Parallels," *Journal of Siberian Federal University. Humanities & Social Sciences* 5, no. 2 (2012): 172-205.

______. "The Quest for Transcendence in Theology and Science," in *The Blackwell Companion to Science and Christianity*, Wiley-Blackwell, 2012, pp. 554-64.

______. "Towards Constituting the Identity of the Universe: Apophaticism and Transcendental Delimiters in Cosmology," *Journal of Siberian Federal University. Humanities & Social Sciences* 5, no. 3 (2012): 358-94.

______. "Cosmology and Teleology: Purposiveness in the Study of the Universe through the Reading of Kant's Third Critique," *Journal of Siberian Federal University. Humanities & Social Sciences* 5, no. 9 (2012): 1304-35.

______. "A 'Participatory Universe' of J. A. Wheeler as an Intentional Correlate of Embodied Subjects and an Example of Purposiveness in Physics," *Journal of Siberian Federal University. Humanities & Social Sciences* 6, no. 3 (2013): 415-37.

______. "The Cosmos of the World and the Cosmos of the Church: St. Maximus the Confessor, Modern Cosmology and the Sense of the Universe." In *Knowing the Purpose of Creation Through the Resurrection*, Proceedings of the Symposium on St. Maximus the Confessor, Belgrade, October 18-21, 2012, ed. Bishop Maxim Vasiljevic, 297-333. Alhambra CA: Sebastian, 2013. (A shortened version of this paper is also published in *The Messenger. Journal of the Deanery of Great Britain and Ireland*, Part 1, no. 22 (March 2013): 13-36; Part 2, no. 23 (July 2013): 16-32).

______. "The Universe as a Construct: Epistemic Beliefs and Coherence of Justification in Modern Cosmology," *Journal of Siberian Federal University. Humanities & Social Sciences* 6, no. 7 (2013): 957-1001.

______. "Coherence of Epistemic Justification versus the Principle of Correspondence in Modern Cosmology," *Izvestiya: Herzen University*

Journal of Humanities and Sciences (St. Petersburg, Russia) no. 156 (2013): 59-67.

______. "The Ecclesial Dimension in the Dialogue between Theology and Science," *International Journal for the Study of the Christian Church* 13, no. 3 (2013): 1-19.

______. "Man and the Universe in Patristic Thought: the Teaching of Maximus the Confessor and Modern Cosmology," *Journal of Siberian Federal University. Humanities & Social Sciences* 7, no. 6 (2014), 959-991.

______. "The Universe as a Saturated Phenomenon: The Christian Concept of Creation in View of Modern Philosophical and Scientific Developments," *Theology and Science* 12, no. 3 (2014), 236-59.

Newton-Smith, W. H. *The Rationality of Science*. London and New York: Routledge, 1996.

Nieman, A. "Welcome to the Neighbourhood: Belonging to the Universe," *Leonardo* 38, no. 5 (2005): 383-88.

Nietzsche, F. *Thus Spoke Zarathustra*, vol. 11 of *The Complete Works of Friedrich Nietzsche*, ed. Oscar Levy. New York: Macmillan, 1919.

O'Murchadha, F. "Nature as Other: Hermeneutical Approach to Science." In *Continental and Postmodern Perspectives in the Philosophy of Science*, ed. B. E. Babich, D. B. Bergoffen, and S. V. Glynn, 189-203. Aldershot, UK: Avebury, 1995.

Ortega y Gasset, J. *Man and Crisis*. London: Allen & Unwin, 1959.

Pabst, A., and C. Schneider, eds. *Encounter Between Eastern Orthodoxy and Radical Orthodoxy: Transfiguring the World Through the Word*. Farnham, UK: Ashgate, 2009.

Pannenberg, W. *Toward a Theology of Nature: Essays on Science and Faith*. Louisville: Westminster John Knox, 1993.

______. *The Historicity of Nature: Essays on Science and Theology*. West Conshohocken, PA: Templeton Foundation, 2008.

Papin, L. "This Is Not a Universe: Metaphor, Language, and Representation," *Proceedings of the Modern Language Association* 107, no. 5 (1992): 1253-65.

Pascal, B. *Pensées. Selections.* Translated and edited by M. Jarret-Kerr. London: SCM, 1959.

______. *Pensées.* Translated by Louis Lafuma. Paris: Éditions du Seuil, 1962.

______. "De l'art de persuader." In *L'espirit géométrique* et *De l'art de persuader.* Paris: Bordas, 2002.

Patton, C. M., and J. A. Wheeler. "Is physics legislated by cosmogony?." In *Encyclopaedia of Ignorance*, ed. R. Duncan and M. Weston-Smith, 19-35. Oxford: Pergamon, 1977.

Pavlenko, A. N. "Universalism and Cosmic Harmony: A Principle of Genetic Similarity," *Skepsis* XV, no. 1 (2004): 389-401.

______. "European Cosmology: Between 'Birth' and 'Creation.'" In *Scientific and Theological Thinking of Ultimate Questions: Cosmology, Creation, Eschatology*, ed. A. A. Grib, 122-30. Moscow: St. Andrew's Biblical Theological Institute, 2008 (in Russian).

______. "A Place of 'Chaos' in the New World 'Order.'" In *Philosophical Problems of Cosmology: The Universe from "Nothing" or the Universe from "Non-Being"?*, 66-87 Moscow: URSS/Librocom, 2012.

______. "The Principle of "Observability": Why the Theory of the Infinite Universe Cannot Be Sustained." In *Philosophical Problems of Cosmology: The Universe from "Nothing" or the Universe from "Non-Being"?*, 131-42. Moscow: URSS/Librocom, 2012.

Penrose, R. "Singularities and Time-Asymmetry." In *General Relativity: An Einstein Centenary Survey*, ed. S. W. Hawking and W. Israel, 581-638. Cambridge: Cambridge University Press, 1979s

______. *The Emperor's New Mind.* Oxford University Press, 1989.

______. *The Road to Reality.* London: Vintage, 2005.

______. *Cycles of Time: An Extraordinary New View of the Universe.* London: Bodley Head, 2010.

Planck, M. "Religion und Naturwissenschaft." Vortrag gehalten im Baltikum Mai 1937 von Dr. Max Planck, 2te unveränd Auflage. Leipzig: Joh. Ambrosius Barth, 1938.

Powell, J. "Perfection as Cosmological Postulate: Aristotle and Bruno," *The Philosophical Review* 44, no. 1 (1935): 57-68.

Prestige, G. L. *God in Patristic Thought.* London: SPCK, 1955.

Primack, J., and N. E. Abrams. *The View from the Centre of the Universe: Discovering Our Extraordinary Place in the Cosmos.* London: Fourth Estate, 2006.

Puhalo, Lazar Archbishop. *Freedom to Believe: Personhood and Freedom on Orthodox Christian Ontology.* Dewdney, BC: Synaxis, 2001.

Ratzinger, J. Kardinal, *Wesen und Auftrag der Theologie.* Freiburg: Johannes Verlag Einsiedeln, 1993.

Razeev, D. N. *Teleology of Immanuel Kant.* St. Petersburg: Nauka, 2010 (in Russian).

Rees, M. *Our Final Century, A Scientist's Warning: How Terror, Error, and Environmental Disaster Threaten Humankind's Future in This Century—On Earth and Beyond.* London: Heinemann, 2003.

______. "Our Complex Cosmos and Its Future." In *The Future of Theoretical Physics and Cosmology: Celebrating Stephen Hawking's 60th Birthday*, ed. G. W. Gibbons, E. P. S. Shellard, and S. J. Rankin, 17-37. Cambridge: Cambridge University Press, 2003.

Rescher, N. *The Coherence Theory of Truth.* New York: University Press of America, 1989.

Resnik, M. *Mathematics as a Science of Patterns.* Oxford: Clarendon, 2005.

Ricoeur, P. *Husserl: An Analysis of His Phenomenology.* Evanston, IL: Northwestern Unversity Press, 1967.

Rizzacasa, A. “The Epistemology of the Sciences of Nature in Relation to the Teleology of Research in the Thought of the Later Husserl.” In *The Teleologies in Husserlian Phenomenology*, ed. A.-T. Tymieniecka, *Analecta Husserliana* 16, 73–84. Dordrecht: Reidel, 1979.

Rodzyanko, B. Bishop. *Theory of the Universe's Decay and Faith of the Fathers: Capadocian Theology—the Key to Apologetics of Our Time.* Moscow: Palomnik, 2003 (in Russian).

Romano, C. *L'événement et le monde.* Paris: Presses Universitaires de France, 1998.

Rosin V. M. *Types and Discourses of Scientific Thinking.* Moscow: Editorial URSS, 2000 (in Russian).

______. “Towards a Problem of Demarcation of the Natural and Human Sciences and Where to One Must Relate Cosmology,” *Epistemology and Philosophy of Science* 11, no. 1 (2007): 111–28 (in Russian).

Rothman, A., and G. F. R. Ellis. “Lost Horizons,” *American Journal of Physics* 61 (1993): 883–93.

Roush, S. “Copernicus, Kant, and the Anthropic Cosmological Principles,” *Studies in History and Philosophy of Modern Physics* 34 (2003): 5–35.

Sakharov, Sophrony Archimandrite. *St. Silouan the Athonite.* Moscow: St. Trinity and Sergius Lavra, 1999 (in Russian).

Scheler, M. *Die Stellung des Menschen im Kosmos.* Moscow: Gnosis, 1994 (in German and Russian).

Schrödinger, E. *What Is Life? The Physical Aspect of the Living Cell* with *Mind and Matter* & *Autobiographical Sketches.* Cambridge: Cambridge University Press, 1992.

Shakhovskoi, Jonn Archbishop. *On the Mystery of Human Life.* Moscow: Lodiya, 2003 (in Russian).

Shapiro, S. *Thinking about Mathematics: The Philosophy of Mathematics.* Oxford: Oxford University Press, 2000.

Sherrard, P. *The Rape of Man and Nature: An Enquiry into the Originsand Consequences of Modern Science*. Ipswich, UK: Golgonooza, 1991.

Shmaliy, V. "The Spirit or/and Spirits in Creation?" In *The Spirit in Creation and New Creation: Science and Theology in Western and Orthodox Realms*, ed. M. Welker, 79-94. Grand Rapids and Cambridge: Eerdmans, 2012.

Smith, J. K. *Speech and Theology: Language and the Logic of Incarnation.* London: Routledge, 2002.

_______. *Radical Orthodoxy: Mapping a Postsecular Theology*. Grand Rapids: Baker Academic, 2004.

Sokolov, S. *The Other World and the Time of the Universe: Time and Eternity*. Moscow: Kovcheg, 2008 (in Russian).

Sokolowski, R. *Husserlian Meditations.* Evanston, IL: Northwestern University Press, 1974.

_______. *Introduction to Phenomenology*. Cambridge: Cambridge University Press, 2000.

Staniloae, D. *Theology and the Church.* Crestwood, NY: St. Vladimir's Seminary Press, 1980.

_______. *Experience of God*, vol. 1 of *Orthodox Dogmatic Theology*. Brookline, MA: Holy Cross Orthodox Press, 1998.

Steinbock, A. J. *Home and Beyond: Generative Phenomenology after Husserl.* Evanston, IL: Northwestern University Press, 1995.

Stenger, V. *God the Failed Hypothesis: How Science Shows That God Does Not Exist*. New York: Prometheus, 2008.

Stoeger, W. "Astronomy's Integrating Impact on Culture: A Ladrièrean Hypothesis," *Leonardo* 29, no. 2 (1996): 151–54.

_______. "Are Anthropic Arguments, Involving Multiverses and Beyond, Legitimate?" In *Universe or Multiverse*, ed. B. J. Carr, 445–57. Cambridge: Cambridge University Press, 2007.

Strasser, S. *Phénoménologie et sciences de l'homme. Vers un nouvel esprit scientifique*. Louvain: Publications Universitaires, 1967.

Ströker, E. *Investigations in Philosophy of Space.* Athens: Ohio University Press, 1965.

Swimme, B. *The Hidden Heart of the Cosmos: Humanity and the New Story.* New York: Orbis, 1996.

Temple, D. "The New Design Argument: What Does It Prove?" In *Science, Technology, and Religious Ideas*, ed. M. H. Shale and G. W. Shields, 127-39. Lanham, MD: University Press of America, 1994.

Tegmark, M. "Parallel Universes." In *Science and Ultimate Reality: From Quantum to Cosmos*, ed. J. D. Barrow, P. C. W. Davies, and C. Harper, 459-91. Cambridge: Cambridge University Press, 2003.

______. *Our Mathematical Universe: My Quest for the Ultimate Nature of Reality.* London: Allen Lane, 2014.

Theokritoff, E. *Living in God's Creation: Orthodox Perspectives on Ecology.* Crestwood, NY: St. Vladimir's Seminary Press, 2009.

Thunberg, L. *Microcosm and Mediator: The Theological Anthropology of Maximus the Confessor.* Chicago: Open Court, 1995.

Tieszen, R. *Phenomenology, Logic, and the Philosophy of Mathematics.* Cambridge: Cambridge University Press, 2005.

Torrance, T. F. "The Relation of the Incarnation to Space in Nicene Theology." In *The Ecumenical World of Orthodox Civilization, Russia and Orthodoxy*, vol. 3, *Essays in Honour of Georges Florovsky*, ed. A. Blane and T. E. Bird, 43-70. The Hague: Mouton, 1974.

______. "The Greek Conception of Space in the Background of Early Christian Theology." In T. F. Torrance, *Divine Meaning: Studies in Patristic Hermeneutics*, 289-342. Edinburgh: T. & T. Clark, 1995.

______. "Ultimate and Penultimate Beliefs in Science." In *Facets of Faith & Science*, vol. 1, *Historiography and Modes of Interaction*, ed. J. M. van der Meer, 151-76. Lanham, MD: University Press of America, 1996.

______. *Theological Science.* Edinburgh: T. & T. Clark, 1996.

______. *Space, Time and Incarnation.* Edinburgh: T. & T. Clark, 1997.

______. *Divine and Contingent Order*. Edinburgh: T. & T. Clark, 1998.

______. *The Grammar of Theology: Consonance between Theology and Science*. Edinburgh: T. & T. Clark, 2001.

Tryon, E. P. "Is the Universe a Quantum Fluctuation?" In *Modern Cosmology and Philosophy*, ed. J. Leslie, 222–25. New York: Prometheus, 1998.

Vialatoux, J. *L'intention philosophique*. Paris: Presses Universitaires de France, 1965.

Walker, A., ed. *Different Gospels*. London: Hodder & Stoughton, 1988.

Walker, R. C. S. *The Coherence Theory of Truth: Realism, Anti-Realism, Idealism*. London: Routledge, 1988.

Ward, K. *Religion and Creation*. Oxford: Clarendon, 1996.

Weinberg, S. *Cosmology*. Oxford: Oxford University Press, 2008.

Weyl, H. *Mind and Nature: Selected Writings on Philosophy, Mathematics and Physics*. Princeton: Princeton University Press, 2009.

______. *Philosophy of Mathematics and Natural Science*. Princeton: Princeton University Press, 2009.

Wheeler, J. A. *At Home in the Universe*. New York: American Institute of Physics, 1994.

______. "From Relativity to Mutability." In *Symposium on the Development of the Physicist's Conception of Nature in the Twentieth Century, Miramare, Italy Trieste, 1972*, ed. J. Mehra, 202–47. Dordrecht: Reidel, 1973.

______. "World as a System Self-Synthesized by Quantum Networking," *IBM Journal of Resarch and Development* 32 (1988): 4–15.

______. "Time Today." In *Physical Origins of Time Asymmetry*, ed. J. J. Halliwell, J. Pérez-Mercader, and W. H. Zurek, 1–29. Cambridge: Cambridge University Press, 1994.

White, A. R. "Coherence Theory of Truth." In *The Encyclopedia of Philosophy*, vol. 2, ed. P. Edwards, 130–33. New York: Macmillan, 1967.

Whittaker, E. "Some Disputed Questions in the Philosophy of the Physical Sciences," *Proceedings of the Royal Society of Edinburgh* 61 (1941): 160–75.

Yannaras, C. *Postmodern Metaphysics.* Brookline, MA: Holy Cross Orthodox Press, 2004.

______. *Person and Eros.* Brookline, MA: Holy Cross Orthodox Press, 2007.

______. *On the Absence and Unknowability of God.* London: T. & T. Clark, 2005.

______. "The Church in Post-Communist Europe," *The Meaning of Reality: Essays on Existence and Communion, Eros and History*, 123-43. Los Angeles: Sebastian Press & Indiktos, 2011.

Zizioulas, J. Metropolitan. *Being as Communion.* Crestwood, NY: St. Vladimir's Seminary Press, 1997.

______. *Communion and Otherness: Further Studies in Personhood and the Church.* London: T&T Clark, 2006.

______. "The Contribution of Western Orthodox Theology," *THE MESSENGER, Journal of the Episcopal Vicariate of Great Britain and Ireland*, no. 6 (May 2008): 33-37.

Glossary Terms

Apodictic (apodicticity)

This term signifies a fact or philosophical truth that is clearly established or beyond dispute. The terms originates from Greek apodeiktikos (ἀποδεικτικός), "capable of demonstration." It is an adjectival expression from Aristotelean logic that refers to propositions that are demonstrable, that are necessary or self-evident. *Apodicticity* is the corresponding abstract noun, referring to logical certainty. Apodictic propositions contrast with assertoric propositions, which merely assert that something is (or is not) the case, and with problematic propositions, which assert only the possibility of something being true. Proof is opposed to probable reasoning. Kant contrasts "apodictic" with "problematic" and "assertoric" in the Critique of Pure Reason (A70/B95).

Apophaticism

This is a general epistemological trend applicable to theology and the sciences, implying that the sense of realities affirmed in terms of their signifiers is never exhausted by the latter, that is, signifiers do not exhaust that which is signified. Historically it has an origin in *apophatic theology* as a way of theologizing which safeguards the absolute transcendence of God against misinterpretations based upon human analogies. However, while one cannot comprehend God in his essence, one can experience his power and love through grace (divine energies). Thus the apophaticism is exactly that freedom which is granted to the faithful to explore experientially their personal way of life in God, the freedom within the horizon of the church's definitions which helps to separate and distinguish truth from its distortion and falsification.

Cosmological principle

The postulate of the spatial, material and nomological uniformity (homogeneity) of the universe. This postulate is required by the methodology of cosmology in order the universe as a whole could be explicable at all. The principle has an interesting history and is philosophical in its nature, being related to the teleology of the human activity of knowledge.

Cosmos noetos

In ancient Greek philosophy this is the intelligible cosmos of divine forms and intellects, embracing the hierarchy of different levels and orders of divine reality.

In Greek patristic context *cosmos noetos*, being an intelligible part of creation, differs ontologically from the sensible creation (*cosmos aestheticos*), so that the difference (*diaphora*) between them represents a constitutive element of creation of the whole world out of nothing. In application to our research *cosmos noetos* simply means the reality of mathematical representations of the physical universe, in particular mathematical models of the universe with no credible empirical evidence. For example the idea of the Big Bang is considered to be a constituted element of the noetic cosmos as if one could "look" at the Big Bang from outside of the physical universe, that is as its intelligible correlate.

Commensurability (with the universe)

In general commensurability implies comparison of things according to their common measure, or their mutual measurability by the same standard. On the basis of this it is obvious that the finite is not commensurable with the infinite. Correspondingly incommensurability of human beings with the universe follows from the simple fact that physically and biologically they occupy an insignificant, infinitely small (on cosmic scales) part of the universe (not mentioning at all the multiverse). However one can speak about the *eidetic or hypostatic* commensurability with the universe as a capacity to intuit the universe in its totality from the event of life and then articulate the universe within a single consciousness. The eidetic *commensurability* with the universe here goes together with the physical *consubstantiality* to the universe as sharing that physical substance that originated in the Big Bang.

Donation

In a modern philosophical usage *donation* is treated as a gift of being able to perceive and reveal things. The very fact of the human existence can be treated as a gift by God, so that humanity exists in a constant givenness of this gift. The characteristic of phenomena as given, that is those which fall under the rubric of "donation," is that they do not subsist in something which causally precedes them. They are given spontaneously from themselves and out of themselves.

Eidetic

The adjective *eidetic* has its origin in the Greek *eidos*, that is form, the external shape, appearance, or configuration of an object, in contradistinction to the matter of which it is composed (in Aristotelian metaphysics this is, the active, determining principle of a thing as distinguished from matter). *Eidetic intuition* implies that one can intuit, or make present to oneself not only individuals with their particular features but also the essences of things. This intuition, together with eidetic variation, and the descriptive science of essences, comprises elements of reasoning which are in constant use in scientific work. In phenomenology eidetic intuition works together with *eidetic reduction* whose purpose is to bracket any considerations concerning the contingent and accidental, and intuit the essential natures or essences of the objects and acts of consciousness. This intuition proceeds from *free variation in imagination*. One imagines variations on an object and retains amid such free variations of an original the invariant, the necessary, universal form, the essential form, without which something of that kind would be inconceivable. Eventually one comes up against something that cannot be varied without destroying that object as an instance of its kind. An example of such a variation and reduction could be the eidetic variation of the physical parameters of this universe through the idea of multiverse with the purpose to assert the unique essential features of this universe.

Enhypostatic, enhypostasis, enhypostasization

The Greek Patristic meaning of the words *enhypostatic* or *enhypostasis* originates in the theology of Leontius of Byzantium in the context of Christological discussions of 6-7c A.D., and whose meaning according to "A Patristic Greek Lexicon" can be described as: "being, existing in an hypostasis or Person," "subsistent in, inherent." *Enhypostasis* points towards something which has its being in the other hypostasis. The "hypostatic inherence" of the universe in the Logos of God can then be interpreted as the Logos' eternal manifestations in different modes of *participation* by created beings in him. This *participation* does not assume any ontological causation; for to participate in the Logos means to be made by the Logos a participating being, that is to be made as a being in the hypostasis of the Logos himself. This implies that existence through participation in the Logos is subsistence in his Personhood, that is the *inherence* in his hypostasis. Another example is a theological assertion that man is hypostasis [personality] of the cosmos, its conscious and personal self-expression; it is he who gives meaning to things and who has to transfigure them. The universe as *the expressed and articulated existence* is possible only in human hypostasis, that is it acquires some qualities of existence if it is reflected in the personality of humanity. The link between the universe as articulated existence and the apprehending being is not ontological, but rather hypostatic or personal. A theologian would say that existence of the universe as the articulated existence is *enhypostatic*. This is the same as to say that the universe, as an articulated system of notions is *enhypostasized* by human beings.

Facticity

In philosophy, the terms facticity has a multiplicity of meanings from "factuality" and "contingency" to the intractable conditions of human existence. In our context the term comes to mean that which resists explanation and interpretation, or that which cannot be presented in the phenomenality of objects and physical causation. M. Heidegger discusses facticity as the "thrownness" of individual existence into the world. He refers not only to a brute fact, or the factuality of a concrete historical situation, but to something that already informs and has been taken up in existence, even if it is unnoticed or left unattended.

Life-world

In a phenomenological context the notion of the life-world has many meanings. However for the purposes of this book one can summarise it as follows. The life-world is that I find constantly present to myself and standing over against me one spatiotemporal fact-world to which I belong, as do all other human beings found in it and related to it in the same way as I am. I find this fact-world to be out there and I take it just as it gives itself to me as something that exists out there. This is the mundane world of immediate experience of living on a sensorial and conscious level. The life world is that underlying structure of consciousness which always accompanies the sense of existence. In comparison with the constructed world of science, the life-world precedes the latter, because the objective world of science always implies those acts of consciousness in and through which the life-world appears as always present and pregiven, as existing in its own right prior to all scientific endeavour. Therefore, for understanding of the world of science, one must return to the life-world and elucidate the role it plays in several respects in the constitution and development of science. An inquiry into the very facticity of science, its very possibility, returns the inquiring mind to the foundation of that same consciousness from within which science is constructed—and this is the life-world.

Multiverse

Theoretical notion reminiscent to the perennial idea of *plurality of worlds*. In modern cosmology this notion has different physical and mathematical interpretations starting with the ensemble of the different physical universes disjoint in space and finishing by the variety of all mathematically possible worlds in a platonic sense. Some physicists advocate the relevance of the theory of the multiverse to physics, some other claim that this notion has a philosophical standing and is not physical.

Noetic-noematic

Noetos in Greek usage is that which belongs to or is characteristic of the intellect.

Intellection in this case is not an abstract concept or a visual image, but the act of the function of the intellect whereby it apprehends realities in a direct manner.

Noetic aspect of the act of knowing is a characteristic relating to mental activity or the intellect as opposite to the noematic pole of perception and consciousness in general. Noema (plural: noemata) derives from the Greek word νόημα meaning thought or what is thought about. In phenomenology noema is used as a technical term to stand for the *object* or *content* of a thought, judgment, or perception. Said simply the noetic is related to human intentional activity, whereas noematic pole represents the content of that which this activity actualises or receives in response.

Nomological uniformity (of the universe)

Constitutes a dimension of the cosmological principle related to the similarity of the laws of physics across the universe. The terms *nomological* (from othe Greek *nomos*) relates to or denotes principles that resemble laws, especially those laws of nature which are neither logically necessary nor theoretically explicable, but just are so.

Phenomenality

In general this term describes the quality or state of a *phenomenon*. For example phenomenality of mundane things corresponds to their being perceptible by the senses or through immediate experience. This constitutes the notion of the phenomenal world, as the world of visible, empirical phenomena. One can talk about phenomenality of objects as entities being constituted according to the rubrics of "I think," so that such a phenomenality can be described in four rubrics: quantity, quality, relation and modality. The phenomenality of objects is different from the event-like manifestations, whose phenomenality cannot be reduced to the stated four rubrics and where there is the excess of intuition over the discursive faculty.

Retrodiction

Retrodiction is the act of making a "prediction" about the past. Predicting the features of the early universe cosmology, in fact, makes an act of retrodiction about the properties of the universe in the past. A retrodiction occurs when already gathered data is accounted for by a later theoretical advance in a more convincing fashion. The advantage of a retrodiction over a prediction is that the already gathered data is more likely to be free of experimenter bias. Another use refers to a process by which one attempts to test a theory whose predictions are too long-term to be tested (or not tested at all) by waiting for a future event to occur. Instead, one speculates about uncertain events in the more distant past, and applies the theory to consider how it would have predicted a known event in the less distant past. This is useful in, for example, the fields of archaeology, climatology, evolutionary biology and cosmology.

Saturated Phenomena

Stand for the group of phenomena which cannot be represented in the phenomenality of objects, that is in rubrics of: quantity, quality, relation and modality. The issue of the saturated phenomenon concerns the possibility that certain phenomena do not manifest themselves in the mode of objects and yet still do manifest themselves. These phenomena undergo saturation by the excess of intuition over the concept or signification in them; the saturated phenomena cannot be constituted because they are saturated. Here such a definition of experience is implied that it cannot be determined by a transcendental subject. On the contrary, it is to the extent that *ego* cannot comprehend the phenomenon that this *ego* is constituted by it. The examples of the saturated phenomena can be found in various fields of the human activity: painting, revelation of the Divine, the givenness of truth, events of life etc. Theory of the saturated phenomena was advanced by J. L. Marion in books listed in the bibliography.

Scientific research as communion

is a conviction that through study of the universe as God's creation scientists commune with the effected words of God through the worldly manifestations. Communion with God through contemplating his effected words, together with the ecclesial sacramental communion and communion through the words of Scripture, contributes to the fullness of communion with God.

Supersensible

Something that is beyond the range of what is perceptible by the senses, not belonging to the experienceable physical world. Presented to the subject without any material reference provided by sensibility. One can associate the supresensible with the intelligible. However, the analogy with the transcendent is very limited, for the supersensible transcends empirical reality towards the realm of ideal forms, but it remains a part of the created realm. Transcendence in an absolute sense, that is beyond creation, can be qualified not only transition to the supersensible but also to the superintelligible. For example, the essence of God is beyond sense-experience as well as beyond any intellection.

Tradition

in theology—historical concreteness of theology as related not only to historical events associated with Christianity, but with the permanent presence of the Holy Spirit in the Church through its liturgical invocation. Tradition as the perseverance of the Church institutions through centuries through direct communion with God.

in (European) science—historical concreteness of the natural sciences as related to the structural path of the European philosophical thought dated back to pre-Christian Hellenic philosophy.

Transcendental sense of physics

Physics is seen not only as a compendium of facts and theories about the observable universe, but as a structural path of humanity in its comprehension of the universe, the path which is historically concrete and includes the conditions of observability and mathematical expressibility of reality.

Transworldly (origin of the universe)

The characteristic of the foundation and origin of the universe as having non-worldly character, as contingent upon the realm beyond the world, that is, the uncreated.

The universe

Whereas the *physical universe*, studied by astronomy, astrophysics and cosmology, comprises the visible totality of the physical reality together with its invisible counterparts, *the universe as a whole* constitutes a philosophical and theological notion. The universe as a whole, or the world, is asserted as creation in contradistinction with the uncreated realm of the Divine. Thus the notion of the universe as a whole acquires some unique features related to the salvation history, namely to its biblically described creation by God out of nothing, the Fall of humanity, the Incarnation of the Son of God, Jesus Christ in it, and the soteriological transfiguration of the universe in the age to come.

Index of Subjects

Index of Scripture Passages